Laboratory Manual

Hole's
Human Anatomy
&Physiology

TENTH EDITION

DAVID SHIER
JACKIE BUTLER
RICKI LEWIS

TERRY R. MARTIN
Kishwaukee College

McGraw Hill **Higher Education**

Boston   Burr Ridge, IL   Dubuque, IA   Madison, WI   New York   San Francisco   St. Louis
Bangkok   Bogotá   Caracas   Kuala Lumpur   Lisbon   London   Madrid   Mexico City
Milan   Montreal   New Delhi   Santiago   Seoul   Singapore   Sydney   Taipei   Toronto

HOLE'S HUMAN ANATOMY & PHYSIOLOGY LABORATORY MANUAL
TENTH EDITION

Published by McGraw-Hill, a business unit of The McGraw-Hill Companies, Inc., 1221 Avenue
of the Americas, New York, NY 10020. Copyright © 2004, 2002, 1999, 1996 by The McGraw-Hill
Companies, Inc. All rights reserved. No part of this publication may be reproduced or distributed in
any form or by any means, or stored in a database or retrieval system, without the prior written
consent of The McGraw-Hill Companies, Inc., including, but not limited to, in any network or other
electronic storage or transmission, or broadcast for distance learning.

Some ancillaries, including electronic and print components, may not be available to customers
outside the United States.

This book is printed on acid-free paper.

1 2 3 4 5 6 7 8 9 0 KGP/KGP 0 9 8 7 6 5 4 3

ISBN 0-07-243891-6

Publisher: *Martin J. Lange*
Sponsoring editor: *Michelle Watnick*
Senior developmental editor: *Patricia Hesse*
Director of development: *Kristine Tibbetts*
Marketing manager: *James F. Connely*
Senior project manager: *Rose Koos*
Lead production supervisor: *Sandy Ludovissy*
Media project manager: *Sandra M. Schnee*
Senior media technology producer: *Barbara R. Block*
Designer: *K. Wayne Harms*
Cover designer: *Christopher Reese*
Cover image: *Hoby Finn/Gettyimages*
Senior photo research coordinator: *John C. Leland*
Photo research: *Billie Porter*
Compositor: *Precision Graphics*
Typeface: *10/12 Garamond Book*
Printer: *Quebecor World Kingsport*

The credits section for this book begins on page 512 and is considered an extension
of the copyright page.

Some of the laboratory experiments included in this text may be hazardous if materials are handled
improperly or if procedures are conducted incorrectly. Safety precautions are necessary when you are
working with chemicals, glass test tubes, hot water baths, sharp instruments, and the like, or for any
procedures that generally require caution. Your school may have set regulations regarding safety
procedures that your instructor will explain to you. Should you have any problems with materials
or procedures, please ask your instructor for help.

www.mhhe.com

CONTENTS

PREFACE

This laboratory manual was prepared to be used with the textbook *Hole's Human Anatomy and Physiology*, 10th edition, by David Shier, Jackie Butler, and Ricki Lewis. As with the textbook, the laboratory manual is designed for students with minimal backgrounds in the physical and biological sciences who are pursuing careers in allied health fields.

The laboratory manual contains sixty-two laboratory exercises and reports, which are closely integrated with the chapters of the textbook. The exercises are planned to illustrate and review anatomical and physiological facts and principles presented in the textbook and to help students investigate some of these ideas in greater detail.

Often the laboratory exercises are short or are divided into several separate procedures. This allows an instructor to select those exercises or parts of exercises that will best meet the needs of a particular program. Also, exercises requiring a minimal amount of laboratory equipment have been included.

The laboratory exercises include a variety of special features that are designed to stimulate interest in the subject matter, to involve students in the learning process, and to guide them through the planned activities. These special features include the following:

MATERIALS NEEDED

This section lists the laboratory materials that are required to complete the exercise and to perform the demonstrations and optional activities.

SAFETY

A list of safety guidelines is included inside the front cover. Each lab session that requires special safety guidelines has a safety section following "Materials Needed." Your instructor might require some modifications of these guidelines.

INTRODUCTION

The introduction briefly describes the subject of the exercise or the ideas that will be investigated.

PURPOSE OF THE EXERCISE

The purpose provides a statement concerning the intent of the exercise—that is, what will be accomplished.

LEARNING OBJECTIVES

The learning objectives list in general terms what a student should be able to do after completing the exercise.

PROCEDURE

The procedure provides a set of detailed instructions for accomplishing the planned laboratory activities. Usually these instructions are presented in outline form so that a student can proceed through the exercise in stepwise fashion. Often the student is referred to particular sections of a textbook for necessary background information or for review of subject matter presented previously.

The procedures include a wide variety of laboratory activities and, from time to time, direct the student to complete various tasks in the laboratory reports.

LABORATORY REPORTS

A laboratory report to be completed by the student immediately follows each exercise. These reports include various types of review activities, spaces for sketches of microscopic objects, tables for recording observations and experimental results, and questions dealing with the analysis of such data.

It is hoped that as a result of these activities, students will develop a better understanding of the structural and functional characteristics of their bodies and will increase their skills in gathering information by observation and experimentation. Some of the exercises also include demonstrations, optional activities, and useful illustrations.

DEMONSTRATIONS

Demonstrations appear in separate boxes. They describe specimens, specialized laboratory equipment, or other materials of interest that an instructor may want to display to enrich the student's laboratory experience.

OPTIONAL ACTIVITIES

Optional activities also appear in separate boxes. They encourage students to extend their laboratory experiences. Some of these activities are open-ended in that they suggest the student plan an investigation or experiment and carry it out after receiving approval from the laboratory instructor.

THE USE OF ANIMALS IN BIOLOGY EDUCATION*

The National Association of Biology Teachers (NABT) believes that the study of organisms, including nonhuman animals, is essential to the understanding of life on Earth. NABT recommends the prudent and responsible use of animals in the life science classroom. NABT believes that biology teachers should foster a respect for life. Biology teachers also should teach about the interrelationship and interdependency of all things.

Classroom experiences that involve nonhuman animals range from observation to dissection. NABT supports these experiences so long as they are conducted within the long-established guidelines of proper care and use of animals, as developed by the scientific and educational community.

As with any instructional activity, the use of nonhuman animals in the biology classroom must have sound educational objectives. Any use of animals, whether for observation or dissection, must convey substantive knowledge of biology. NABT believes that biology teachers are in the best position to make this determination for their students.

NABT acknowledges that no alternative can substitute for the actual experience of dissection or other use of animals and urges teachers to be aware of the limitations of alternatives. When the teacher determines that the most effective means to meet the objectives of the class do not require dissection, NABT accepts the use of alternatives to dissection, including models and the various forms of multimedia. The Association encourages teachers to be sensitive to substantive student objections to dissection and to consider providing appropriate lessons for those students where necessary.

To implement this policy, NABT endorses and adopts the "Principles and Guidelines for the Use of Animals in Precollege Education" of the Institute of Laboratory Animals Resources (National Research Council). Copies of the "Principles and Guidelines" may be obtained from the ILAR (2101 Constitution Avenue, NW, Washington, DC 20418; 202 334-2590).

---

* Adopted by the Board of Directors in October 1995. This policy supersedes and replaces all previous NABT statements regarding animals in biology education.

ILLUSTRATIONS

Diagrams from the textbook and diagrams similar to those in the textbook often are used as aids for reviewing subject matter. Other illustrations provide visual instructions for performing steps in procedures or are used to identify parts of instruments or specimens. Micrographs are included to help students identify microscopic structures or to evaluate student understanding of tissues.

In some exercises, the figures include line drawings that are suitable for students to color with colored pencils. This activity may motivate students to observe the illustrations more carefully and help them to locate the special features represented in the figures. Students can check their work by referring to the corresponding full-color illustrations in the textbook.

It should be noted that frequent variations exist in anatomical structures among humans. The illustrations in the textbook and the laboratory manual represent normal (normal means the most common variation) anatomy.

FEATURES OF THIS EDITION

This new edition of the laboratory manual has been made more user-friendly. Many of the changes are a result of evaluations and suggestions from anatomy and physiology students. Many suggestions from users and reviewers of the ninth edition have been incorporated. Some features include the following:

1. To meet the need for clearer and more definite safety guidelines, a safety list is located inside the front cover, and safety sections are found in appropriate labs.
2. A section called "Study Skills for Anatomy and Physiology" is located in the front material. This section was written by students enrolled in a Human Anatomy and Physiology course.
3. The "Materials Needed" section is located at the beginning of the laboratory exercise for greater ease in laboratory preparations.
4. New art has been created for the textbook and the lab manual. Additional color has been included for many figures.
5. To clarify whether a figure label refers to a general area or a specific structure, "clue" words in parentheses have been added to some figures to direct students in their answers. The first example is figure 2.1.

6. A list of terms is provided to assist the labeling of many figures. The first example is figure 11.3.
7. Computer literacy is integrated with relevant laboratory exercises. "Web Quest" activities are found at the end of most exercises. Discover the answers to many scientific questions at www.mhhe.com/shier10 Click on Student Edition and click on Martin Lab Manual, Web Quest. Here you'll find links to help you with your quest.
8. "Critical Thinking Applications" are 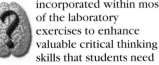 incorporated within most of the laboratory exercises to enhance valuable critical thinking skills that students need throughout their lives.
9. Two assessment tools (rubrics) for laboratory reports are included in Appendix 2.
10. The *Instructor's Manual for the Laboratory Manual to Accompany Hole's Human Anatomy and Physiology* has been revised and updated. It is found in the Online Learning Center for Hole's Human Anatomy & Physiology, 10th edition, www.mhhe.com/shier10 and www.mhhe.com/biosci/ap/labcentral
11. A supplement of four computerized physiology labs with laboratory reports using Intelitool products is available. The title is *Intelitool Supplementary Lab Exercises to Accompany the Laboratory Manual for Hole's Human Anatomy and Physiology* (0-697-27976-6).

REVIEWERS

I would like to express my sincere gratitude to all users of the laboratory manual who provided suggestions for its improvement. I am especially grateful for the contributions of the reviewers who kindly reviewed the manual and examined the manuscript of the new edition. Their thoughtful comments and valuable suggestions are greatly appreciated. They include the following:

Mahmoud Bishr
Labette Community College

Jennifer L. Borash
Horry-Georgetown Technical College

Jennifer Carr Burtwistle
Northeast Community College

Peter I. Ekechukwu
Horry-Georgetown Technical College

Lori K. Garrett
Danville Area Community College

Louis Giacinti
Milwaukee Area Technical College

Michael J. Harman
North Harris College

June A. Kinsley
Harcum College

Randy Lankford
Galveston College

Barbara Wineinger
Vincennes University Jasper

ACKNOWLEDGMENTS

I value all the support and encouragement by the staff at McGraw-Hill, including Michael Lange, Marty Lange, Michelle Watnick, Pat Hesse, Kristine Tibbetts, Rose Koos, and Darlene Schueller. Special recognition is granted to Colin Wheatley for his insight, confidence, wisdom, warmth, and friendship.

I am grateful for the professional talent of David Shier, Jackie Butler, and Ricki Lewis for all of their coordination efforts and guidance with the laboratory manual. The deepest gratitude is extended to John W. Hole, Jr. for his years of dedicated effort in previous editions of this classic work and for the opportunity to revise his established laboratory manual.

I am particularly thankful to Dr. Norman Jenkins, retired president of Kishwaukee College, and Dr. David Louis, president of Kishwaukee College, for their support and confidence in my endeavors. I am appreciative for the expertise of J. Womack Photography. The professional reviews of the nursing procedures were provided by Kathy Schnier. I am also grateful to Michele Dukes, Troy Hanke, Jenifer Holtzclaw, Angele Myska, Sparkle Neal, Robert Stockley, Shatina Thompson, Jana Voorhis, and DeKalb Clinic Chartered for their contributions. There have been valuable contributions from my students who have supplied thoughtful suggestions and assisted in clarification of details.

To my son Ross, I owe gratitude for his keen eye and creative suggestions. Foremost, I am appreciative to Sherrie Martin, my spouse and best friend, for advice, understanding, and devotion throughout the writing and revising.

Terry R. Martin
Kishwaukee College
21193 Malta Road
Malta, IL 60150

ABOUT THE AUTHOR

This tenth edition is the fourth revision by Terry R. Martin of Kishwaukee College. Terry's teaching experience of over thirty years, his interest in students and love for college instruction, and his innovative attitude and use of technology-based learning enhance the solid tradition of John Hole's laboratory manual. Among Terry's awards are the 1972 Kishwaukee College Outstanding Educator, 1977 Phi Theta Kappa Outstanding Instructor Award, 1989 Kishwaukee College ICCTA Outstanding Educator Award, 1996 Who's Who Among America's Teachers, 1996 Kishwaukee College Faculty Board of Trustees Award of Excellence, and 1998 Continued Excellence Award for Phi Theta Kappa Advisors. Terry's professional memberships include the National Association of Biology Teachers, Illinois Association of Community College Biologists, Human Anatomy and Physiology Society, Chicago Area Anatomy and Physiology Society (founding member), Phi Theta Kappa (honorary member), DeKalb County Prairie Stewards, and Nature Conservancy. In addition to writing many publications, he co-produced with Hassan Rastegar a videotape entitled *Introduction to the Human Cadaver and Prosection,* published by Wm. C. Brown Publishers in 1989. Terry revised the *Laboratory Manual to Accompany Hole's Essentials of Human Anatomy and Physiology,* 8th edition, and authored *Human Anatomy and Physiol-*

ogy Laboratory Manual, Fetal Pig Dissection, 2nd edition. Terry serves as a Faculty Consultant for the Advanced Placement Biology examination readings. During 1994, Terry was a faculty exchange member in Ireland. He has also been involved in his community,

most notably as a Commissioner for Boy Scouts of America. We are pleased to have Terry continue the tradition of John Hole's laboratory manual.

The Editor

TO THE STUDENT

The exercises in this laboratory manual will provide you with opportunities to observe various anatomical parts and to investigate certain physiological phenomena. Such experiences should help you relate specimens, models, microscope slides, and your own body to what you have learned in the lecture and read about in the textbook.

The following list of suggestions may help to make your laboratory activities more effective and profitable.

1. Prepare yourself before attending the laboratory session by reading the assigned exercise and reviewing the related sections of the textbook. It is important to have some understanding of what will be done in the laboratory before you come to class.

2. Bring your laboratory manual and textbook to each laboratory session. These books are closely integrated and will help you complete most of the exercises.

3. Be on time. During the first few minutes of the laboratory meeting, the instructor often will provide verbal instructions. Make special note of any changes in materials to be used or procedures to be followed. Also listen carefully for information about special techniques to be used and precautions to be taken.

4. Keep your work area clean and your materials neatly arranged so that you can locate needed items quickly. This will enable you to proceed efficiently and will reduce the chances of making mistakes.

5. Pay particular attention to the purpose of the exercise, which states what you are to accomplish in general terms, and to the learning objectives, which list what you should be able to do as a result of the laboratory experience. Then, before you leave the class, review the objectives and make sure that you can meet them.

6. Precisely follow the directions in the procedure and proceed only when you understand them clearly. Do not improvise procedures unless you have the approval of the laboratory instructor. Ask questions if you do not understand exactly what you are supposed to do and why you are doing it.

7. Handle all laboratory materials with care. These materials often are

fragile and expensive to replace. Whenever you have questions about the proper treatment of equipment, ask the instructor.

8. Treat all living specimens humanely and try to minimize any discomfort they might experience.

9. Although at times you might work with a laboratory partner or a small group, try to remain independent when you are making observations, drawing conclusions, and completing the activities in the laboratory reports.

10. Record your observations immediately after making them. In most cases, such data can be entered in spaces provided in the laboratory reports.

11. Read the instructions for each section of the laboratory report before you begin to complete it. Think about the questions before you answer them. Your responses should be based on logical reasoning and phrased in clear and concise language.

12. At the end of each laboratory period, clean your work area and the instruments you have used. Return all materials to their proper places and dispose of wastes, including glassware or microscope slides that have become contaminated with human blood or body fluids, as directed by the laboratory instructor. Wash your hands thoroughly before leaving the laboratory.

STUDY SKILLS FOR ANATOMY AND PHYSIOLOGY

My students have found that certain study skills worked well for them while enrolled in Human Anatomy and Physiology. Although everyone has their own learning style, there are techniques that work well for most students. Using some of the skills listed here could make your course more enjoyable and rewarding.

1. **Note taking:** Look for the main ideas and briefly express them in your own words. Organize, edit, and review your notes soon after the lecture. Add textbook information to your notes as you reorganize them. Underline or highlight with different colors the important points, major headings, and key terms. Study your notes daily, as they provide sequential building blocks of the course content.

2. **Chunking:** Organize information into logical groups or categories. Study and master one chunk of information at a time. For example, study the bones of the upper limb, lower limb, trunk, and head as separate study tasks.

3. **Mnemonic devices:** An *acrostic* is a combination of association and imagery to aid your memory. It is often in the form of a poem, rhyme, or jingle in which the first letter of each word corresponds to the first letters of the words you need to remember. **S**o **L**ong **T**op **P**art, **H**ere **C**omes **T**he **T**humb is an example of such a mnemonic device to remember the eight carpals in the correct sequence. *Acronyms* are words that are formed by the first letters of the items to remember. *IPMAT* is an example of this type of mnemonic device to help you remember the phases of the cell cycle in the correct sequence. Try creating some of your own.

4. **Study groups:** Small study groups that meet periodically to review course material and compare notes have helped and encouraged many students. However, keep the group on the task at hand. Work as a team and alternate leaders. This group often becomes a support group.

5. **Recording and recitation:** An auditory learner can benefit by recording lectures and review sessions with a cassette recorder. Many students listen to the taped sessions as they drive or just before going to bed. Reading your notes aloud can help also. Explain the material to anyone (even if there are no listeners). Talk about anatomy and physiology in everyday conversations.

6. **Note cards/flash cards:** Make your own. Add labels and colors to enhance the material. Keep them with you in your pocket or purse. Study them often and for short periods of time. Concentrate on a small number of cards at one time. Shuffle your cards and have someone quiz you on their content. As you become familiar with the material, you can set aside cards that don't require additional mastery.

7. **Time management:** Prepare monthly, weekly, and daily schedules. Include dates of quizzes, exams, and projects on the calendar. On your daily schedule, budget several short study periods. Daily repetition alleviates cramming for exams. Prioritize your time so that you still have time for work and leisure activities. Find an appropriate study atmosphere with minimum distractions.

Best wishes on your anatomy and physiology endeavor.

CORRELATION OF TEXTBOOK CHAPTERS AND LABORATORY EXERCISES

CORRELATION OF TEXTBOOK CHAPTERS AND SUPPLEMENTAL LABORATORY EXERCISES

SCIENTIFIC METHOD AND MEASUREMENTS

MATERIALS NEEDED

Meterstick
Calculator
Human skeleton

Scientific investigation involves a series of logical steps to arrive at explanations for various biological phenomena. This technique, called the *scientific method,* is used in all disciplines of science. It allows scientists to draw logical and reliable conclusions about phenomena.

The scientific method begins with *observations* related to the topic under investigation. This step commonly involves the accumulation of previously acquired information and/or your own observations of the phenomenon. These observations are used to formulate a tentative explanation known as the *hypothesis.* An important attribute of a hypothesis is that it must be testable. The testing of the hypothesis involves performing a carefully controlled experiment to obtain data that can be used to support, disprove, or modify the hypothesis. An *analysis of data* is conducted using all of the information collected during the experiment. Data analysis may include organization and presentation of data as tables, graphs, and drawings. From the interpretation of the data analysis, *conclusions* are drawn. (If the data do not support the hypothesis, you must reexamine the experimental design and the data, and if needed, develop a new hypothesis.) The final presentation of the information is made from the conclusions. Results and conclusions are presented to the scientific community for evaluation through peer reviews, presentations at professional meetings, and published articles. If many investigators working independently can validate the hypothesis by arriving at the same conclusions, the explanation becomes a **theory.** A theory that is verified continuously over a period of time and accepted by the scientific community becomes known as a **scientific law** or **principle.** A scientific law serves as the standard explanation for an observation unless it is disproved by new information. The five components of the scientific method are summarized as

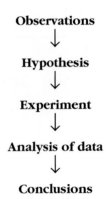

Observations
↓
Hypothesis
↓
Experiment
↓
Analysis of data
↓
Conclusions

Metric measurements are characteristic tools of scientific investigations. Because the English system of measurements is often used in the United States, the investigator must make conversions from the English system to the metric system. A reference table for the conversion of English units of measure to metric units for length, mass, volume, time, and temperature is located inside the back cover of the laboratory manual.

PURPOSE OF THE EXERCISE

To become familiar with the scientific method of investigation, to learn how to formulate sound conclusions, and to provide opportunities to use the metric system of measurements.

LEARNING OBJECTIVES

After completing this exercise, you should be able to

1. List in the correct order and describe all steps of the scientific method.
2. Use the scientific method to test the validity of a hypothesis concerning the direct, linear relationship between human height and upper limb length.
3. Make conversions from English measurements to the metric system, and vice versa.
4. Formulate a hypothesis and test it using the scientific method.

Figure 1.1 Measurement of upper limb length.

PROCEDURE A—USING THE STEPS OF THE SCIENTIFIC METHOD

1. Many people have observed a correlation between the length of the upper and lower limbs and the height (height for this lab means overall height of the subject) of an individual. For example, a person who has long upper limbs (the arm, forearm, and hand combined) tends to be tall. Make some visual observations of other people in your class to observe a possible correlation.
2. From such observations, the following hypothesis is formulated: The length of a person's upper limb is equal to 0.4 (40%) of the height of the person. Test this hypothesis by performing the following experiment.
3. In this experiment, use a meterstick to measure an upper limb length of ten subjects. For each measurement, place the meterstick in the axilla (armpit) and record the length in centimeters to the end of the longest finger (fig. 1.1). Obtain the height of each person in centimeters by measuring them without shoes against a wall (fig. 1.2). The height of each person can be calculated by multiplying each individual's height in inches by 2.54 to obtain his/her height in centimeters. Record all your measurements in Part A of Laboratory Report 1.

Figure 1.2 Measurement of height.

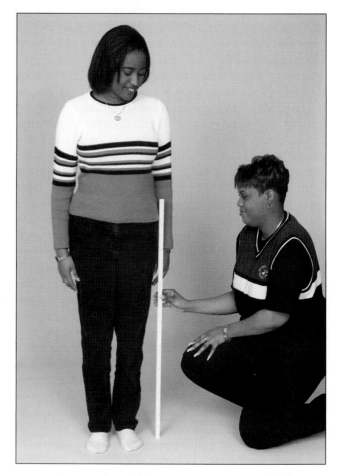

4. The data collected from all of the measurements can now be analyzed. The expected (predicted) correlation between upper limb length and height is determined using the following equation:

Height × 0.4 = expected upper limb length

The observed (actual) correlation to be used to test the hypothesis is determined by

**Length of upper limb/height
= actual % of height**

5. A graph is an excellent way to display a visual representation of the data. Plot the subjects' data in Part A of the laboratory report. Plot the upper limb length of each subject on the x-axis and the height of each person on the y-axis. A line is already located on the graph that represents a hypothetical relationship of 0.4 (40%) upper limb length compared to height. This is a graphic representation of the original hypothesis.

6. Compare the distribution of all of the points (actual height and upper limb length) that you placed on the graph with the distribution of the expected correlation represented by the hypothesis.

7. Complete Part A of the laboratory report.

PROCEDURE B—DESIGN AN EXPERIMENT

 ### Critical Thinking Application

You have probably concluded that there is some correlation of the length of body parts to height. Often when a skeleton is found, it is not complete, especially when paleontologists discover a skeleton. It is occasionally feasible to use the length of a single bone to estimate the height of an individual. Observe human skeletons and locate the radius bone in the forearm. Use your observations to identify a mathematical relationship between the length of the radius and height. Formulate a hypothesis that can be tested. Make measurements, analyze data, and develop a conclusion from your experiment. Complete Part B of the laboratory report.

 Web Quest

Examine additional information about the scientific method at
www.mhhe.com/shier10

Click on Student Edition. Click on Martin Lab Manual, Web Quest.

SCIENTIFIC METHOD AND MEASUREMENTS

PART A

1. Record measurements for height and the upper limb length of ten subjects. Use a calculator to determine the expected upper limb length and the actual percentage (as a decimal or a percentage) of the height for the ten subjects. Record your results in the following table.

Subject	Height (cm)	Measured Upper Limb Length (cm)	Height × 0.4 = Expected Upper Limb Length (cm)	Actual % of Height = Upper Limb Length (cm)/ Height (cm)
1.				
2.				
3.				
4.				
5.				
6.				
7.				
8.				
9.				
10.				

2. Plot the distribution of data (upper limb length and height) collected for the ten subjects on the following graph. The line located on the graph represents the **expected** 0.4 (40%) ratio of upper limb length to measured height (the original hypothesis). (Note that the x-axis represents upper limb length and the y-axis represents height.) Draw a line of *best fit* through the distribution of points of the plotted data of the ten subjects. Compare the two distributions (expected line and the distribution line drawn for the ten subjects).

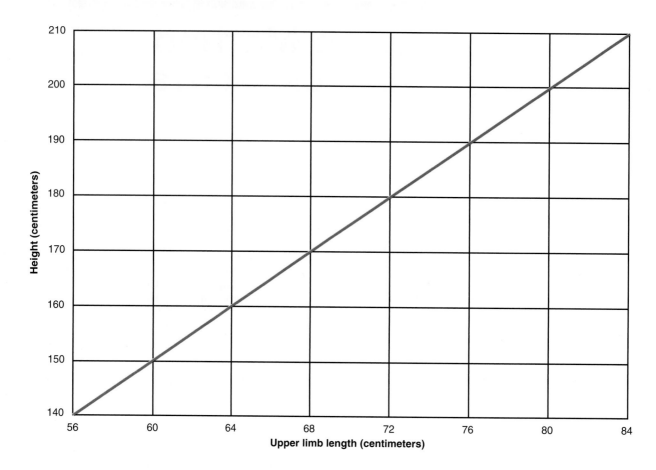

3. Does the distribution of the ten subjects' measured upper limb lengths support or disprove the original hypothesis? _____ Explain your answer.

PART B

1. Describe your observations of a possible correlation between the radius length and height.

2. Write a hypothesis based on your observations.

3. Describe the design of the experiment that you devised to test your hypothesis.

4. Place your analysis of the data in this space in the form of a table and a graph.

5. Based on an analysis of your data, what conclusions can you make? Did these conclusions confirm or refute your original hypothesis?

6. Discuss your results and conclusions with other classmates. What common conclusion can the class formulate about the correlation between radius length and height?

LABORATORY EXERCISE 2

BODY ORGANIZATION AND TERMINOLOGY

MATERIALS NEEDED

Textbook
Dissectible torso (manikin)
Variety of specimens or models sectioned along various
    planes

For Optional Activity:

Colored pencils

The major features of the human body include certain cavities, a set of membranes associated with these cavities, and a group of organ systems composed of related organs. In order to communicate effectively with each other about the body, scientists have devised names to describe these body features. They also have developed terms to represent the relative positions of body parts, imaginary planes passing through these parts, and body regions.

PURPOSE OF THE EXERCISE

To review the organizational pattern of the human body, to review its organ systems and the organs included in each system, and to become acquainted with the terms used to describe the relative position of body parts, body sections, and body regions.

LEARNING OBJECTIVES

After completing this exercise, you should be able to

1. Locate and name the major body cavities and identify the membranes associated with each cavity.
2. Name the organ systems of the human organism.
3. List the organs included within each system and locate the organs in a dissectible torso.
4. Describe the general functions of each system.
5. Define the terms used to describe the relative positions of body parts.
6. Define the terms used to identify body sections and identify the plane along which a particular specimen is cut.
7. Define the terms used to identify body regions.

PROCEDURE A—BODY CAVITIES AND MEMBRANES

1. Review the sections entitled "Body Cavities" and "Thoracic and Abdominopelvic Membranes" in chapter 1 of the textbook.
2. As a review activity, label figures 2.1, 2.2, and 2.3.
3. Locate the following features on the reference plates near the end of chapter 1 of the textbook and on the dissectible torso:

dorsal cavity

    cranial cavity

    vertebral canal (spinal cavity)

ventral cavity

    thoracic cavity

        mediastinum

        pleural cavity

    abdominopelvic cavity

        abdominal cavity

        pelvic cavity

diaphragm

smaller cavities within the head

    oral cavity

    nasal cavity with connected sinuses

    orbital cavity

    middle ear cavity

    membranes and cavities

    pleural cavity

        parietal pleura

        visceral pleura

    pericardial cavity

        parietal pericardium (covered by fibrous pericardium)

        visceral pericardium (epicardium)

Figure 2.1 Label the major body cavities.

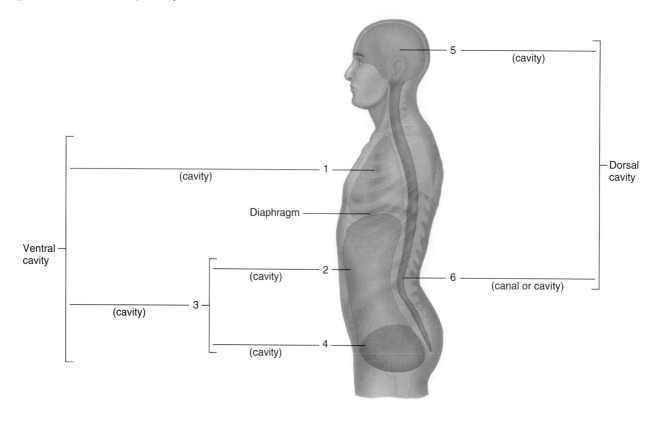

5 _____ (cavity)

Dorsal cavity

1 _____ (cavity)

Diaphragm _____

Ventral cavity

2 _____ (cavity)

6 _____ (canal or cavity)

3 _____ (cavity)

4 _____ (cavity)

Figure 2.2 Label the smaller cavities and sinuses within the head.

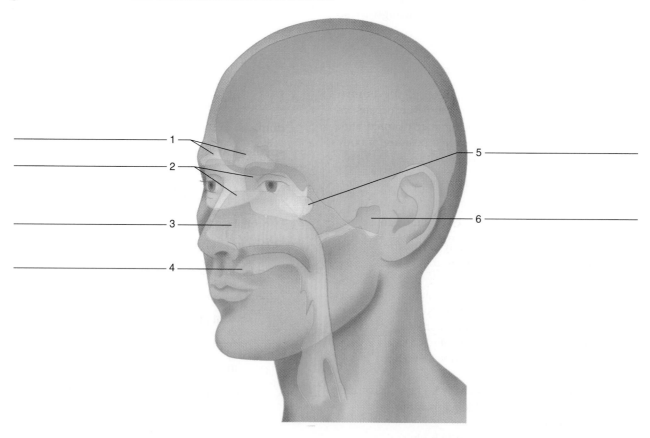

1

2

3

4

5

6

Figure 2.3 Label the thoracic membranes and cavities in (*a*) and the abdominopelvic membranes and cavity in (*b*) as shown in these superior views of transverse sections.

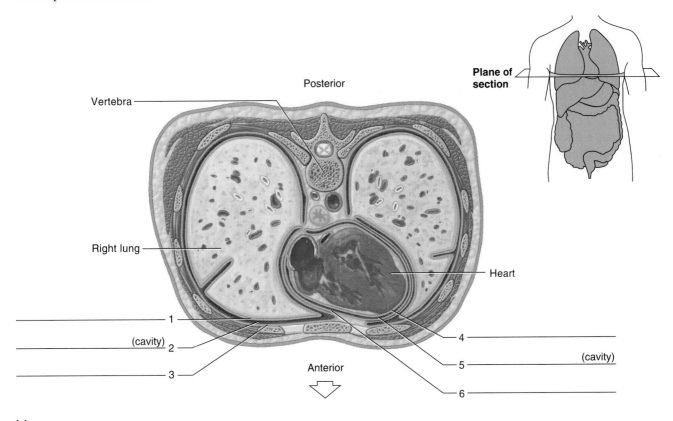

(a)

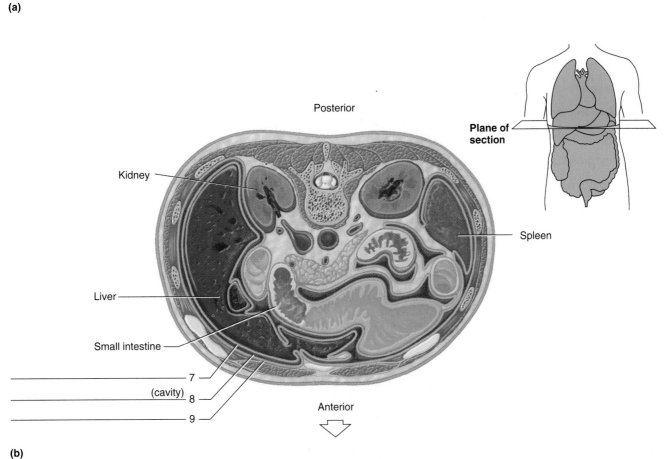

(b)

            peritoneal cavity

                parietal peritoneum

                visceral peritoneum

4. Complete Parts A and B of Laboratory Report 2.

PROCEDURE B—ORGAN SYSTEMS

1. Review the section entitled "Organ Systems" in chapter 1 of the textbook.
2. Use the reference plates on pages 29–36 of the textbook and the dissectible torso to locate the following organs:

integumentary system

skin

accessory organs such as hair and nails

skeletal system

bones

ligaments

cartilages

muscular system

skeletal muscles

tendons

nervous system

brain

spinal cord

nerves

endocrine system

pituitary gland

thyroid gland

parathyroid glands

adrenal glands

pancreas

ovaries

testes

pineal gland

thymus gland

cardiovascular system

heart

arteries

veins

lymphatic system

lymphatic vessels

lymph nodes

thymus gland

spleen

tonsils

digestive system

mouth

tongue

teeth

salivary glands

pharynx

esophagus

stomach

liver

gallbladder

pancreas

small intestine

large intestine

respiratory system

nasal cavity

pharynx

larynx

trachea

bronchi

lungs

urinary system

kidneys

ureters

urinary bladder

urethra

male reproductive system

scrotum

testes

penis

urethra

female reproductive system

>ovaries

>uterine tubes (oviducts; fallopian tubes)

>uterus

>vagina

3. Complete Parts C and D of the laboratory report.

PROCEDURE C—RELATIVE POSITIONS, PLANES, SECTIONS, AND REGIONS

1. Review the section entitled "Anatomical Terminology" in chapter 1 of the textbook.
2. As a review activity, label figures 2.4, 2.5, and 2.6.
3. Examine the sectioned specimens on the demonstration table and identify the plane along which each is cut.
4. Complete Parts E, F, G, H, and I of the laboratory report.

Figure 2.4   Label the planes represented in this illustration.

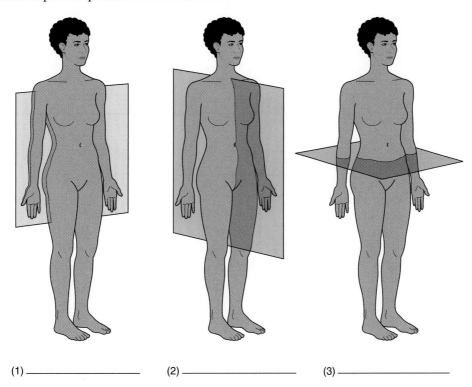

(1) _____   (2) _____   (3) _____

Figure 2.5 Label (*a*) the regions and (*b*) the quadrants of the abdominal area.

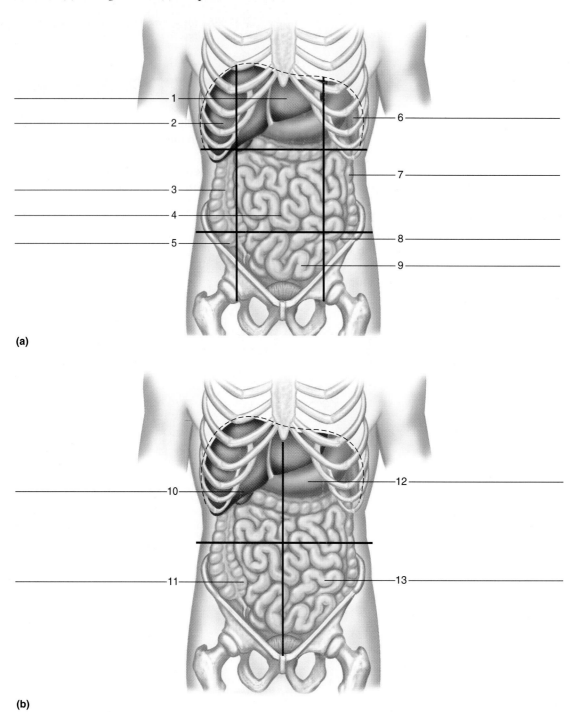

(a)

(b)

Figure 2.6 Label these diagrams with terms used to describe body regions: (*a*) anterior regions; (*b*) posterior regions.

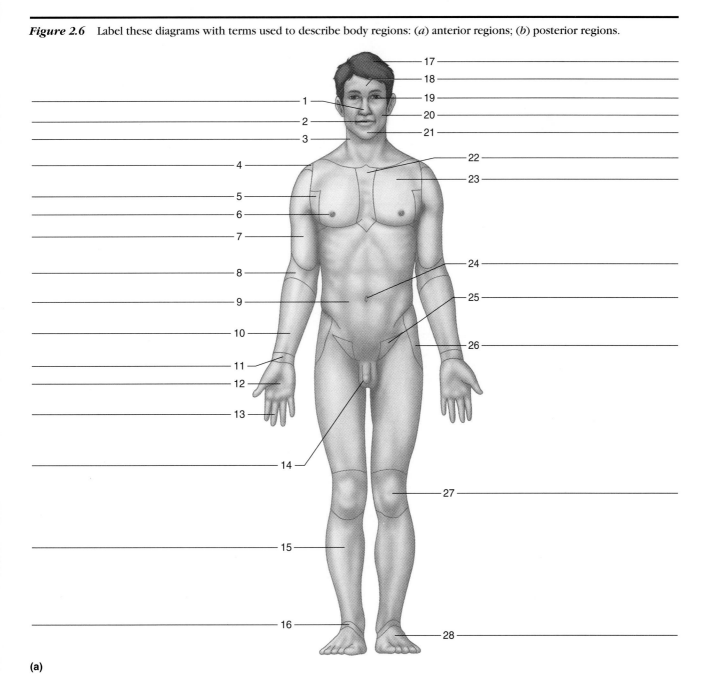

(a)

Figure 2.6 *Continued.*

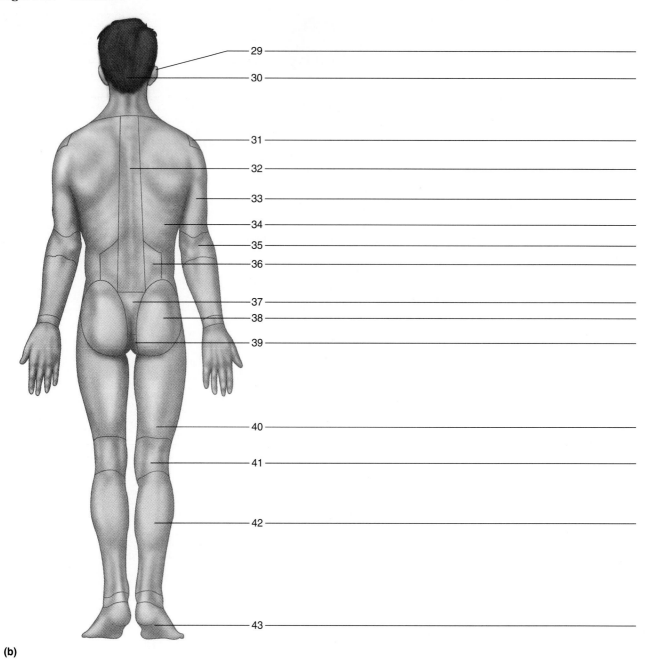

29 _____

30 _____

31 _____

32 _____

33 _____

34 _____

35 _____

36 _____

37 _____

38 _____

39 _____

40 _____

41 _____

42 _____

43 _____

(b)

BODY ORGANIZATION AND TERMINOLOGY

PART A

Match the body cavities in column A with the organs contained in the cavities in column B. Place the letter of your choice in the space provided.

Column A		Column B
a. Abdominal cavity	_____ 1.	Liver
b. Cranial cavity	_____ 2.	Lungs
c. Middle ear cavity	_____ 3.	Spleen
d. Oral cavity	_____ 4.	Stomach
e. Orbital cavity	_____ 5.	Brain
f. Pelvic cavity	_____ 6.	Teeth
g. Thoracic cavity	_____ 7.	Gallbladder
h. Vertebral canal (spinal cavity)	_____ 8.	Urinary bladder
	_____ 9.	Eyes
	_____ 10.	Spinal cord
	_____ 11.	Rectum
	_____ 12.	Ear bones
	_____ 13.	Heart
	_____ 14.	Esophagus

PART B

Complete the following statements:

1. The membrane on the surface of the lung is called the _____.

2. The membrane on the surface of the heart is called the _____.

3. The membrane that lines the wall of the abdominopelvic cavity is called the_____.

4. The membrane on the surface of the stomach is called the_____.

5. The thin, watery fluid located between the pleural membranes is called _____.

6. Epicardium is another name for _____.

7. The region of the thoracic cavity between the two lungs is called the_____.

8. The muscular structure that separates the thoracic and abdominopelvic cavities is called the _____.

PART C

Match the organ systems in column A with the functions in column B. Place the letter of your choice in the space provided.

Column A		**Column B**
a. Cardiovascular system	_____ 1.	Main system that secretes hormones
b. Digestive system	_____ 2.	Provides an outer covering of the body
c. Endocrine system	_____ 3.	Produces a new organism
d. Integumentary system	_____ 4.	Stimulates muscles to contract and interprets information from sensory units
e. Lymphatic system		
f. Muscular system	_____ 5.	Provides a framework for soft tissues and produces blood cells in red marrow
g. Nervous system		
h. Reproductive system	_____ 6.	Exchanges gases between air and blood
i. Respiratory system	_____ 7.	Transports excess fluid from tissues to blood
j. Skeletal system	_____ 8.	Maintains posture and generates most body heat
k. Urinary system	_____ 9.	Removes liquid and wastes from blood and transports them to the outside
	_____ 10.	Converts food molecules into forms that are absorbed
	_____ 11.	Transports nutrients, wastes, and gases throughout the body

PART D

Match the organ systems in column A with the organs in column B. Place the letter of your choice in the space provided. (In some cases, there may be more than one correct answer.)

Column A		**Column B**
a. Cardiovascular system	_____ 1.	Adrenal and parathyroid glands
b. Digestive system	_____ 2.	Arteries and veins
c. Endocrine system	_____ 3.	Brain and spinal cord
d. Integumentary system	_____ 4.	Gallbladder and esophagus
e. Lymphatic system	_____ 5.	Kidneys and ureters
f. Muscular system	_____ 6.	Larynx and lungs
g. Nervous system	_____ 7.	Ligaments
h. Reproductive system (female)	_____ 8.	Ovaries and uterus
i. Reproductive system (male)	_____ 9.	Prostate gland and testes
j. Respiratory system	_____ 10.	Skin
k. Skeletal system	_____ 11.	Spleen and tonsils
l. Urinary system	_____ 12.	Tendons

PART E

Indicate whether each of the following sentences makes correct or incorrect usage of the word in boldface type (assume that the body is in the anatomical position). If the sentence is incorrect, supply a term that will make it correct in the space provided.

1. The mouth is **superior** to the nose. _____

2. The stomach is **inferior** to the diaphragm. _____

3. The trachea is **anterior** to the spinal cord. _____

4. The larynx is **posterior** to the esophagus. _____

5. The heart is **medial** to the lungs. _____

6. The kidneys are **inferior** to the adrenal glands. _____

7. The hand is **proximal** to the elbow. _____

8. The knee is **proximal** to the ankle. _____

9. Blood in **deep** blood vessels gives color to the skin. _____

10. A **peripheral** nerve passes from the spinal cord into the limbs. _____

11. The spleen and gallbladder are **ipsilateral.** _____

12. The dermis is the **superficial** layer of the skin. _____

PART F

Name each of the planes represented in figure 2.7*a* and the sections represented in figure 2.7*b*.

1. _____   4. _____

2. _____   5. _____

3. _____   6. _____

Figure 2.7   Name (*a*) the planes and (*b*) the sections represented in these diagrams.

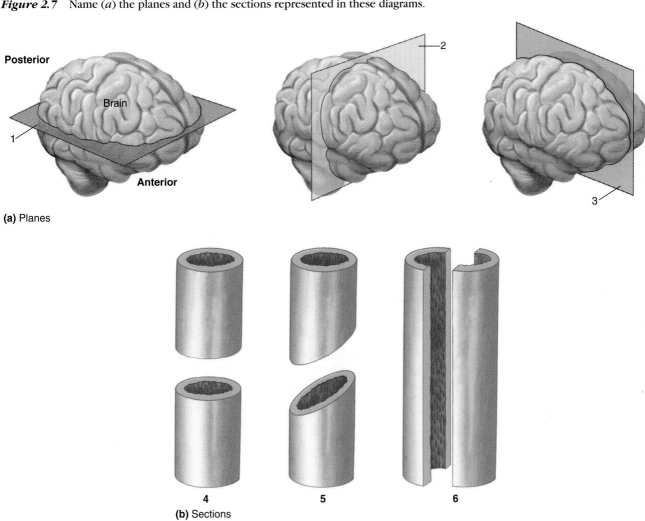

(a) Planes

(b) Sections

PART G

Match the body regions in column A with the locations in column B. Place the letter of your choice in the space provided.

Column A	Column B
a. Antebrachial	_____ 1. Wrist
b. Antecubital	_____ 2. Ribs
c. Axillary	_____ 3. Reproductive organs
d. Brachial	_____ 4. Armpit
e. Buccal	_____ 5. Elbow
f. Carpal	_____ 6. Forehead
g. Cephalic	_____ 7. Buttocks
h. Cervical	_____ 8. Forearm
i. Costal	_____ 9. Back
j. Crural	_____ 10. Neck
k. Cubital	_____ 11. Arm
l. Dorsal	_____ 12. Cheek
m. Frontal	_____ 13. Leg
n. Genital	_____ 14. Head
o. Gluteal	_____ 15. Space in front of elbow

PART H

Match the body regions in column A with the locations in column B. Place the letter of your choice in the space provided.

Column A	Column B
a. Inguinal	_____ 1. Pelvis
b. Lumbar	_____ 2. Breasts
c. Mammary	_____ 3. Ear
d. Mental	_____ 4. Between anus and reproductive organs
e. Occipital	_____ 5. Sole
f. Otic	_____ 6. Middle of thorax
g. Palmar	_____ 7. Chest
h. Pectoral	_____ 8. Navel
i. Pedal	_____ 9. Chin
j. Pelvic	_____ 10. Behind knee
k. Perineal	_____ 11. Foot
l. Plantar	_____ 12. Lower posterior region of head
m. Popliteal	_____ 13. Abdominal wall near thigh
n. Sternal	_____ 14. Lower back
o. Umbilical	_____ 15. Palm

Critical Thinking Application

State the quadrant of the abdominopelvic cavity where the pain or sound would be located for each of the six common conditions listed. In some cases, there may be more than one correct answer, and pain is sometimes referred to another region. This phenomenon, called *referred pain,* occurs when pain is interpreted as originating from some area other than the parts being stimulated.

1. Stomach ulcer   _____

2. Appendicitis   _____

3. Bowel sounds   _____

4. Gallbladder attack   _____

5. Kidney stone in left ureter _____

6. Ruptured spleen   _____

CARE AND USE OF THE MICROSCOPE

Because the human eye cannot perceive objects less than 0.1 mm in diameter, a microscope is an essential tool for the study of small structures such as cells. The microscope usually used for this purpose is the *compound light microscope.* It is called compound because it uses two sets of lenses: an eyepiece lens system and an objective lens system. The eyepiece lens system magnifies, or compounds, the image reaching it after the image is magnified by the objective lens system. Such an instrument can magnify images of small objects up to about one thousand times.

PURPOSE OF THE EXERCISE

To become familiar with the major parts of a compound light microscope and their functions and to make use of the microscope to observe small objects.

LEARNING OBJECTIVES

After completing this exercise, you should be able to

1. Locate and identify the major parts of a compound light microscope.
2. Describe the functions of these parts.
3. Calculate the total magnification produced by various combinations of eyepiece and objective lenses.
4. Prepare a simple microscope slide.
5. Make proper use of the microscope to observe small objects.

PROCEDURE

1. Familiarize yourself with the following list of rules for care of the microscope:
   a. Keep the microscope under its *dustcover* and in a cabinet when it is not being used.
   b. Handle the microscope with great care. It is an expensive and delicate instrument. To move it or carry it, hold it by its *arm* with one hand and support its *base* with the other hand (fig. 3.1).
   c. To clean the lenses, rub them gently with *lens paper* or a high-quality cotton swab. If the lenses need additional cleaning, follow the directions in the "Lens-Cleaning Technique" section that follows.
   d. If the microscope has a substage lamp, be sure the electric cord does not hang off the laboratory table where someone might trip over it. The bulb life can be extended if the lamp is cool before the microscope is moved.
   e. Never drag the microscope across the laboratory table after you have placed it.
   f. Never remove parts of the microscope or try to disassemble the eyepiece or objective lenses.
   g. If your microscope is not functioning properly, report the problem to your laboratory instructor immediately.

Figure 3.1 Major parts of a compound light microscope with a monocular body and a mechanical stage. Some microscopes are equipped with a binocular body.

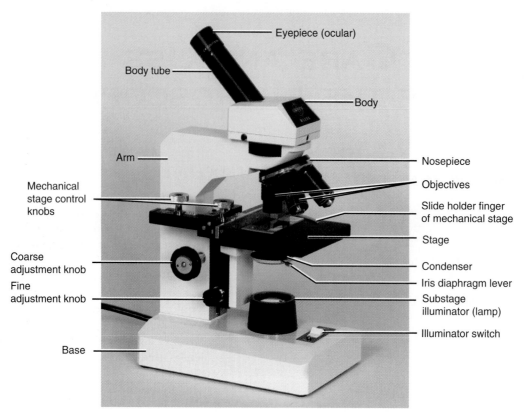

2. Observe a compound light microscope and study figure 3.1 to learn the names of its major parts. Note that the lens system of a compound microscope includes three parts—the condenser, objective lens, and eyepiece.

Light enters this system from a *substage illuminator (lamp)* or *mirror* and is concentrated and focused by a *condenser* onto a microscope slide or specimen placed on the *stage.* The condenser, which contains a set of lenses, usually is kept in its highest position possible.

The *iris diaphragm,* which is located between the light source and the condenser, can be used to increase or decrease the intensity of the light entering the condenser. Locate the lever that operates the iris diaphragm beneath the stage and move it back and forth. Note how this movement causes the size of the opening in the diaphragm to change. (Some microscopes have a revolving plate called a disc diaphragm beneath the stage instead of an iris diaphragm. Disc diaphragms have different-sized holes to admit varying amounts of light.) Which way do you move the diaphragm to increase the light intensity? _____ Which way to decrease it? _____

After light passes through a specimen mounted on a microscope slide, it enters an *objective lens system.* This lens projects the light upward into the *body tube,* where it produces a magnified image of the object being viewed.

The *eyepiece (ocular) lens* system then magnifies this image to produce another image, which is seen by the eye. Typically, the eyepiece lens magnifies the image ten times (10×). Look for the number in the metal of the eyepiece that indicates its power (fig. 3.2). What is the eyepiece power of your microscope? _____

The objective lenses are mounted in a revolving *nosepiece* so that different magnifications can be achieved by rotating any one of several objective lenses into position above the specimen. Commonly, this set of lenses includes a scanning objective (4×), a low-power objective (10×), and a high-power objective, also called a high-dry-power objective (about 40×). Sometimes an oil immersion objective (about 100×) is present. Look for the number printed on each objective that indicates its power. What are the objective lens powers of your microscope? _____

To calculate the *total magnification* achieved when using a particular objective, multiply the power of the eyepiece by the power of the objective used. Thus, the 10× eyepiece and the 40× objective produce a total magnification of 10 × 40, or 400×.

24

Figure 3.2 The powers of this 10× eyepiece (*a*) and this 40× objective (*b*) are marked in the metal. DIN is an international optical standard on quality optics. The 0.65 on the 40× objective is the numerical aperture, which is a measure of the light-gathering capabilities.

(a)

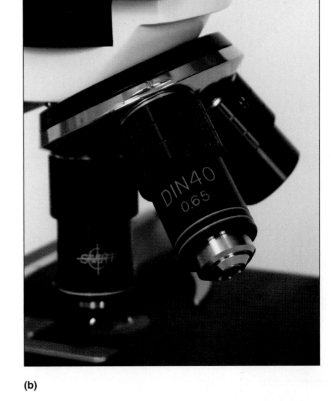

(b)

3. Complete Part A of Laboratory Report 3.
4. Turn on the substage illuminator and look through the eyepiece. You will see a lighted circular area called the *field of view.*

    You can measure the diameter of this field of view by focusing the lenses on the millimeter scale of a transparent plastic ruler. To do this, follow these steps:
    a. Place the ruler on the microscope stage in the spring clamp of a slide holder finger on a mechanical stage or under the stage (slide) clips. (*Note:* If your microscope is equipped with a mechanical stage, it may be necessary to use a short section cut from a transparent plastic ruler. The section should be several millimeters long and can be mounted on a microscope slide for viewing.)
    b. Center the millimeter scale in the beam of light coming up through the condenser and rotate the scanning objective into position.
    c. While you watch from the side to prevent the lens from touching anything, lower the objective until it is as close to the ruler as possible, using the *coarse adjustment knob* and then using the *fine adjustment knob* (fig. 3.3). (*Note:* The adjustment knobs on some microscopes move the stage upward and downward for focusing.)

    d. Look into the eyepiece and use the coarse adjustment knob to raise the objective lens until the lines of the millimeter scale come into sharp focus.
    e. Adjust the light intensity by moving the *iris diaphragm lever* so that the field of view is brightly illuminated but comfortable to your eye. At the same time, take care not to overilluminate the field because transparent objects tend to disappear in very bright light.
    f. Position the millimeter ruler so that its scale crosses the greatest diameter of the field of view. Also, move the ruler so that one of the millimeter marks is against the edge of the field of view.
    g. Measure the distance across the field of view in millimeters.
5. Complete Part B of the laboratory report.
6. Most microscopes are designed to be *parfocal.* This means that when a specimen is in focus with a lower-power objective, it will be in focus (or nearly so) when a higher-power objective is rotated into position. Always center the specimen in the field of view before changing to higher objectives.

    Rotate the low-power objective into position, and then look at the millimeter scale of the transparent plastic ruler. If you need to move the low-power objective to sharpen the focus, use the *fine adjustment knob.*

Figure 3.3 When you focus using a particular objective, you can prevent it from touching the specimen by watching from the side.

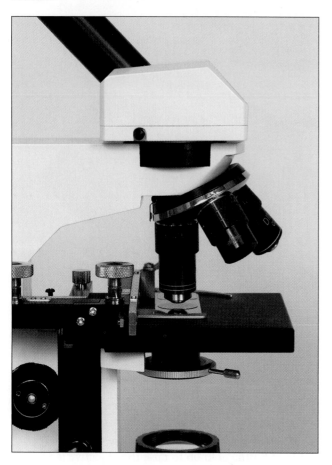

Adjust the iris diaphragm so that the field of view is properly illuminated. Once again, adjust the millimeter ruler so that the scale crosses the field of view through its greater diameter, and position the ruler so that a millimeter mark is against one edge of the field.

Try to measure the distance across the field of view in millimeters.

7. Rotate the high-power objective into position while you watch from the side, and then observe the millimeter scale on the plastic ruler. All focusing using high-power magnification should be done only with the fine adjustment knob. If you use the coarse adjustment knob with the high-power objective, you can accidently force the objective into the coverslip. This is because the *working distance* (the distance from the objective lens to the slide on the stage) is much shorter when using higher magnifications.

Adjust the iris diaphragm for proper illumination. Usually more illumination when using higher magnifications will help you to view the objects more clearly. Try to measure the distance across the field of view in millimeters.

8. Locate the numeral 4 (or 9) on the plastic ruler and focus on it using the scanning objective. Note how the number appears in the field of view. Move the plastic ruler to the right and note which way the image moves. Slide the ruler away from you and again note how the image moves.

9. Examine the slide of the three colored threads using the low-power objective and then the high-power objective. Focus on the location where the three threads cross. By using the fine adjustment knob, determine the order from top to bottom by noting which color is in focus at different depths. The other colored threads will still be visible, but they will be blurred. Be sure to notice whether the stage or the body tube moves up and down with the adjustment knobs of the microscope that is being used for this depth determination. The vertical depth of the specimen that is clearly in focus is called the *depth of field* (*focus*). Whenever specimens are examined, continue to use the fine adjustment focusing knob to determine relative depths of structures that are clearly in focus within cells, giving a three-dimensional perspective. It should be noted that the depth of field is less at higher magnifications.

Critical Thinking Application

What was the sequence of the three colored threads from top to bottom?

_____

_____

_____

10. Complete Parts C and D of the laboratory report.

DEMONSTRATION

A compound light microscope is sometimes equipped with a micrometer scale mounted in the eyepiece. Such a scale is subdivided into fifty to one hundred equal divisions (fig. 3.4). These arbitrary divisions can be calibrated against the known divisions of a micrometer slide placed on the microscope stage. Once the values of the divisions are known, the length and width of a microscopic object can be measured by superimposing the scale over the magnified image of the object.

Observe the micrometer scale in the eyepiece of the demonstration microscope. Focus the low-power objective on the millimeter scale of a micrometer slide (or a plastic ruler) and measure the distance between the divisions on the micrometer scale in the eyepiece. What is the distance between the finest divisions of the scale in micrometers? _____

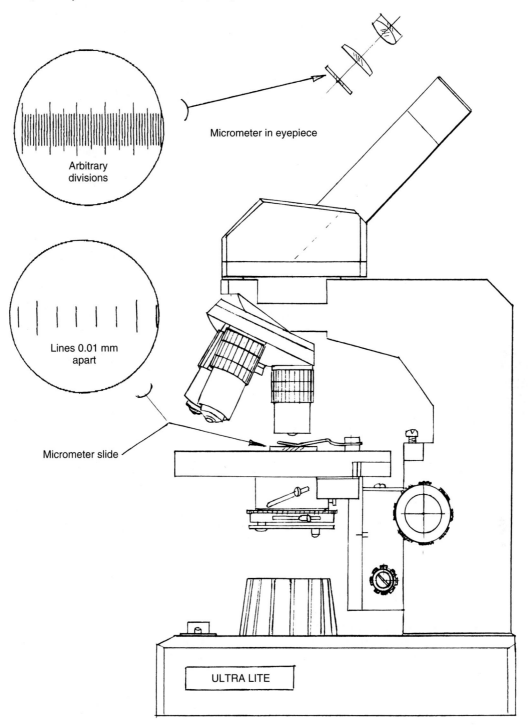

Figure 3.4 The divisions of a micrometer scale in an eyepiece can be calibrated against the known divisions of a micrometer slide. (Courtesy of Swift Instruments, Inc., San Jose, California)

Micrometer in eyepiece

Arbitrary divisions

Lines 0.01 mm apart

Micrometer slide

ULTRA LITE

11. Prepare several temporary *wet mounts* using any small, transparent objects of interest, and examine the specimens using the low-power objective and then a high-power objective to observe their details. To prepare a wet mount, follow these steps (fig. 3.5):
   a. Obtain a precleaned microscope slide.
   b. Place a tiny, thin piece of the specimen you want to observe in the center of the slide, and use a medicine dropper to put a drop of water over it. Consult with your instructor if a drop of stain might enhance the image of any cellular structures of your specimen. If the specimen is solid, you might want to tease some of it apart with dissecting needles. In any case, the specimen must be thin enough so that light can

Figure 3.5 Steps in the preparation of a wet mount.

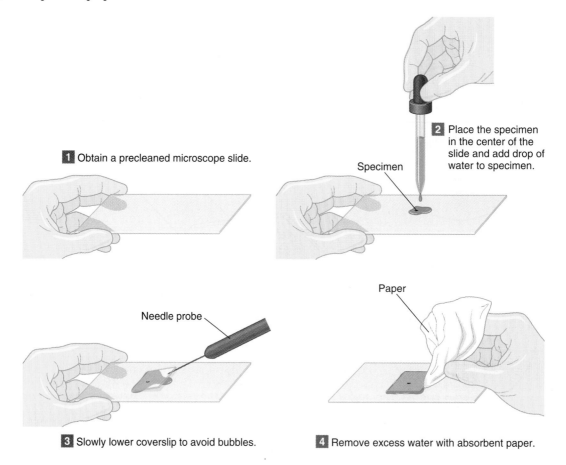

1 Obtain a precleaned microscope slide.

2 Place the specimen in the center of the slide and add drop of water to specimen.

Specimen

Needle probe

Paper

3 Slowly lower coverslip to avoid bubbles.

4 Remove excess water with absorbent paper.

pass through it. Why is it necessary for the specimen to be so thin?

_____

_____

c. Cover the specimen with a coverslip. Try to avoid trapping bubbles of air beneath the coverslip by slowly lowering it at an angle into the drop of water.

d. Remove any excess water from the edge of the coverslip with absorbent paper. If your microscope has an inclination joint, do not tilt the microscope while observing wet mounts because the fluid will flow.

e. Place the slide under the stage (slide) clips or in the slide holder on a mechanical stage, and position the slide so that the specimen is centered in the light beam passing up through the condenser.

f. Focus on the specimen using the scanning objective first. Next focus using the low-power objective, and then examine it with the high-power objective.

12. If an oil immersion objective is available, use it to examine the specimen. To use the oil immersion objective, follow these steps:

a. Center the object you want to study under the high-power field of view.

b. Rotate the high-power objective away from the microscope slide, place a small drop of immersion oil on the coverslip, and swing the oil immersion objective into position. To achieve sharp focus, use the fine adjustment knob only.

c. You will need to open the iris diaphragm more fully for proper illumination. More light is needed because the oil immersion objective covers a very small lighted area of the microscope slide.

d. Because the oil immersion objective must be very close to the coverslip to achieve sharp focus, care must be taken to avoid breaking the coverslip or damaging the objective lens. For this reason, never lower the objective when you are looking into the eyepiece. Instead, always raise the objective to achieve focus, or prevent the objective from touching the coverslip by watching the microscope slide and coverslip from the side if the objective needs to be lowered.

13. When you have finished working with the microscope, remove the microscope slide from the

Figure 3.6 A stereomicroscope, which is also called a dissecting microscope.

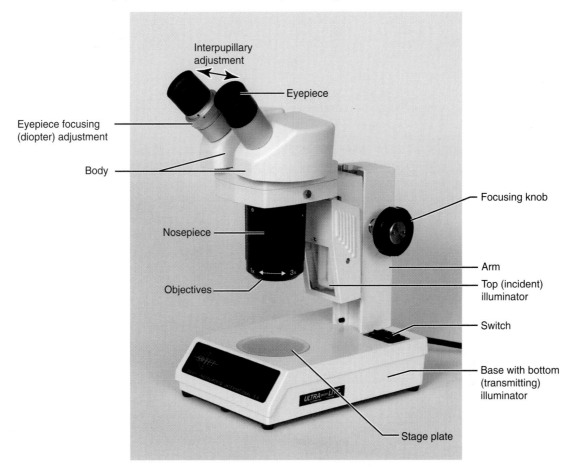

stage and wipe any oil from the objective lens with lens paper or a high-quality cotton swab. Swing the scanning objective or the low-power objective into position. Wrap the electric cord around the base of the microscope and replace the dustcover.

14. Complete Part E of the laboratory report.

DEMONSTRATION

A stereomicroscope (dissecting microscope) (fig. 3.6) is useful for observing the details of relatively large, opaque specimens. Although this type of microscope achieves less magnification than a compound microscope, it has the advantage of producing a three-dimensional image rather than the flat, two-dimensional image of the compound light microscope. In addition, the image produced by the stereomicroscope is positioned in the same manner as the specimen, rather than being reversed and inverted as it is by the compound light microscope.

Observe the stereomicroscope. Note that the eyepieces can be pushed apart or together to fit the distance between your eyes. Focus the microscope on the end of your finger. Which way does the image move when you move your finger to the right? _____

When you move it away? _____

If the instrument has more than one objective, change the magnification to higher power. Use the instrument to examine various small, opaque objects available in the laboratory.

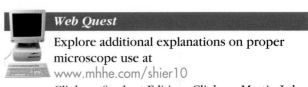

Web Quest

Explore additional explanations on proper microscope use at
www.mhhe.com/shier10

Click on Student Edition. Click on Martin Lab Manual, Web Quest.

CARE AND USE OF THE MICROSCOPE

PART A

Complete the following:

1. What total magnification will be achieved if the 10× eyepiece and the 10× objective are used? _____

2. What total magnification will be achieved if the 10× eyepiece and the 100× objective are used? _____

PART B

Complete the following:

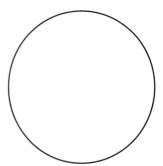

1. Sketch the millimeter scale as it appears under the scanning objective magnification. (The circle represents the field of view through the microscope.)

2. What is the diameter of the scanning field of view in millimeters? _____

3. Microscopic objects often are measured in *micrometers.* A micrometer equals 1/1,000 of a millimeter and is symbolized by μm. What is the diameter of the scanning power field of view in micrometers? _____

4. If a circular object or specimen extends halfway across the scanning field, what is its diameter in millimeters? _____

5. What is its diameter in micrometers? _____

PART C

Complete the following:

1. Sketch the millimeter scale as it appears using the low-power objective.

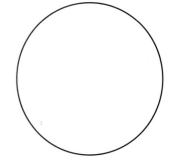

2. What do you estimate the diameter of this field of view to be in millimeters? _____

3. How does the diameter of the scanning power field of view compare with that of the low-power field?

_____

4. Why is it more difficult to measure the diameter of the high-power field of view than the low-power field?

_____

_____

5. What change occurred in the light intensity of the field of view when you exchanged the low-power objective for the high-power objective? _____

6. Sketch the numeral 4 (or 9) as it appears through the scanning objective of the compound microscope.

7. What has the lens system done to the image of the numeral? (Is it right side up, upside down, or what?) _____

8. When you moved the ruler to the right, which way did the image move? _____

9. When you moved the ruler away from you, which way did the image move? _____

PART D

Match the names of the microscope parts in column A with the descriptions in column B. Place the letter of your choice in the space provided.

Column A	Column B
a. Adjustment knob	_____ 1. Increases or decreases the light intensity
b. Arm	
c. Condenser	_____ 2. Platform that supports a microscope slide
d. Eyepiece (ocular)	_____ 3. Concentrates light onto the specimen
e. Field of view	
f. Iris diaphragm	_____ 4. Causes objective lens (or stage) to move upward or downward
g. Nosepiece	
h. Objective lens system	_____ 5. After light passes through the specimen, it next enters this lens system
i. Stage	
j. Stage (slide) clip	_____ 6. Holds a microscope slide in position
	_____ 7. Contains a lens at the top of the body tube
	_____ 8. Serves as a handle for carrying the microscope
	_____ 9. Part to which the objective lenses are attached
	_____ 10. Circular area seen through the eyepiece

PART E

Prepare sketches of the objects you observed using the microscope. For each sketch, include the name of the object, the magnification you used to observe it, and its estimated dimensions in millimeters and micrometers.

CELL STRUCTURE AND FUNCTION

SAFETY

- Review all the safety guidelines inside the front cover.
- Clean laboratory surfaces before and after laboratory procedures.
- Wear disposable gloves for the wet-mount procedures of the cells lining the inside of the cheek.
- Work only with your own materials when preparing the slide of cheek cells. Observe the same precautions as with all body fluids.
- Dispose of laboratory gloves, slides, coverslips, and toothpicks as instructed.
- Precautions should be taken to avoid cellular stains from contacting your clothes and skin.
- Wash your hands before leaving the laboratory.

Cells are the "building blocks" from which all parts of the human body are formed. They account for the shape, organization, and construction of the body and are responsible for carrying on its life processes. Under the light microscope, with a properly applied stain to make structures visible, the **cell (plasma) membrane,** the **cytoplasm,** and a **nucleus** are easily seen. The cytoplasm is composed of a clear fluid, the *cytosol,* and numerous *cytoplasmic organelles* suspended in the cytosol.

PURPOSE OF THE EXERCISE

To review the structure and functions of major cellular components and to observe examples of human cells.

LEARNING OBJECTIVES

After completing this exercise, you should be able to

1. Name and locate the components of a cell on a model or diagram and describe the general functions of these components.
2. Prepare a wet mount of cells lining the inside of the cheek, stain the cells, and identify the major components of these cells.
3. Locate cells on prepared slides of human tissues and identify their major components.
4. Identify major cellular components in a transmission electron micrograph.

PROCEDURE

1. Review the section entitled "A Composite Cell" in chapter 3 of the textbook.
2. Observe the animal cell model and identify its structures.
3. As a review activity, label figure 4.1 and study figure 4.2.
4. Complete Part A of Laboratory Report 4.
5. Prepare a wet mount of cells lining the inside of the cheek. To do this, follow these steps:
   a. Gently scrape (force is not necessary and should be avoided) the inner lining of your cheek with the broad end of a flat toothpick.
   b. Stir the toothpick in a drop of water on a clean microscope slide and dispose of the toothpick as directed by your instructor.
   c. Cover the drop with a coverslip.
   d. Observe the cheek cells by using the microscope. Compare your image with figure 4.3. To report what you observe, sketch a single cell in the space provided in Part B of the laboratory report.
6. Prepare a second wet mount of cheek cells, but this time, add a drop of dilute methylene blue or iodine-potassium-iodide stain to the cells. Cover the liquid with a coverslip and observe the cells with the microscope. Add to your sketch any additional structures you observe in the stained cells.
7. Answer the questions in Part B of the laboratory report.

Figure 4.1 Label the structures of this composite cell.

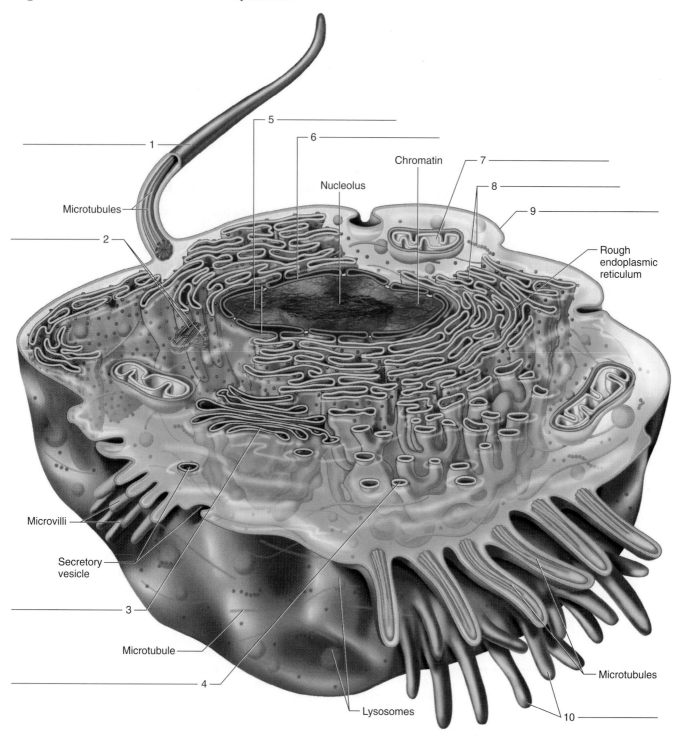

Microtubules

Nucleolus

Chromatin

Rough endoplasmic reticulum

Microvilli

Secretory vesicle

Microtubule

Lysosomes

Microtubules

8. Using the microscope, observe each of the prepared slides of human tissues. To report what you observe, sketch a single cell of each type in the space provided in Part C of the laboratory report.

9. Answer the questions in Part C of the laboratory report.

10. Study figure 4.4 *a–c.* These electron micrographs represent extremely thin slices of cells. Note that each micrograph contains a section of a nucleus and some cytoplasm. Compare the organelles shown in these micrographs with the organelles included in the electron micrographs in chapter 3 of the textbook.

11. Complete Parts D and E of the laboratory report.

Figure 4.2 The structures of the cell membrane.

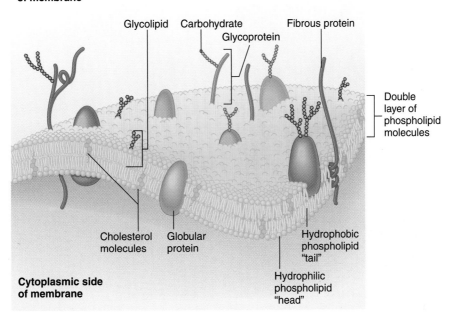

Extracellular side
of membrane

Glycolipid Carbohydrate Fibrous protein

Glycoprotein

Double
layer of
phospholipid
molecules

Cholesterol
molecules

Globular
protein

Hydrophobic
phospholipid
"tail"

Cytoplasmic side
of membrane

Hydrophilic
phospholipid
"head"

Figure 4.3 Stained cell lining the inside of the cheek as viewed through the compound light microscope using the high-power objective (400×).

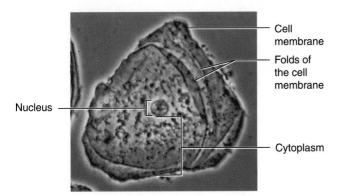

Cell
membrane

Folds of
the cell
membrane

Nucleus

Cytoplasm

Critical Thinking Application

The cells lining the inside of the cheek are frequently removed for making observations of basic cell structure. The cells are from stratified squamous epithelium. Explain why these cells are used instead of outer body surface tissue. Why was the removal of inside cheek cells painless and lacking in any blood loss?

_____

_____

_____

_____

_____

_____

_____

Figure 4.4 Transmission electron micrographs of cellular components. All of the views are only portions of a cell. Magnifications: (*a*) 26,000×; (*b*)10,000×; (*c*) 30,000×. Identify the numbered cellular structures, using the terms provided.

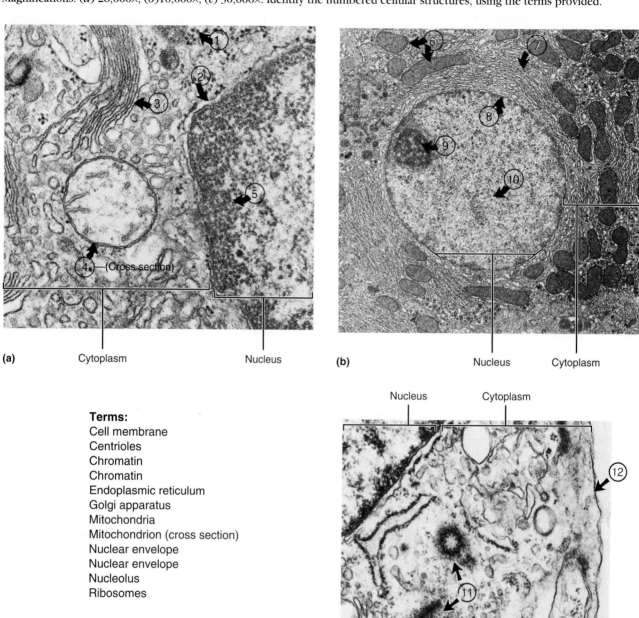

(a)   Cytoplasm   Nucleus

(b)   Nucleus   Cytoplasm

(Cross section)

Nucleus   Cytoplasm

(c)

Terms:
Cell membrane
Centrioles
Chromatin
Chromatin
Endoplasmic reticulum
Golgi apparatus
Mitochondria
Mitochondrion (cross section)
Nuclear envelope
Nuclear envelope
Nucleolus
Ribosomes

OPTIONAL ACTIVITY

Investigate the microscopic structure of various plant materials. To do this, prepare very tiny, thin slices of plant specimens, using a single-edged razor blade. (*Take care not to injure yourself with the blade.*) Keep the slices in a container of water until you are ready to observe them. To observe a specimen, place it into a drop of water on a clean microscope slide and cover it with a coverslip. Use the microscope and view the specimen using low- and high-power magnifications. Observe near the edges where your section of tissue is most likely to be one cell thick. Add a drop of dilute methylene blue or iodine-potassium-iodide stain, and note if any additional structures become visible. How are the microscopic structures of the plant specimens similar to the human tissues you observed?

_____

How are they different?

_____

OPTIONAL ACTIVITY

Prepare a wet mount of the *Amoeba* and *Paramecium* by putting a drop of culture on a clean glass slide. Gently cover with a clean coverslip. Observe the movements of the *Amoeba* with pseudopodia and the *Paramecium* with cilia. Try to locate cellular components such as the cell membrane, nuclear envelope, nucleus, mitochondria, and contractile vacuoles. Describe the movement of the *Amoeba*.

_____

_____

Describe the movement of the *Paramecium*.

_____

_____

Web Quest

Review cell structure and function at
www.mhhe.com/shier10
Click on Student Edition. Click on Martin Lab Manual, Web Quest.

CELL STRUCTURE AND FUNCTION

PART A

Match the cellular components in column A with the descriptions in column B. Place the letter of your choice in the space provided.

Column A	Column B
a. Cell (plasma) membrane	_____ 1. Loosely coiled fibers containing protein and DNA within nucleus
b. Centrosome	
c. Chromatin	_____ 2. Cellular product, such as a pigment melanin in skin
d. Cilia	_____ 3. Location of the energy release from food molecules
e. Cytoplasm	
f. Endoplasmic reticulum	_____ 4. Nonmembranous structure that contains the centrioles
g. Golgi apparatus	_____ 5. Small RNA-containing particles for the synthesis of proteins
h. Inclusion	
i. Lysosome	_____ 6. Membranous sac formed by the pinching off of pieces of cell membrane
j. Microfilament	
k. Microtubule	_____ 7. Dense body of RNA within the nucleus
l. Mitochondrion	_____ 8. Slender tubes that provide movement in cilia and flagella
m. Nuclear envelope	
n. Nucleolus	_____ 9. Organelles composed of membrane-bound sacs, canals, and vesicles
o. Nucleus	
p. Peroxisome	_____ 10. Outside boundary of the cell
q. Ribosome	_____ 11. Occupies space between cell membrane and nucleus
r. Vesicle/vacuole	_____ 12. Flattened membranous sacs that package a secretion
	_____ 13. Motile processes that are numerous and short and are associated with some cells
	_____ 14. Tiny rods in meshworks or bundles that help cell to shorten
	_____ 15. Membranous sac that contains digestive enzymes
	_____ 16. Contains enzymes that decompose hydrogen peroxide
	_____ 17. Separates nuclear contents from cytoplasm
	_____ 18. Spherical organelle that contains chromatin and nucleolus

PART B

Complete the following:

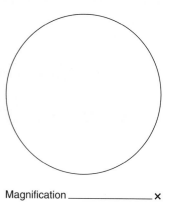

1. Sketch a single cheek cell that has been stained. Label the cellular components you recognize. (The circle represents the field of view through the microscope.)

Magnification _____ ✕

2. After comparing the wet mount and the stained cheek cells, state the advantage that was gained by staining cells.

_____

_____

3. Are the stained cheek cells nearly the same size and shape? _____ Propose an explanation for your answer. _____

_____

PART C

Complete the following:

1. Sketch a single cell of each kind you observed in the prepared slides of human tissues. Name the tissue and label the cellular components you recognize.

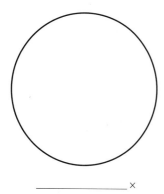

_____ ✕

Tissue _____

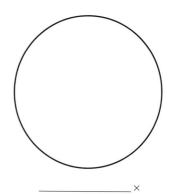

_____ ✕

Tissue _____

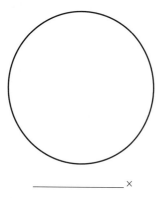

_____ ✕

Tissue _____

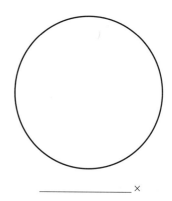

_____ ✕

Tissue _____

40

2. What do the various kinds of cells in these tissues have in common? _____

_____

3. What are the main differences you observed among these cells? _____

_____

PART D

Identify the structures indicated by the arrows in figure 4.4.

1. _____   7. _____

2. _____   8. _____

3. _____   9. _____

4. _____   10. _____

5. _____   11. _____

6. _____   12. _____

PART E

Answer the following questions after observing the transmission electron micrographs in figure 4.4.

1. What cellular structures were visible in the transmission electron micrographs that were not apparent in the cells you observed using the microscope? _____

_____

2. Before they can be observed by using a transmission electron microscope, cells are sliced into very thin sections. What disadvantage does this procedure present in the study of cellular parts? _____

_____

MOVEMENTS THROUGH CELL MEMBRANES

MATERIALS NEEDED

For Procedure A—Diffusion:

Textbook
Petri dish
White paper
Forceps
Potassium permanganate crystals
Millimeter ruler (thin and transparent)

For Procedure B—Osmosis:

Textbook
Thistle tube
Molasses (or Karo dark corn syrup)
Selectively permeable (semipermeable) membrane
    (presoaked dialysis tubing of 1 5/16″ or greater
    diameter)
Ring stand and clamp
Beaker
Rubber band
Millimeter ruler

*For Procedure C—Hypertonic, Hypotonic,
and Isotonic Solutions:*

Textbook
Test tubes
Marking pen
Test-tube rack
10-mL graduated cylinder
Medicine dropper
Uncoagulated animal blood
Distilled water
0.9% NaCl (aqueous solution)
3.0% NaCl (aqueous solution)
Clean microscope slides
Coverslips
Compound light microscope

For Procedure D—Filtration:

Textbook
Glass funnel
Filter paper
Ring stand and ring
Beaker
Powdered charcoal
1% glucose (aqueous solution)
1% starch (aqueous solution)
Test tubes
10-mL graduated cylinder
Water bath (boiling water)
Benedict's solution

Iodine-potassium-iodide solution
Medicine dropper

For Alternative Osmosis Activity:

Fresh chicken egg
Beaker
Laboratory balance
Spoon
Vinegar
Corn syrup (Karo)

SAFETY

- Clean laboratory surfaces before and after laboratory procedures.
- Wear disposable gloves when handling chemicals and animal blood.
- Wear safety glasses when using chemicals.
- Dispose of laboratory gloves and blood-contaminated items as instructed.
- Wash your hands before leaving the laboratory.

A cell membrane functions as a gateway through which chemical substances and small particles may enter or leave a cell. These substances move through the membrane by physical processes such as diffusion, osmosis, and filtration, or by physiological processes such as active transport, phagocytosis, or pinocytosis.

PURPOSE OF THE EXERCISE

To demonstrate some of the physical processes by which substances move through cell membranes.

LEARNING OBJECTIVES

After completing this exercise, you should be able to

1. Define *diffusion* and identify examples of diffusion.
2. Define *osmosis* and identify examples of osmosis.
3. Distinguish among hypertonic, hypotonic, and isotonic solutions and observe the effects of these solutions on animal cells.
4. Define *filtration* and identify examples of filtration.

Figure 5.1 To demonstrate diffusion, place one crystal of potassium permanganate in the center of a petri dish containing water. Place the crystal near the millimeter ruler (positioned under the petri dish).

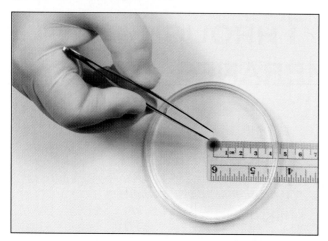

PROCEDURE A—DIFFUSION

1. Review the section entitled "Diffusion" in chapter 3 of the textbook.
2. To demonstrate diffusion, follow these steps:
   a. Place a petri dish half filled with water on a piece of white paper that has a millimeter ruler positioned on the paper. Wait until the water surface is still. Allow approximately 3 minutes.
   b. Using forceps, place one crystal of potassium permanganate near the center of the petri dish and near the millimeter ruler (fig. 5.1).
   c. Measure the radius of the purple circle at 1-minute intervals for 10 minutes.
3. Complete Part A of Laboratory Report 5.

OPTIONAL ACTIVITY

Repeat the demonstration of diffusion using a petri dish filled with ice-cold water and a second dish filled with very hot water. At the same moment, add a crystal of potassium permanganate to each dish and observe the circle as before. What difference do you note in the rate of diffusion in the two dishes? How do you explain this difference?

_____

_____

_____

PROCEDURE B—OSMOSIS

1. Review the section entitled "Osmosis" in chapter 3 of the textbook.
2. To demonstrate osmosis, refer to figure 5.2 as you follow these steps:
   a. One person plugs the tube end of a thistle tube with a finger.

   b. Another person then fills the bulb with molasses until it is about to overflow at the top of the bulb. Note that air remains trapped in the stem.
   c. Cover the bulb opening with a single-thickness piece of moist selectively permeable (semipermeable) membrane. Dialysis tubing that has been soaked for 30 minutes can easily be cut open because it becomes pliable.
   d. Tightly secure the membrane in place with several wrappings of a rubber band.
   e. Immerse the bulb end of the tube in a beaker of water. If leaks are noted, repeat the procedures.
   f. Support the upright portion of the tube with a clamp on a ring stand. Folded paper under the clamp will protect the thistle tube stem from breakage.
   g. Mark the meniscus level of the molasses in the tube. *Note:* The best results will occur if the mark of the molasses is a short distance up the stem of the thistle tube when the experiment starts.
   h. Measure the level changes after 10 minutes and 30 minutes.
3. Complete Part B of the laboratory report.

ALTERNATIVE PROCEDURE

Eggshell membranes possess selectively permeable properties. To demonstrate osmosis using a natural membrane, soak a fresh chicken egg in vinegar for about 24 hours to remove the shell. Use a spoon to carefully handle the delicate egg. Place the egg in a hypertonic solution (corn syrup) for about 24 hours. Remove the egg, rinse it, and using a laboratory balance, weigh the egg to establish a baseline weight. Place the egg in a hypotonic solution (distilled water). Remove the egg and weigh it every 15 minutes for an elapsed time of 75 minutes. Explain any weight changes that were noted during this experiment.

_____

_____

_____

_____

_____

_____

PROCEDURE C—HYPERTONIC, HYPOTONIC, AND ISOTONIC SOLUTIONS

1. Review the section entitled "Osmosis" in chapter 3 of the textbook.
2. To demonstrate the effect of hypertonic, hypotonic, and isotonic solutions on animal cells, follow these steps:
   a. Place three test tubes in a rack and mark them as *tube 1, tube 2,* and *tube 3.* (*Note:* One set of tubes can be used to supply test samples for the entire class.)
   b. Using 10-mL graduated cylinders, add 3 mL of distilled water to tube 1; add 3 mL of 0.9% NaCl to tube 2; and add 3 mL of 3.0% NaCl to tube 3.

Figure 5.2 (*a*) Fill the bulb of the thistle tube with molasses; (*b*) tightly secure a piece of selectively permeable (semipermeable) membrane over the bulb opening; and (*c*) immerse the bulb in a beaker of water. *Note:* These procedures require the participation of two people.

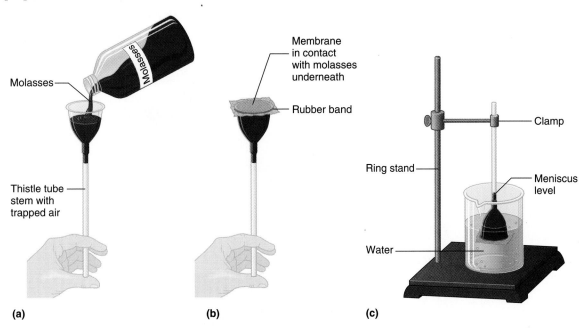

c. Place three drops of fresh uncoagulated animal blood into each of the tubes and gently mix the blood with the solutions. Wait 5 minutes.

d. Using three separate medicine droppers, remove a drop from each tube and place the drops on three separate microscope slides marked *1, 2,* and *3.*

e. Cover the drops with coverslips and observe the blood cells, using the high power of the microscope.

3. Complete Part C of the laboratory report.

ALTERNATIVE PROCEDURE

Various substitutes for blood can be used for Procedure C. Onion, cucumber, or cells lining the inside of the cheek represent three possible options.

PROCEDURE D—FILTRATION

1. Review the section entitled "Filtration" in chapter 3 of the textbook.

2. To demonstrate filtration, follow these steps:

a. Place a glass funnel in the ring of a ring stand over an empty beaker. Fold a piece of filter paper in half and then in half again. Open one thickness of the filter paper to form a cone, wet the cone, and place it in the funnel. The filter paper is used to demonstrate how movement

across membranes is limited by the size of the molecules, but it does not represent a working model of biological membranes.

b. Prepare a mixture of 5 cc (approximately 1 teaspoon) powdered charcoal and equal amounts of 1% glucose solution and 1% starch solution in a beaker. Pour some of the mixture into the funnel until it nearly reaches the top of the filter-paper cone. Care should be taken to prevent the mixture from spilling over the top of the filter paper. Collect the filtrate in the beaker below the funnel (fig. 5.3).

c. Test some of the filtrate in the beaker for the presence of glucose. To do this, place 1 mL of filtrate in a clean test tube and add 1 mL of Benedict's solution. Place the test tube in a water bath of boiling water for 2 minutes and then allow the liquid to cool slowly. If the color of the solution changes to green, yellow, or red, glucose is present (fig. 5.4).

d. Test some of the filtrate in the beaker for the presence of starch. To do this, place a few drops of filtrate in a test tube and add one drop of iodine-potassium-iodide solution. If the color of the solution changes to blue-black, starch is present.

e. Observe any charcoal in the filtrate.

3. Complete Part D of the laboratory report.

Figure 5.3 Apparatus used to illustrate filtration.

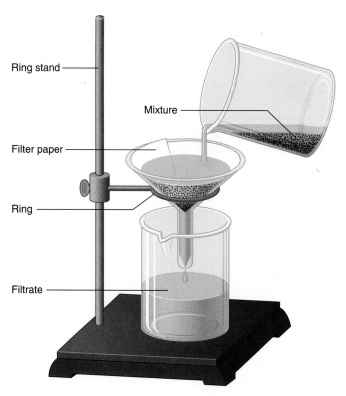

Ring stand

Mixture

Filter paper

Ring

Filtrate

Figure 5.4 Heat the filtrate and Benedict's solution in a boiling water bath for 2 minutes.

Water bath

Filtrate and Benedict's solution

Hot plate

Name _____

Date _____

Section _____

MOVEMENTS THROUGH CELL MEMBRANES

PART A

Complete the following:

1. Enter data for changes in the movement of the potassium permanganate.

Elapsed Time	Radius of Purple Circle in Millimeters
Initial	_____
1 minute	_____
2 minutes	_____
3 minutes	_____
4 minutes	_____
5 minutes	_____
6 minutes	_____
7 minutes	_____
8 minutes	_____
9 minutes	_____
10 minutes	_____

2. Prepare a graph that illustrates the diffusion distance of potassium permanganate in 10 minutes.

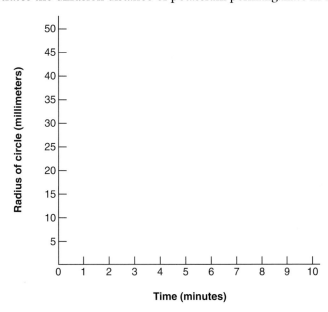

3. Explain your graph. _____

_____

4. Briefly define *diffusion.* _____

_____

Critical Thinking Application

By answering yes or no, indicate which of the following provides an example of diffusion.

1. A perfume bottle is opened, and soon the odor can be sensed in all parts of the room. _____
2. A sugar cube is dropped into a cup of hot water, and, without being stirred, all of the liquid becomes sweet tasting. _____
3. Water molecules move from a faucet through a garden hose when the faucet is turned on. _____
4. A person blows air molecules into a balloon by exhaling forcefully. _____
5. A crystal of blue copper sulfate is placed in a test tube of water. The next day, the solid is gone, but the water is evenly colored. _____

PART B

Complete the following:

1. What was the change in the level of molasses in 10 minutes? _____
2. What was the change in the level of molasses in 30 minutes? _____
3. How do you explain this change? _____

_____

4. Briefly define *osmosis.* _____

_____

Critical Thinking Application

By answering yes or no, indicate which of the following involves osmosis.

1. A fresh potato is peeled, weighed, and soaked in a strong salt solution. The next day, it is discovered that the potato has lost weight. _____
2. Garden grass wilts after being exposed to dry chemical fertilizer. _____
3. Air molecules escape from a punctured tire as a result of high pressure inside. _____
4. Plant seeds soaked in water swell and become several times as large as before soaking. _____
5. When the bulb of a thistle tube filled with water is sealed by a selectively permeable membrane and submerged in a beaker of molasses, the water level in the tube falls. _____

48

PART C

Complete the following:

1. In the spaces, sketch a few blood cells from each of the test tubes, and indicate the magnification.

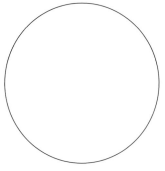

Tube 1
(distilled water)

_____ ×

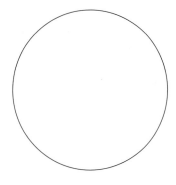

Tube 2
(0.9% NaCl)

_____ ×

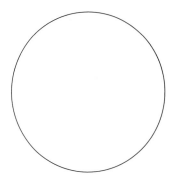

Tube 3
(3.0% NaCl)

_____ ×

2. Based on your results, which tube contained a solution that was hypertonic to the blood cells? _____
   Give the reason for your answer. _____
   _____

3. Which tube contained a solution that was hypotonic to the blood cells? _____
   Give the reason for your answer. _____
   _____

4. Which tube contained a solution that was isotonic to the blood cells? _____
   Give the reason for your answer. _____
   _____

PART D

Complete the following:

1. Which of the substances in the mixture you prepared passed through the filter paper into the filtrate? _____
   _____

2. What evidence do you have for your answer to question 1?_____
   _____
   _____

3. What force was responsible for the movement of substances through the filter paper? _____
   _____

4. What substances did not pass through the filter paper? _____
   _____

5. What factor prevented these substances from passing through? _____
   _____
   _____

6. Briefly define *filtration.* _____
   _____
   _____

Critical Thinking Application

By answering yes or no, indicate which of the following involves filtration.

1. Oxygen molecules move into a cell and carbon dioxide molecules leave a cell because of differences in the concentrations of these substances on either side of the cell membrane. _____

2. Blood pressure forces water molecules from the blood outward through the thin wall of a blood capillary. _____

3. Urine is forced from the urinary bladder through the tubular urethra by muscular contractions. _____

4. Air molecules enter the lungs through the airways when air pressure is greater outside these organs than inside. _____

5. Coffee is made using a coffeemaker (not instant). _____

LABORATORY EXERCISE 6

THE CELL CYCLE

The cell cycle consists of the series of changes a cell undergoes from the time it is formed until it divides. Typically, a newly formed cell grows to a certain size and then divides to form two new cells (*daughter cells*). This cell division process involves two major steps: (1) division of the cell's nuclear parts, which is called *mitosis* (karyokinesis) and (2) division of the cell's cytoplasm (*cytokinesis*). Before the cell divides, it must synthesize biochemicals and other contents. This period of preparation is called *interphase.* The extensive period of interphase is divided into three phases. The S phase, when DNA synthesis occurs, is between two gap phases (G_1 and G_2), when cell growth occurs and cytoplasmic organelles duplicate.

PURPOSE OF THE EXERCISE

To review the stages in the cell cycle and to observe cells in various stages of their life cycles.

LEARNING OBJECTIVES

After completing this exercise, you should be able to

1. Describe the cell cycle.
2. Identify the stages in the life cycle of a particular cell.
3. Arrange into a correct sequence a set of models or drawings of cells in various stages of their life cycles.

PROCEDURE

1. Review the section entitled "The Cell Cycle" in chapter 3 of the textbook.

2. As a review activity, study the various stages of the cell's life cycle represented in figure 6.1 and label the structures indicated in figure 6.2.
3. Observe the animal mitosis models and review the major events in a cell's life cycle represented by each of them. Be sure you can arrange these models in correct sequence if their positions are changed. The acronym IPMAT can help you arrange the correct order of phases in the cell cycle. This includes interphase followed by the four phases of mitosis. Cytokinesis overlaps anaphase and telophase.
4. Complete Part A of Laboratory Report 6.
5. Obtain a slide of the whitefish mitosis (blastula).
   a. Examine the slide using the high-power objective of a microscope. The tissue on this slide was obtained from a developing embryo (blastula) of a fish, and many of the embryonic cells are undergoing mitosis. Note that the chromosomes of these dividing cells are darkly stained (fig. 6.3).
   b. Search the tissue for cells in various stages of cell division. Note that there are several sections on the slide. If you cannot locate different stages in one section, examine the cells of another section since the stages occur at random.
   c. Each time you locate a cell in a different stage, sketch it in an appropriate circle in Part B of the laboratory report.

 Critical Thinking Application

Which stage (phase) of the cell cycle was the most numerous in the blastula?_____

Explain your answer._____
_____
_____

6. Complete Parts C, D, and E of the laboratory report.

Figure 6.1 The cell cycle: interphase, mitosis, and cytokinesis.

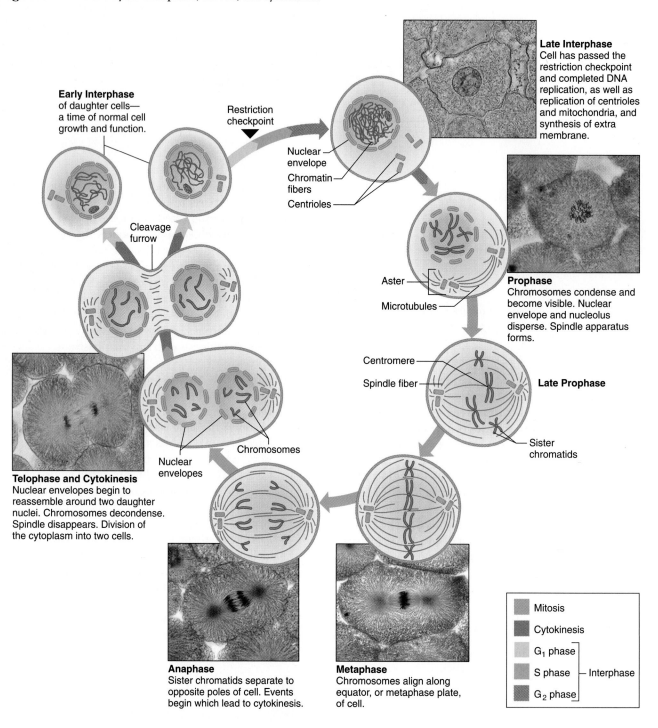

Late Interphase
Cell has passed the restriction checkpoint and completed DNA replication, as well as replication of centrioles and mitochondria, and synthesis of extra membrane.

Early Interphase
of daughter cells—a time of normal cell growth and function.

Restriction checkpoint

Nuclear envelope
Chromatin fibers
Centrioles

Prophase
Chromosomes condense and become visible. Nuclear envelope and nucleolus disperse. Spindle apparatus forms.

Cleavage furrow

Aster
Microtubules

Centromere
Spindle fiber

Late Prophase

Sister chromatids

Telophase and Cytokinesis
Nuclear envelopes begin to reassemble around two daughter nuclei. Chromosomes decondense. Spindle disappears. Division of the cytoplasm into two cells.

Nuclear envelopes
Chromosomes

Anaphase
Sister chromatids separate to opposite poles of cell. Events begin which lead to cytokinesis.

Metaphase
Chromosomes align along equator, or metaphase plate, of cell.

Mitosis
Cytokinesis
G_1 phase
S phase — Interphase
G_2 phase

Figure 6.2 Label the structures indicated in the dividing cell.

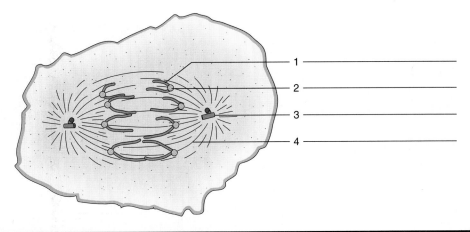

1 _____

2 _____

3 _____

4 _____

Figure 6.3 Cell in prophase (250× micrograph enlarged to 1,000×).

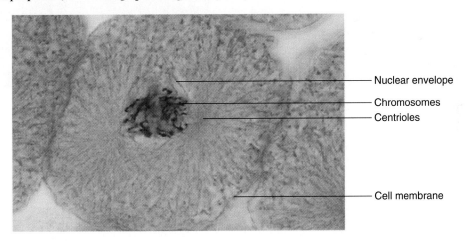

Nuclear envelope

Chromosomes

Centrioles

Cell membrane

DEMONSTRATION

Using the oil immersion objective of a microscope, see if you can locate some human chromosomes by examining a prepared slide of human chromosomes from leukocytes. The cells on this slide were cultured in a special medium and were stimulated to undergo mitosis. The mitotic process was arrested in metaphase by exposing the cells to a chemical called colchicine, and the cells were caused to swell osmotically. As a result of this treatment, the chromosomes were spread apart. A complement of human chromosomes should be visible when they are magnified about 1,000×. Note that each chromosome is double-stranded and consists of two chromatids joined by a common centromere (fig. 6.4).

Web Quest

Review the phases of the cell cycle and the events during each phase. Search these at

www.mhhe.com/shier10

Click on Student Edition. Click on Martin Lab Manual, Web Quest.

Figure 6.4 (*a*) A complement of human chromosomes (2,700×). (*b*) A *karyotype* can be constructed by arranging the homologous chromosome pairs together in a chart. The completed karyotype indicates a normal male.

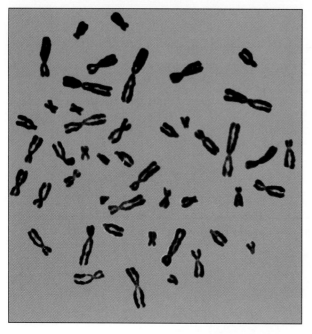

(a)

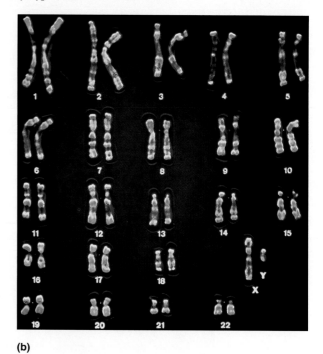

(b)

THE CELL CYCLE

PART A

Complete the table.

Stage	Major Events Occurring
Interphase (G₁, S, and G₂)	
Mitosis (karyokinesis) Prophase	
Metaphase	
Anaphase	
Telophase	
Cytoplasmic division (cytokinesis)	

PART B

Sketch an interphase cell and cells in different stages of mitosis to illustrate the whitefish cell's life cycle. Label the major cellular structures represented in the sketches and indicate cytokinesis locations. (The circles represent fields of view through the microscope.)

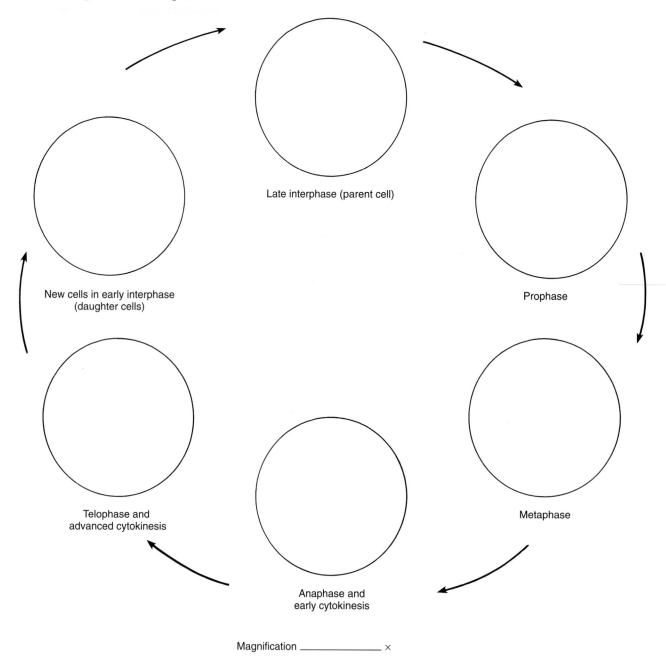

Late interphase (parent cell)

New cells in early interphase
(daughter cells)

Prophase

Telophase and
advanced cytokinesis

Metaphase

Anaphase and
early cytokinesis

Magnification _____ ×

PART C

Complete the following:

1. In what ways are the new cells (daughter cells), which result from a cell cycle, similar? _____

_____

_____

2. How do the new cells differ slightly? _____

_____

_____

3. Distinguish between mitosis (karyokinesis) and cytoplasmic division (cytokinesis). _____

_____

_____

PART D

Identify the mitotic stage represented by each of the micrographs in figure 6.5(*a–d*).

a. _____

b. _____

c. _____

d. _____

PART E

Identify the structures indicated by numbers in figure 6.5.

1. _____

2. _____

3. _____

4. _____

5. _____

6. _____

Figure 6.5 Identify the mitotic stage of the cell in each of these micrographs of the whitefish blastula (250×).

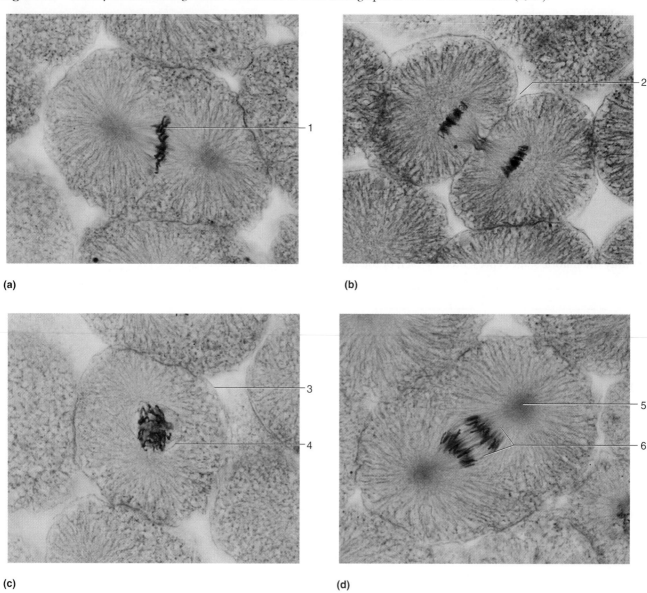

(a)

(b)

(c)

(d)

EPITHELIAL TISSUES

MATERIALS NEEDED

Textbook
Compound light microscope
Prepared slides of the following epithelial tissues:
  Simple squamous epithelium (lung)
  Simple cuboidal epithelium (kidney)
  Simple columnar epithelium (small intestine)
  Pseudostratified (ciliated) columnar epithelium
   (trachea)
  Stratified squamous epithelium (esophagus)
  Transitional epithelium (urinary bladder)

For Optional Activity:

Colored pencils

A tissue is composed of a layer or group of cells—cells that are similar in size, shape, and function. Within the human body, there are four major types of tissues: (1) epithelial, which cover the body's external and internal surfaces; (2) connective, which bind and support parts; (3) muscle, which make movement possible; and (4) nervous, which conduct impulses from one part of the body to another and help to control and coordinate body activities.

Epithelial tissues are tightly packed single (simple) to multiple (stratified) layers of cells that provide protective barriers. The underside of this tissue layer contains a basement membrane layer to which the epithelial cells anchor. Epithelial cells always have a free surface that is exposed to the outside or to an open space internally. Many shapes of the cells exist that are used to name and identify the variations. Many of the prepared slides contain more than the tissue to be studied, so be certain that your view matches the correct tissue. Also be aware that stained colors of all tissues might vary.

PURPOSE OF THE EXERCISE

To review the characteristics of epithelial tissues and to observe examples.

LEARNING OBJECTIVES

After completing this exercise, you should be able to

1. Describe the general characteristics of epithelial tissues.
2. List six types of epithelial tissues.
3. Describe the special characteristics of each type of epithelial tissue.
4. Indicate a location and function of each type of epithelial tissue.
5. Identify examples of epithelial tissues.

PROCEDURE

1. Review the section entitled "Epithelial Tissues" in chapter 5 of the textbook.
2. Complete Part A of Laboratory Report 7.
3. Use the microscope to observe the prepared slides of types of epithelial tissues. As you observe each tissue, look for its special distinguishing features as described in the textbook, such as cell size, shape, and arrangement. Compare your prepared slides of epithelial tissues to the micrographs in figure 7.1.
4. Complete Part B of the laboratory report.
5. Test your ability to recognize each type of epithelial tissue. To do this, have a laboratory partner select one of the prepared slides, cover its label, and focus the microscope on the tissue. Then see if you can correctly identify the tissue.

Web Quest

Identify tissues from micrographs and examine the structural components of tissues. Search these at

www.mhhe.com/shier10

Click on Student Edition. Click on Martin Lab Manual, Web Quest.

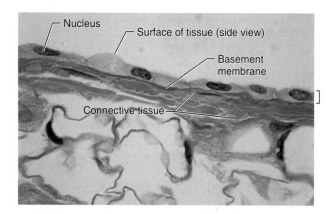

(a) Simple squamous epithelium (side view) (1,100×)

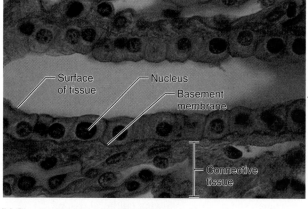

(b) Simple cuboidal epithelium (800×)

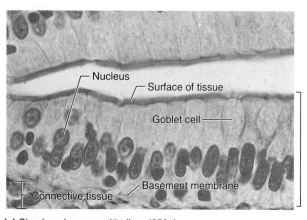

(c) Simple columnar epithelium (650×)

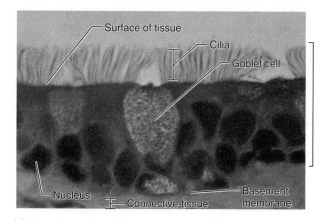

(d) Pseudostratified columnar epithelium with cilia (1,500×)

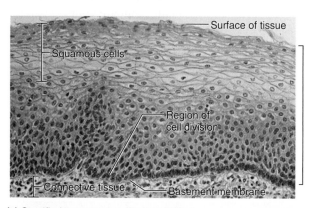

(e) Stratified squamous epithelium (370×)

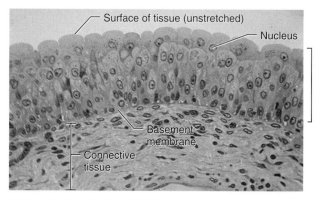

(f) Transitional epithelium (190×)

EPITHELIAL TISSUES

PART A

Match the tissues in column A with the characteristics in column B. Place the letter of your choice in the space provided. (Some answers may be used more than once.)

Column A

a.   Simple columnar epithelium
b.   Simple cuboidal epithelium
c.   Simple squamous epithelium
d.   Pseudostratified columnar epithelium
e.   Stratified squamous epithelium
f.   Transitional epithelium

Column B

_____ 1.   Consists of several layers of cube-shaped, elongated, and irregular cells

_____ 2.   Commonly possesses cilia that move sex cells and mucus

_____ 3.   Single layer of flattened cells

_____ 4.   Nuclei located at different levels within cells

_____ 5.   Forms walls of capillaries and air sacs of lungs

_____ 6.   Forms linings of trachea and bronchi

_____ 7.   Younger cells cuboidal, older cells flattened

_____ 8.   Forms inner lining of urinary bladder

_____ 9.   Lines kidney tubules and ducts of salivary glands

_____ 10. Forms lining of stomach and intestines

_____ 11. Nuclei located near basement membrane

_____ 12. Forms lining of oral cavity, anal canal, and vagina

PART B

In the space that follows, sketch a few cells of each type of epithelium you observed. For each sketch, label the major characteristics, indicate the magnification used, write an example of a location in the body, and provide a function.

Simple squamous epithelium (____×)
Location _____
Function _____

Simple cuboidal epithelium (____×)
Location _____
Function _____

Simple columnar epithelium (____×)
Location _____
Function _____

Pseudostratified columnar epithelium
with cilia (____×)
Location _____
Function _____

Stratified squamous epithelium (____×)
Location _____
Function _____

Transitional epithelium (____×)
Location _____
Function _____

Critical Thinking Application

As a result of your observations of epithelial tissues, which one(s) provide(s) the best protection? Explain your answer.

_____

_____

_____

_____

LABORATORY EXERCISE 8

CONNECTIVE TISSUES

MATERIALS NEEDED

Textbook
Compound light microscope
Prepared slides of the following:
   Loose (areolar) connective tissue
   Adipose tissue
   Dense connective tissue (regular type)
   Elastic connective tissue
   Reticular connective tissue
   Hyaline cartilage
   Elastic cartilage
   Fibrocartilage
   Bone (compact, ground, cross section)
   Blood (human smear)

For Optional Activity:

Colored pencils

Connective tissues contain a variety of cell types and occur in all regions of the body. They bind structures together, provide support and protection, fill spaces, store fat, and produce blood cells.

Connective tissue cells are often widely scattered in an abundance of intercellular matrix. The matrix consists of fibers and a ground substance of various densities. Many of the prepared slides contain more than the tissue to be studied, so be certain that your view matches the correct tissue. Additional study of bone and blood will be found in Laboratory Exercises 11 and 38.

PURPOSE OF THE EXERCISE

To review the characteristics of connective tissues and to observe examples of the major types.

LEARNING OBJECTIVES

After completing this exercise, you should be able to

1. Describe the general characteristics of connective tissues.
2. List the major types of connective tissues.
3. Describe the special characteristics of each of the major types of connective tissue.
4. Indicate a location and function of each type of connective tissue.
5. Identify the major types of connective tissues on microscope slides.

PROCEDURE

1. Review the section entitled "Connective Tissues" in chapter 5 of the textbook.
2. Complete Part A of Laboratory Report 8.
3. Use a microscope to observe the prepared slides of various connective tissues. As you observe each tissue, look for its special distinguishing features as described in the textbook. Compare your prepared slides of connective tissues to the micrographs in figure 8.1.
4. Complete Part B of the laboratory report.
5. Test your ability to recognize each of these connective tissues by having a laboratory partner select a slide, cover its label, and focus the microscope on this tissue. Then see if you correctly identify the tissue.

Web Quest

Identify tissues from micrographs and examine the structural components of tissues. Search these at
   www.mhhe.com/shier10
Click on Student Edition. Click on Martin Lab Manual, Web Quest.

Figure 8.1 Micrographs of connective tissues.

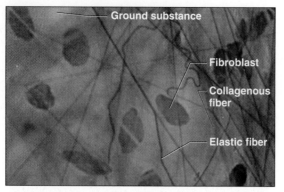

(a) Loose (areolar) connective (400×)

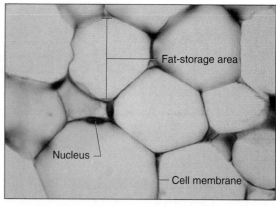

(b) Adipose (265×)

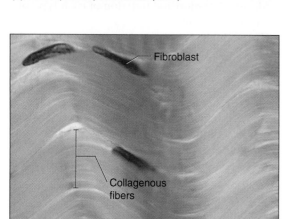

(c) Dense connective (regular) (1,000×)

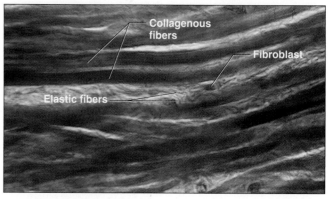

(d) Elastic connective (680×)

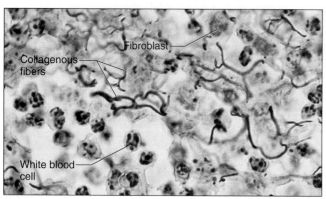

(e) Reticular connective (1,000×)

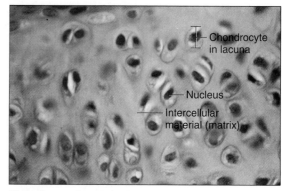

(f) Hyaline cartilage (500×)

Figure 8.1 *Continued.*

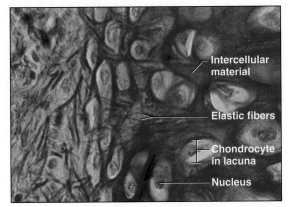

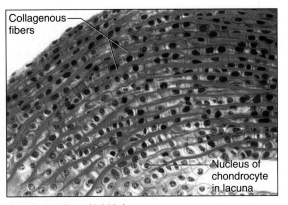

(g) Elastic cartilage (1,200×)

(h) Fibrocartilage (1,800×)

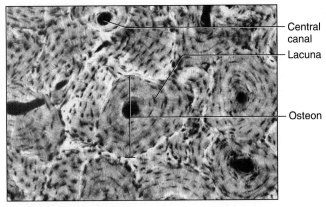

(i) Bone (100×)

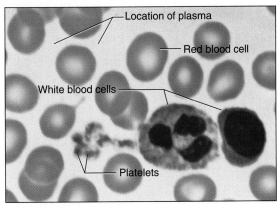

(j) Blood (1,200×)

CONNECTIVE TISSUES

PART A

Match the tissues in column A with the characteristics in column B. Place the letter of your choice in the space provided. (Some answers may be used more than once.)

Column A	Column B
a. Adipose tissue	_____ 1. Forms framework of outer ear
b. Blood	_____ 2. Functions as heat insulator beneath skin
c. Bone	_____ 3. Contains large amounts of fluid and lacks fibers
d. Dense connective tissue	_____ 4. Cells arranged around central canal
e. Elastic cartilage	_____ 5. Binds skin to underlying organs
f. Elastic connective tissue	_____ 6. Main tissue of tendons and ligaments
g. Fibrocartilage	_____ 7. Provides stored energy supply in fat vacuoles
h. Hyaline cartilage	_____ 8. Forms the flexible part of the nasal septum
i. Loose (areolar) connective tissue	_____ 9. Pads between vertebrae that are shock absorbers
j. Reticular connective tissue	_____ 10. Forms supporting rings of respiratory passages
	_____ 11. Cells greatly enlarged with nuclei pushed to sides
	_____ 12. Matrix contains collagen fibers and mineral salts
	_____ 13. Occurs in ligament attachments between vertebrae and artery walls
	_____ 14. Forms supporting tissue in walls of liver and spleen

PART B

In the space that follows, sketch a small section of each of the types of connective tissues you observed. For each sketch, label the major characteristics, indicate the magnification used, write an example of a location in the body, and provide a function.

Loose (areolar) connective tissue (___×)
Location _____
Function _____

Adipose tissue (___×)
Location _____
Function _____

Dense connective tissue (___×)
Location _____
Function _____

Elastic connective tissue (___×)
Location _____
Function _____

Reticular connective tissue (___×)
Location _____
Function _____

Hyaline cartilage (___×)
Location _____
Function _____

Elastic cartilage (____×)
Location _____
Function _____

Fibrocartilage (____×)
Location _____
Function _____

Bone (____×)
Location _____
Function _____

Blood (____×)
Location _____
Function _____

Critical Thinking Application

Abdominal impact injuries often involve the spleen. Explain the structural tissue characteristics that make the spleen so vulnerable to serious injury.

_____

_____

_____

_____

OPTIONAL ACTIVITY

Use colored pencils to differentiate various cellular structures in Part B. Select a different color for the cells, fibers, and ground substance whenever visible.

MUSCLE AND NERVOUS TISSUES

Muscle tissues are characterized by the presence of elongated cells or muscle fibers that can contract. As they shorten, these fibers pull at their attached ends and cause body parts to move. The three types of muscle tissues are skeletal, smooth, and cardiac.

Nervous tissues occur in the brain, spinal cord, and nerves. They consist of neurons (nerve cells), which are the impulse-conducting cells of the nervous system, and neuroglial cells, which perform supportive and protective functions.

PURPOSE OF THE EXERCISE

To review the characteristics of muscle and nervous tissues and to observe examples of these tissues.

LEARNING OBJECTIVES

After completing this exercise, you should be able to

1. List the three types of muscle tissues.
2. Describe the general characteristics of muscle tissues.
3. Describe the special characteristics of each type of muscle tissue.
4. Indicate a location and function of each type of muscle tissue.
5. Identify examples of muscle tissues.
6. Describe the general characteristics of nervous tissues.
7. Identify nervous tissue.

PROCEDURE

1. Review the sections entitled "Muscle Tissues" and "Nervous Tissues" in chapter 5 of the textbook.
2. Complete Part A of Laboratory Report 9.
3. Using the microscope, observe each of the types of muscle tissues on the prepared slides. Look for the special features of each type, as described in the textbook. Compare your prepared slides of muscle tissues to the micrographs in figure 9.1 and the muscle tissue characteristics in table 9.1.
4. Observe the prepared slide of nervous tissue and identify neurons (nerve cells), neuron processes, and neuroglial cells. Compare your prepared slide of nervous tissue to the micrograph in figure 9.1.
5. Complete Part B of the laboratory report.
6. Test your ability to recognize each of these muscle and nervous tissues by having your laboratory partner select a slide, cover its label, and focus the microscope on this tissue. Then see if you correctly identify the tissue.

Web Quest

Identify tissues from micrographs, and examine the structural components of tissues. Search these at
www.mhhe.com/shier10
Click on Student Edition. Click on Martin Lab Manual, Web Quest.

Figure 9.1 Micrographs of muscle and nervous tissues.

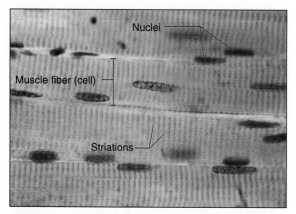

(a) Skeletal muscle (700×)

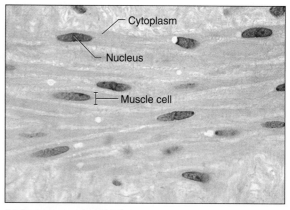

(b) Smooth muscle (900×)

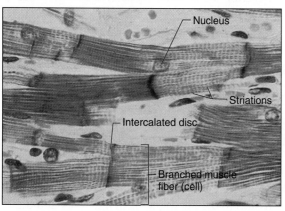

(c) Cardiac muscle (400×)

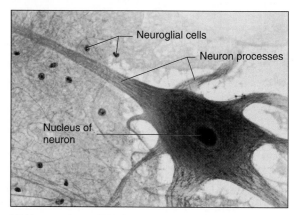

(d) Nervous tissue (300×)

Table 9.1 Muscle Tissue Characteristics

Characteristic	Skeletal Muscle	Cardiac Muscle	Smooth Muscle
Appearance of cells	Unbranched and relatively parallel	Branched and connected in complex networks	Spindle-shaped
Striations	Present and obvious	Present	Absent
Nucleus	Multinucleated	Uninucleated (usually)	Uninucleated
Intercalated discs	Absent	Present	Absent
Control	Voluntary	Involuntary	Involuntary

MUSCLE AND NERVOUS TISSUES

PART A

Match the tissues in column A with the characteristics in column B. Place the letter of your choice in the space provided. (Some answers may be used more than once.)

Column A	Column B
a. Cardiac muscle	_____ 1. Coordinates, regulates, and integrates body functions
b. Nervous tissue	_____ 2. Contains intercalated discs
c. Skeletal muscle	_____ 3. Muscle that lacks striations
d. Smooth muscle	_____ 4. Striated and involuntary
	_____ 5. Striated and voluntary
	_____ 6. Contains neurons and neuroglial cells
	_____ 7. Muscle attached to bones
	_____ 8. Muscle that composes heart
	_____ 9. Moves food through the digestive tract
	_____ 10. Transmits impulses along cytoplasmic extensions

PART B

In the space that follows, sketch a few cells or fibers of each of the three types of muscle tissues and of nervous tissue as they appear through the microscope. For each sketch, label the major structures of the cells or fibers, indicate the magnification used, write an example of a location in the body, and provide a function.

Skeletal muscle tissue (____×)
Location _____
Function _____

Smooth muscle tissue (____×)
Location _____
Function _____

Cardiac muscle tissue (____×)
Location _____
Function _____

Nervous tissue (____×)
Location _____
Function _____

OPTIONAL ACTIVITY

Use colored pencils to differentiate various cellular structures in Part B.

Laboratory Exercise 10

Integumentary System

MATERIALS NEEDED

Textbook
Skin model
Hand magnifier or dissecting microscope
Forceps
Microscope slide and coverslip
Compound light microscope
Prepared microscope slide of human scalp or axilla
Prepared slide of heavily pigmented human skin
Prepared slide of thick skin (plantar or palmar)

For Optional Activity:

Tattoo slide

The integumentary system includes the skin, hair, nails, sebaceous glands, and sweat glands. These organs provide a protective covering for deeper tissues, aid in regulating body temperature, retard water loss, house sensory receptors, synthesize various chemicals, and excrete small quantities of wastes.

PURPOSE OF THE EXERCISE

To observe the organs and tissues of the integumentary system and to review the functions of these parts.

LEARNING OBJECTIVES

After completing this exercise, you should be able to

1. Name the organs of the integumentary system.
2. Describe the major functions of these organs.
3. Distinguish among epidermis, dermis, and the subcutaneous layer.
4. Identify the layers of the skin, a hair follicle, an arrector pili muscle, a sebaceous gland, and a sweat gland on a microscope slide, diagram, or model.

PROCEDURE

1. Review the sections entitled "Skin and Its Tissues" and "Accessory Organs of the Skin" in chapter 6 of the textbook.

2. As a review activity, label figures 10.1, 10.2, and 10.3. Locate as many of these structures as possible on a skin model.
3. Complete Part A of Laboratory Report 10.
4. Use the hand magnifier or dissecting microscope and proceed as follows:
   a. Observe the skin, hair, and nails on your hand.
   b. Compare the type and distribution of hairs on the front and back of your forearm.
5. Use low-power magnification of the compound light microscope and proceed as follows:
   a. Pull out a single hair with forceps and mount it on a microscope slide under a coverslip.
   b. Observe the root and shaft of the hair and note the scalelike parts that make up the shaft.
6. Complete Part B of the laboratory report.
7. As vertical sections of human skin are observed, remember that the lenses of the microscope invert and reverse images. It is important to orient the position of the epidermis, dermis, and subcutaneous (hypodermis) layers using scan magnification before continuing with additional observations. Compare all of your skin observations to figure 10.4. Use low-power magnification of the compound light microscope and proceed as follows:
   a. Observe the prepared slide of human scalp or axilla.
   b. Locate the epidermis, dermis, and subcutaneous layer, a hair follicle, an arrector pili muscle, a sebaceous gland, and a sweat gland.
   c. Focus on the epidermis with high power and locate the stratum corneum and stratum basale (stratum germinativum). Note how the shapes of the cells in these two layers differ.
   d. Observe the dense connective tissue (irregular type) that makes up the bulk of the dermis.
   e. Observe the adipose tissue that composes most of the subcutaneous layer.

Figure 10.1 Label this vertical section of skin.

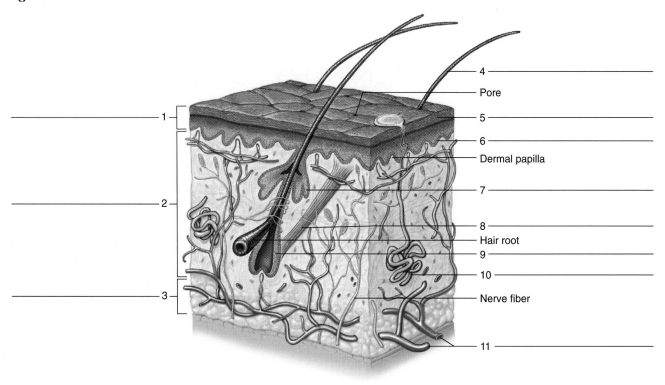

Pore

Dermal papilla

Hair root

Nerve fiber

Figure 10.2 Label the epidermal layers in this section of thick skin from the palm of the hand.

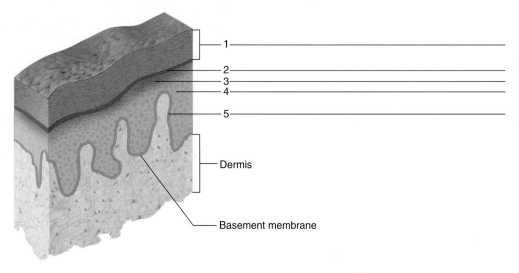

Dermis

Basement membrane

Figure 10.3 Label the features associated with this hair follicle.

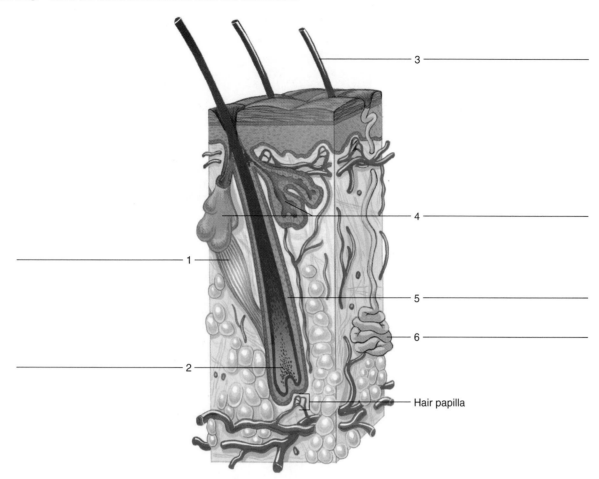

8. Observe the prepared slide of heavily pigmented human skin with low-power magnification. Note that the pigment is most abundant in the epidermis. Focus on this region with the high-power objective. The pigment-producing cells, or melanocytes, are located among the deeper layers of epidermal cells. Differences in skin color are primarily due to the amount of pigment (melanin) produced by these cells. The number of melanocytes in the skin is about the same for members of all racial groups.

Critical Thinking Application

Explain the advantage for melanin granules of being located in the deep layer of the epidermis.

_____

_____

_____

_____

9. Observe the prepared slide of thick skin from the palm of a hand or the sole of a foot. Locate the stratum lucidum. Note how the stratum corneum compares to your observation of human scalp.
10. Complete Part C of the laboratory report.
11. Using low-power magnification, locate a hair follicle that has been sectioned longitudinally through its bulblike base. Also locate a sebaceous gland close to the follicle and find a sweat gland. Observe the detailed structure of these parts with high-power magnification.
12. Complete Parts D and E of the laboratory report.

Figure 10.4 Features of human skin are indicated in these micrographs. Magnifications: (*a*) 290×; (*b*) 30× micrograph enlarged to 280×; (*c*) 45×; (*d*) 110×.

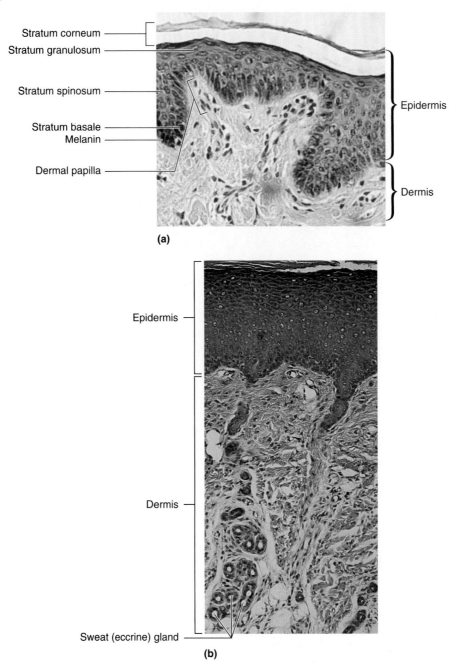

Stratum corneum

Stratum granulosum

Stratum spinosum

Stratum basale

Melanin

Dermal papilla

Epidermis

Dermis

(a)

Epidermis

Dermis

Sweat (eccrine) gland

(b)

Figure 10.4   *Continued.*

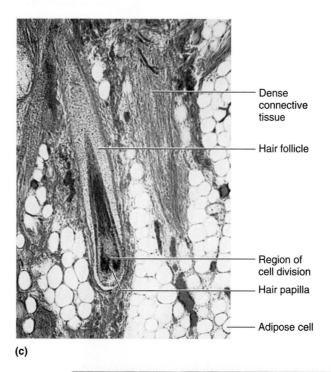

Dense connective tissue

Hair follicle

Region of cell division

Hair papilla

Adipose cell

(c)

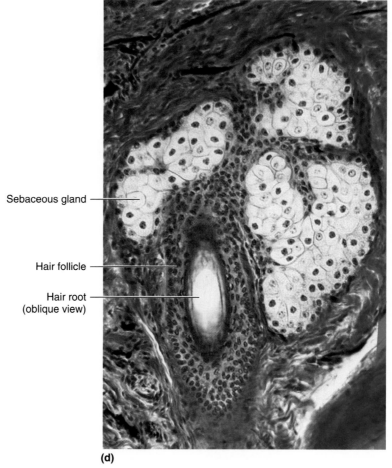

Sebaceous gland

Hair follicle

Hair root (oblique view)

(d)

OPTIONAL ACTIVITY

Observe a vertical section of human skin through a tattoo, using low-power magnification. Note the location of the dispersed ink granules within the upper portion of the dermis. From a thin vertical section of a tattoo, it is not possible to determine the figure or word of the entire tattoo as seen on the surface of the skin. Compare this to the location of melanin granules found in heavily pigmented skin. Describe why a tattoo is permanent and a suntan is not.

_____

_____

_____

Web Quest

Identify skin layers from micrographs and review the functions of the skin structures. Search these at

www.mhhe.com/shier10

Click on Student Edition. Click on Martin Lab Manual, Web Quest.

INTEGUMENTARY SYSTEM

PART A

Match the structures in column A with the description and functions in column B. Place the letter of your choice in the space provided.

Column A		Column B
a. Apocrine sweat gland	_____ 1.	An oily secretion that helps to waterproof body surface
b. Arrector pili muscle	_____ 2.	Outermost layer of epidermis
c. Dermis	_____ 3.	Become active at puberty
d. Eccrine sweat gland	_____ 4.	Epidermal pigment
e. Epidermis	_____ 5.	Inner layer of skin
f. Hair follicle	_____ 6.	Responds to elevated body temperature
g. Keratin	_____ 7.	Pigment-producing cell
h. Melanin	_____ 8.	General name of entire superficial layer of the skin
i. Melanocyte	_____ 9.	Gland that secretes an oily substance
j. Sebaceous gland	_____ 10.	Hard protein of nails and hair
k. Sebum	_____ 11.	Binds skin to underlying organs
l. Stratum basale	_____ 12.	Cell division and deepest layer of epidermis
m. Stratum corneum	_____ 13.	Tubelike part that contains the root of the hair
n. Subcutaneous layer	_____ 14.	Causes hair to stand on end and goose bumps to appear

PART B

Complete the following:

1. How does the skin of your palm differ from that on the back (posterior) of your hand? _____

_____

_____

2. Describe the differences you observed in the type and distribution of hair on the front (anterior) and back (posterior) of your forearm. _____

_____

3. Explain how a hair is formed. _____

_____

4. What cells produce the pigment in hair? _____

_____

PART C

Complete the following:

1. Distinguish among epidermis, dermis, and subcutaneous layer._____

   _____

   _____

2. How do the cells of stratum corneum and stratum basale differ?_____

   _____

   _____

3. State the specific location of melanin observed in heavily pigmented skin. _____

   _____

4. What special qualities does the connective tissue of the dermis have? _____

   _____

PART D

Complete the following:

1. What part of the hair extends from the hair papilla to the body surface? _____

2. In which layer of skin are sebaceous glands found?_____

3. How are sebaceous glands associated with hair follicles? _____

   _____

4. In which layer of skin are sweat glands usually located? _____

PART E

Sketch and label a vertical section of human skin, using the scanning objective.

STRUCTURE AND CLASSIFICATION OF BONE

MATERIALS NEEDED

Textbook
Prepared microscope slide of ground compact bone
Human bone specimens including long, short, flat, and
    irregular types
Human long bone, sectioned longitudinally
Fresh animal bones, sectioned longitudinally and
    transversely
Compound light microscope
Dissecting microscope

For Demonstration:

Fresh chicken bones (radius and ulna from wings)
Vinegar or dilute hydrochloric acid

SAFETY

- Wear disposable gloves for handling fresh bones and
  for the demonstration of a bone soaked in vinegar or
  dilute hydrochloric acid.
- Wash your hands before leaving the laboratory.

A bone represents an organ of the skeletal system. As such, it is composed of a variety of tissues including bone tissue, cartilage, dense connective tissue, blood, and nervous tissue.

Although various bones of the skeleton vary greatly in size and shape, they have much in common structurally and functionally.

PURPOSE OF THE EXERCISE

To review the way bones are classified and to examine the structure of a long bone.

LEARNING OBJECTIVES

After completing this exercise, you should be able to

1. Name four groups of bones based on their shapes and give an example for each group.
2. Locate and name the major structures of a long bone.
3. Describe the functions of various structures of a bone.
4. Distinguish between compact and spongy bone.

PROCEDURE

1. Review the section entitled "Bone Structure" in chapter 5 of the textbook.
2. Reexamine the microscopic structure of bone tissue by observing a prepared microscope slide of ground compact bone. Use the figures of bone tissue in chapter 5 of the textbook to locate the following features:

   osteon (Haversian   lacuna (small chamber
       system)   for an osteocyte)

   central canal   canaliculus
       (Haversian canal)

   lamella

3. Review the section entitled "Bone Structure" in chapter 7 of the textbook.
4. As a review activity, label figures 11.1 and 11.2.
5. Observe the individual bone specimens and arrange them into groups, according to the following shapes:

   long   irregular

   short   sesamoid (round)

   flat

6. Complete Part A of Laboratory Report 11.
7. Examine the sectioned long bones and locate the following:

   epiphysis (proximal and distal)

   epiphyseal plate

   articular cartilage

   diaphysis

   periosteum

   compact bone

   spongy bone

   trabeculae

   medullary cavity

   endosteum

   yellow marrow

   red marrow

Figure 11.1 Label the major structures of this long bone (femur).

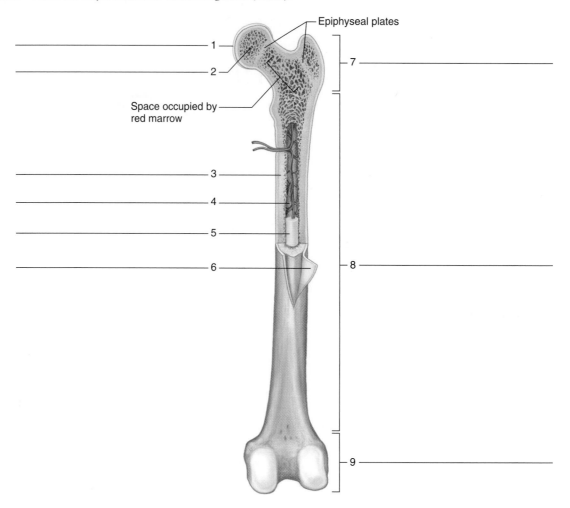

Epiphyseal plates

Space occupied by red marrow

8. Use the dissecting microscope to observe the compact bone and spongy bone of the sectioned specimens. Also examine the marrow in the medullary cavity and the spaces within the spongy bone of the fresh specimen.
9. Complete Parts B and C of the laboratory report.

Critical Thinking Application

Explain how bone cells embedded in a solid ground substance obtain nutrients and eliminate wastes.

_____

_____

_____

_____

DEMONSTRATION

Examine a fresh chicken bone and a chicken bone that has been soaked for several days in vinegar or overnight in dilute hydrochloric acid. Wear disposable gloves for handling these bones. This acid treatment removes the inorganic salts from the bone matrix. Rinse the bones in water and note the texture and flexibility of each. Based on your observations, what quality of the fresh bone seems to be due to the inorganic salts that were removed by the acid treatment?

_____

Examine the specimen of chicken bone that has been exposed to high temperature (baked at 121°C/250°F for 2 hours). This treatment removes the protein and other organic substances from the bone matrix. What quality of the fresh bone seems to be due to these organic materials?

_____

_____

Figure 11.2 Label the features associated with the microscopic structure of bone.

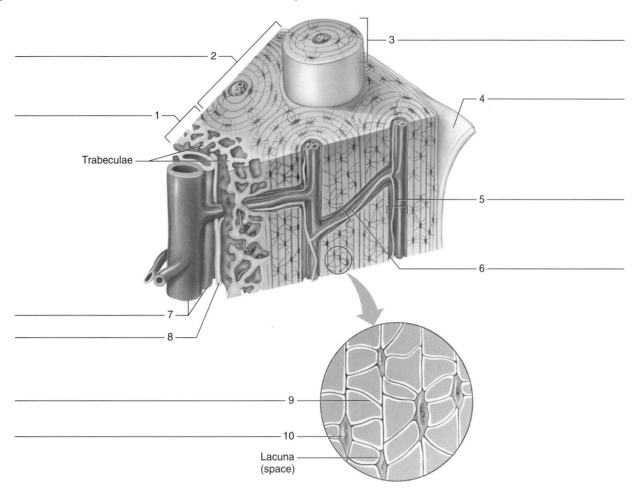

Trabeculae

Lacuna
(space)

STRUCTURE AND CLASSIFICATION OF BONE

PART A

Complete the following statements. (*Note:* Questions 1-6 pertain to bone classification by shape.)

1. A bone that is platelike is classified as a(an) _____ bone.

2. The bones of the wrist are examples of _____ bones.

3. The bone of the thigh is an example of a(an) _____ bone.

4. Vertebrae are examples of _____ bones.

5. The patella (kneecap) is an example of a large _____ bone.

6. The bones of the skull that form a protective covering of the brain are examples of _____ bones.

7. Distinguish between the epiphysis and the diaphysis of a long bone. _____

_____

8. Describe where cartilage is found on the surface of a long bone. _____

_____

9. Describe where dense connective tissue is found on the surface of a long bone. _____

_____

10. Distinguish between the periosteum and the endosteum. _____

_____

_____

PART B

Complete the following:

1. What differences did you note between the structure of compact bone and spongy bone?_____

_____

_____

_____

2. How are these structural differences related to the locations and functions of these two types of bone?_____

_____

_____

3. From your observations, how does the marrow in the medullary cavity compare with the marrow in the spaces of the spongy bone? _____

_____

_____

Identify the structures indicated in figure 11.3.

Figure 11.3 Identify the structures indicated in (*a*) the unsectioned long bone (fifth metatarsal) and (*b*) the partially sectioned long bone, using the terms provided.

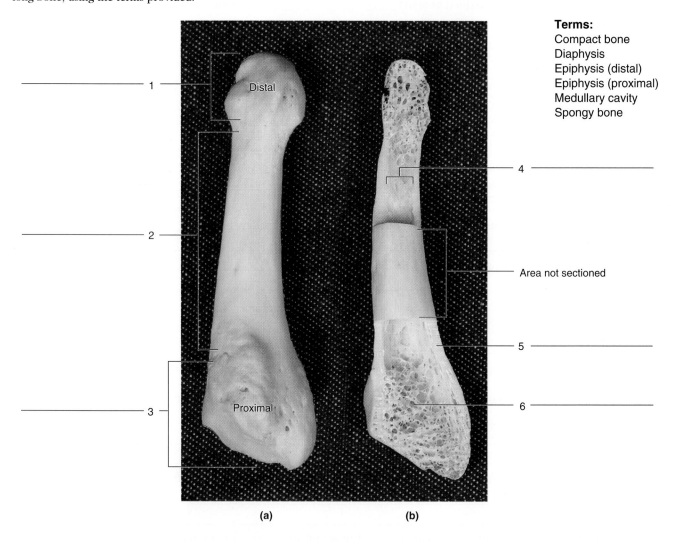

Terms:
Compact bone
Diaphysis
Epiphysis (distal)
Epiphysis (proximal)
Medullary cavity
Spongy bone

ORGANIZATION OF THE SKELETON

MATERIALS NEEDED

Textbook
Articulated human skeleton

For Optional Activity:
Colored pencils

For Demonstration:
Radiographs (X rays) of skeletal structures

The skeleton can be divided into two major portions: (1) the axial skeleton, which consists of the bones and cartilages of the head, neck, and trunk, and (2) the appendicular skeleton, which consists of the bones of the limbs and those that anchor the limbs to the axial skeleton.

PURPOSE OF THE EXERCISE

To review the organization of the skeleton, the major bones of the skeleton, and the terms used to describe skeletal structures.

LEARNING OBJECTIVES

After completing this exercise, you should be able to

1. Distinguish between the axial skeleton and the appendicular skeleton.
2. Locate and name the major bones of the human skeleton.
3. Define the terms used to describe skeletal structures and locate examples of such structures on the human skeleton.

PROCEDURE

1. Review the section entitled "Skeletal Organization" in chapter 7 of the textbook. (Note that pronunciations of the names for major skeletal structures are included within the narrative of chapter 7.)
2. As a review activity, label figure 12.1.
3. Examine the human skeleton and locate the following parts. As you locate the following bones, note the number of each in the skeleton. Palpate as many of the corresponding bones in your own skeleton as possible.

axial skeleton

  skull

cranium	8
face	14
middle ear bone	6

  hyoid bone   1

  vertebral column

vertebra	24
sacrum	1
coccyx	1

  thoracic cage

rib	24
sternum	1

appendicular skeleton

  pectoral girdle

scapula	2
clavicle	2

  upper limbs

humerus	2
radius	2
ulna	2
carpal	16
metacarpal	10
phalanx	28

  pelvic girdle

coxa (hipbone; innominate)	2

Figure 12.1 Label the major bones of the skeleton: (*a*) anterior view; (*b*) posterior view.

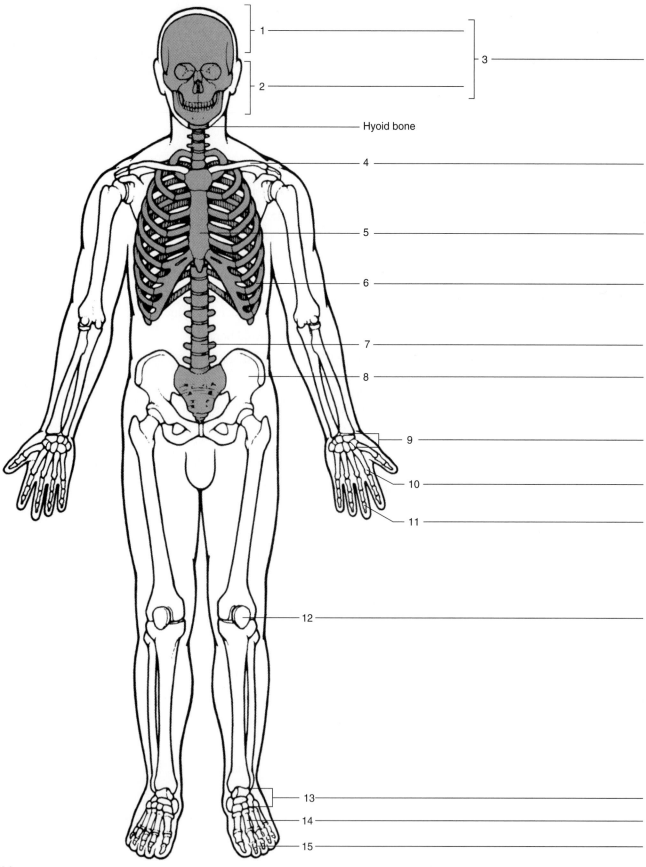

1 _____

2 _____

3 _____

Hyoid bone

4 _____

5 _____

6 _____

7 _____

8 _____

9 _____

10 _____

11 _____

12 _____

13 _____

14 _____

15 _____

(a)

Figure 12.1 Continued.

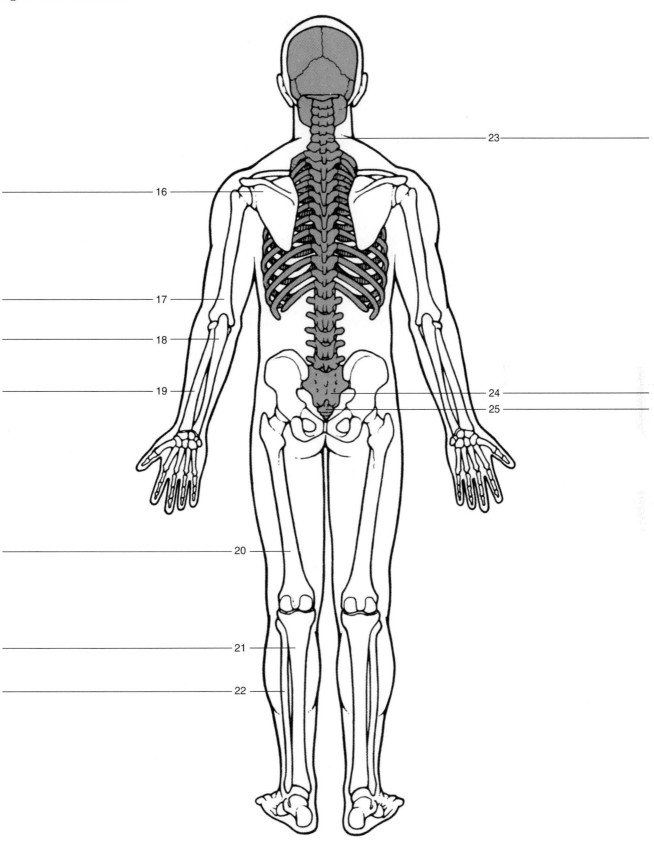

(b)

lower limbs

femur	2
tibia	2
fibula	2
patella	2
tarsal	14
metatarsal	10
phalanx	28

Total   206 bones

OPTIONAL ACTIVITY

Use colored pencils to distinguish the individual bones in figure 12.1.

4. Study table 7.4 in chapter 7 of the textbook. Locate an example of each of the following features (bone markings) on the bone listed, noting the size, shape, and location in the human skeleton:

condyle—occipital

crest—coxa (hipbone)

epicondyle—femur

facet—vertebra

fontanel—skull

foramen—vertebra

fossa—humerus

fovea—femur

head—humerus

meatus—temporal

process—temporal

sinus—frontal

spine—scapula

suture—skull

trochanter—femur

tubercle—humerus

tuberosity—tibia

Critical Thinking Application

Locate and name the largest foramen in the skull.

_____

_____

Locate and name the largest foramen in the skeleton.

_____

_____

5. Complete Parts A, B, C, and D of Laboratory Report 12.

DEMONSTRATION

Images on radiographs (X rays) are produced by allowing X rays from an X-ray tube to pass through a body part and to expose photographic film that is positioned on the opposite side of the part. The image that appears on the film after it is developed reveals the presence of parts with different densities. Bone, for example, is very dense tissue and is a good absorber of X rays. Thus, bone generally appears light on the film. Air-filled spaces, on the other hand, absorb almost no X rays and appear as dark areas on the film. Liquids and soft tissues absorb intermediate quantities of X rays, so they usually appear in various shades of gray.

Examine the available radiographs (X rays) of skeletal structures by holding each film in front of a light source. Identify as many of the bones and features as you can.

Name _____

Date _____

Section _____

ORGANIZATION OF THE SKELETON

PART A

Complete the following statements:

1. The extra bones that sometimes develop between the flat bones of the skull are called _____.

2. Small bones occurring in some tendons are called _____ bones.

3. The cranium and facial bones compose the _____.

4. The _____ bone supports the tongue.

5. The _____ at the inferior end of the sacrum is composed of several fused vertebrae.

6. Most ribs are attached anteriorly to the _____.

7. The thoracic cage is composed of _____ pairs of ribs.

8. The scapulae and clavicles together form the _____.

9. The humerus, radius, and _____ articulate to form the elbow joint.

10. The wrist is composed of eight bones called _____.

11. The coxae (hipbones) are attached posteriorly to the _____.

12. The pelvic girdle (coxae), sacrum, and coccyx together form the _____.

13. The _____ covers the anterior surface of the knee.

14. The bones that articulate with the distal ends of the tibia and fibula are called _____.

15. All finger and toe bones are called _____ .

PART B

Match the terms in column A with the definitions in column B. Place the letter of your choice in the space provided.

Column A	Column B
a. Condyle	_____ 1. Small, nearly flat articular surface
b. Crest	_____ 2. Deep depression
c. Facet	_____ 3. Rounded process
d. Fontanel	_____ 4. Opening or passageway
e. Foramen	_____ 5. Interlocking line of union
f. Fossa	_____ 6. Narrow, ridgelike projection
g. Suture	_____ 7. Soft region between bones of skull

PART C

Match the terms in column A with the definitions in column B. Place the letter of your choice in the space provided.

Column A

a. Fovea
b. Head
c. Meatus
d. Sinus
e. Spine
f. Trochanter
g. Tubercle

Column B

_____ 1. Tubelike passageway

_____ 2. Tiny pit or depression

_____ 3. Small, knoblike process

_____ 4. Thornlike projection

_____ 5. Rounded enlargement at end of bone

_____ 6. Air-filled cavity within bone

_____ 7. Relatively large process

PART D

Identify the bones indicated in figure 12.2.

Figure 12.2 Identify the bones in this random arrangement, using the terms provided.

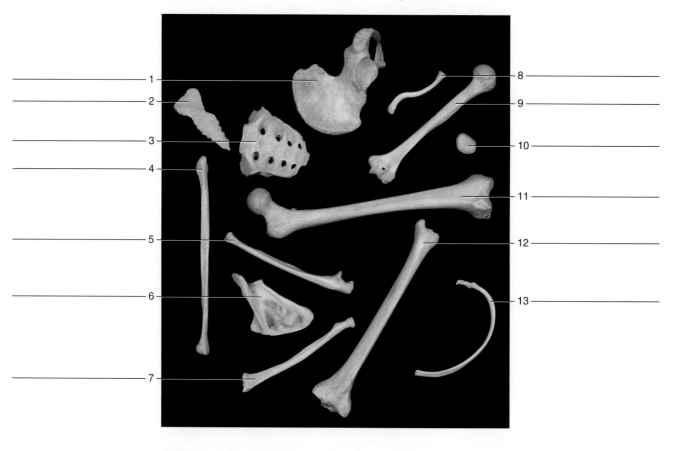

Terms:

Clavicle	Patella	Sternum
Coxa (hipbone)	Radius	Tibia
Femur	Rib	Ulna
Fibula	Sacrum	
Humerus	Scapula	

THE SKULL

MATERIALS NEEDED

Textbook
Human skull, articulated
Human skull, disarticulated (Beauchene)
Human skull, sagittal section

For Optional Activity:

Colored pencils

For Demonstration:

Fetal skull

A human skull consists of twenty-two bones that, except for the lower jaw, are firmly interlocked along sutures. Eight of these immovable bones make up the braincase, or cranium, and thirteen more immovable bones and the mandible form the facial skeleton.

PURPOSE OF THE EXERCISE

To examine the structure of the human skull and to identify the bones and major features of the skull.

LEARNING OBJECTIVES

After completing this exercise, you should be able to

1. Distinguish between the cranium and the facial skeleton.
2. Locate and name the bones of the skull and their major features.
3. Locate and name the major sutures of the cranium.
4. Locate and name the sinuses of the skull.

PROCEDURE

1. Review the section entitled "Skull" in chapter 7 of the textbook.
2. As a review activity, label figures 13.1, 13.2, 13.3, 13.4, and 13.5.
3. Examine the **cranial bones** of the articulated human skull and the sectioned skull. Also observe the corresponding disarticulated bones. Locate the following bones and features in the laboratory specimens and, at the same time, palpate as many of these bones and features in your own skull as possible.

frontal bone   1
   supraorbital foramen
   frontal sinus

parietal bone   2
   sagittal suture
   coronal suture

occipital bone   1
   lambdoidal suture
   foramen magnum
   occipital condyle

temporal bone   2
   squamosal suture
   external auditory meatus
   mandibular fossa
   mastoid process
   styloid process
   carotid canal
   jugular foramen
   internal acoustic meatus
   zygomatic process

sphenoid bone   1
   sella turcica
   greater and lesser wings
   sphenoidal sinus

ethmoid bone   1
   cribriform plate
   perpendicular plate
   superior nasal concha
   middle nasal concha
   ethmoidal sinus
   crista galli

Figure 13.1 Label the anterior bones and features of the skull. (If the line lacks the word *bone*, label the particular feature of that bone.)

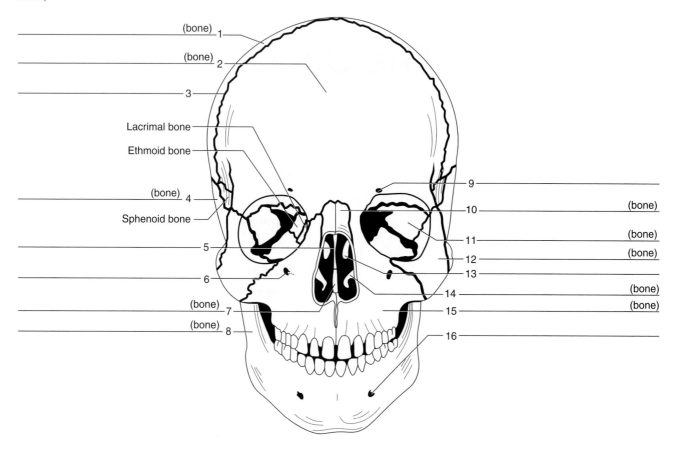

4. Complete Parts A and B of Laboratory Report 13.
5. Examine the **facial bones** of the articulated and sectioned skulls and the corresponding disarticulated bones. Locate the following:

maxilla   2

   maxillary sinus

   palatine process

   alveolar process

   alveolar arch

palatine bone   2

zygomatic bone   2

   temporal process

   zygomatic arch

lacrimal bone   2

nasal bone   2

vomer bone   1

inferior nasal concha 2

mandible   1

   ramus

   mandibular condyle

   coronoid process

   alveolar border

   mandibular foramen

   mental foramen

6. Complete Part C of the laboratory report.

OPTIONAL ACTIVITY

Use colored pencils to differentiate the bones illustrated in figures 13.1–13.5. Select a different color for each bone in the series. This activity should help you locate various bones that are shown in different views in the figures. You can check your work by referring to the corresponding figures in the textbook, which are presented in full color.

Figure 13.2 Label the lateral bones and features of the skull.

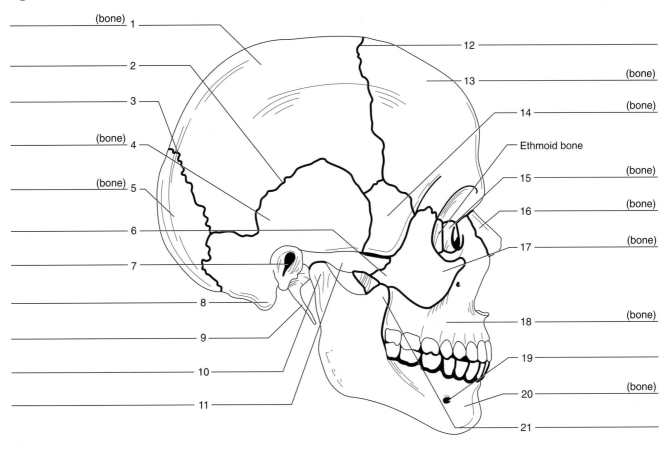

_____ (bone) 1
_____ 2
_____ 3
_____ (bone) 4
_____ (bone) 5
_____ 6
_____ 7
_____ 8
_____ 9
_____ 10
_____ 11

12 _____
13 _____ (bone)
14 _____ (bone)
Ethmoid bone
15 _____ (bone)
16 _____ (bone)
17 _____ (bone)
18 _____ (bone)
19 _____
20 _____ (bone)
21 _____

Figure 13.3 Label the inferior bones and features of the skull.

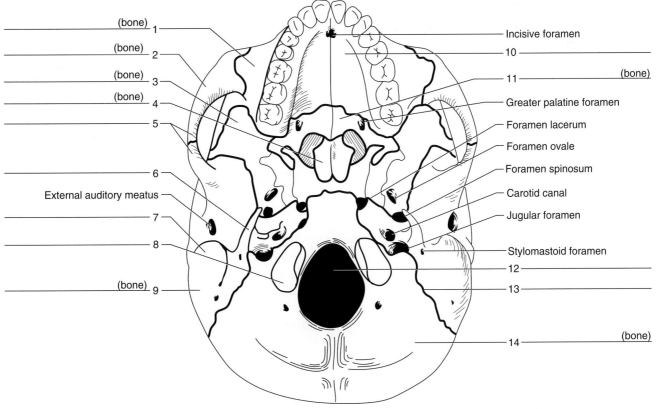

_____ (bone) 1
_____ (bone) 2
_____ (bone) 3
_____ (bone) 4
_____ 5
_____ 6
External auditory meatus
_____ 7
_____ 8
_____ (bone) 9

Incisive foramen
10 _____
11 _____ (bone)
Greater palatine foramen
Foramen lacerum
Foramen ovale
Foramen spinosum
Carotid canal
Jugular foramen
Stylomastoid foramen
12 _____
13 _____
14 _____ (bone)

Figure 13.4 Label the bones and features of the floor of the cranial cavity as viewed from above.

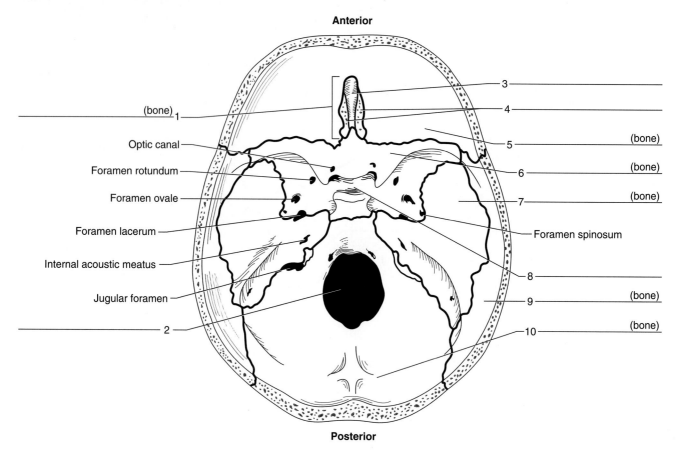

Anterior

(bone)₁

Optic canal

Foramen rotundum

Foramen ovale

Foramen lacerum

Internal acoustic meatus

Jugular foramen

2

3

4

5 (bone)

6 (bone)

7 (bone)

Foramen spinosum

8

9 (bone)

10 (bone)

Posterior

7. Study table 7.8 of chapter 7 in the textbook. Locate the following features of the human skull:

carotid canal

foramen lacerum

foramen magnum

foramen ovale

foramen rotundum

foramen spinosum

greater palatine foramen

hypoglossal canal

incisive foramen

inferior orbital fissure

infraorbital foramen

internal acoustic meatus

jugular foramen

mandibular foramen

mental foramen

optic canal

stylomastoid foramen

superior orbital fissure

supraorbital foramen

8. Complete Parts D and E of the laboratory report.

DEMONSTRATION

Examine the fetal skull (fig. 13.6). Note that the skull is incompletely developed and that the cranial bones are separated by fibrous membranes. These membranous areas are called fontanels, or "soft spots." The fontanels close as the cranial bones grow together. The posterior fontanel usually closes within a few months after birth, whereas the anterior fontanel may not close until the middle or end of the second year. What other features characterize the fetal skull?

Figure 13.5 Label the bones and features of the sagittal section of the skull.

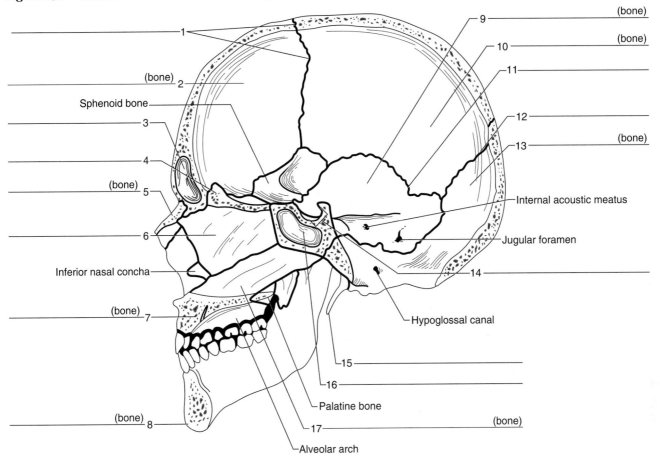

1 _____

(bone) 2 _____

Sphenoid bone _____

3 _____

4 _____

(bone) 5 _____

6 _____

Inferior nasal concha _____

(bone) 7 _____

(bone) 8 _____

9 _____ (bone)

10 _____ (bone)

11 _____

12 _____ (bone)

13 _____

Internal acoustic meatus

Jugular foramen

14 _____

Hypoglossal canal

15 _____

16 _____

Palatine bone

17 _____ (bone)

Alveolar arch

Figure 13.6 Human fetal skeleton.

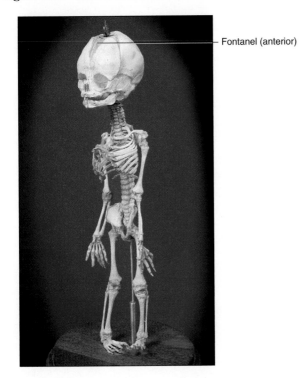

Fontanel (anterior)

Critical Thinking Application

Examine the inside of the cranium on a sectioned skull. What area appears to be the weakest area? Explain your answer.

_____

_____

_____

_____

Web Quest

What are the functions of individual bones and features? Search these and review the anatomy of the skeleton at
www.mhhe.com/shier10
Click on Student Edition. Click on Martin Lab Manual, Web Quest.

Laboratory Report **13**

Name _____

Date _____

Section _____

THE SKULL

PART A

Match the bones in column A with the features in column B. Place the letter of your choice in the space provided. (Some answers are used more than once.)

Column A

a. Ethmoid bone
b. Frontal bone
c. Occipital bone
d. Parietal bone
e. Sphenoid bone
f. Temporal bone

Column B

_____ 1. Forms sagittal, coronal, squamosal, and lambdoidal sutures

_____ 2. Cribriform plate

_____ 3. Crista galli

_____ 4. External auditory meatus

_____ 5. Foramen magnum

_____ 6. Mandibular fossa

_____ 7. Mastoid process

_____ 8. Middle nasal concha

_____ 9. Occipital condyle

_____ 10. Sella turcica

_____ 11. Styloid process

_____ 12. Supraorbital foramen

PART B

Complete the following statements:

1. The _____ suture joins the frontal bone to the parietal bones.

2. The parietal bones are firmly interlocked along the midline by the _____ suture.

3. The _____ suture joins the parietal bones to the occipital bone.

4. The temporal bones are joined to the parietal bones along the _____ sutures.

5. Name the three cranial bones that contain sinuses. _____

_____

6. Name a facial bone that contains a sinus. _____

PART C

Match the bones in column A with the characteristics in column B. Place the letter of your choice in the space provided.

Column A

a. Inferior nasal concha
b. Lacrimal bone
c. Mandible
d. Maxilla
e. Nasal bone
f. Palatine bone
g. Vomer bone
h. Zygomatic bone

Column B

_____ 1. Forms bridge of nose

_____ 2. Only movable bone in the facial skeleton

_____ 3. Contains coronoid process

_____ 4. Creates prominence of cheek inferior and lateral to the eye

_____ 5. Contains sockets of upper teeth

_____ 6. Forms inferior portion of nasal septum

_____ 7. Forms anterior portion of zygomatic arch

_____ 8. Scroll-shaped bone

_____ 9. Forms anterior roof of mouth

_____ 10. Contains mental foramen

_____ 11. Forms posterior roof of mouth

_____ 12. Scalelike part in medial wall of orbit

PART D

Match the passageways in column A with the structures transmitted through them in column B. Place the letter of your choice in the space provided.

Column A

a. Foramen magnum
b. Foramen ovale
c. Foramen rotundum
d. Incisive foramen
e. Internal acoustic meatus
f. Jugular foramen
g. Optic canal

Column B

_____ 1. Maxillary division of trigeminal nerve

_____ 2. Nerve fibers of spinal cord

_____ 3. Optic nerve

_____ 4. Vagus and accessory nerves

_____ 5. Nasopalatine nerves

_____ 6. Mandibular division of trigeminal nerve

_____ 7. Vestibular and cochlear nerves

PART E

Identify the numbered bones and features of the skulls indicated in figures 13.7, 13.8, 13.9, 13.10, and 13.11.

Figure 13.7 Identify the bones and features indicated on this anterior view of the skull, using the terms provided.

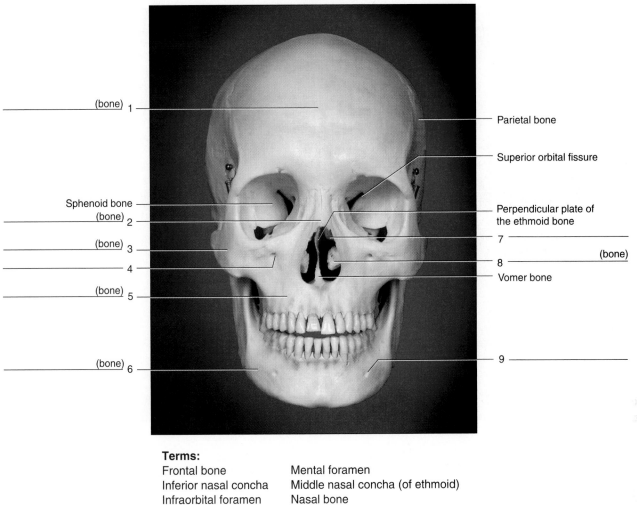

(bone) 1

Parietal bone

Superior orbital fissure

Sphenoid bone

(bone) 2

Perpendicular plate of the ethmoid bone

7

(bone) 3

(bone)

4

8

(bone) 5

Vomer bone

(bone) 6

9

Terms:

Frontal bone
Inferior nasal concha
Infraorbital foramen
Mandible
Maxilla

Mental foramen
Middle nasal concha (of ethmoid)
Nasal bone
Zygomatic bone

Figure 13.8 Identify the bones and features indicated on this lateral view of the skull, using the terms provided.

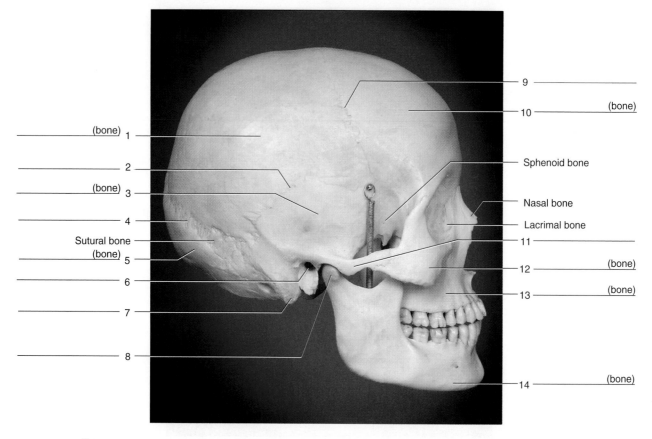

Terms:

Coronal suture	Mandible	Occipital bone	Temporal bone
External auditory meatus	Mandibular condyle	Parietal bone	Zygomatic bone
Frontal bone	Mastoid process	Squamosal suture	Zygomatic process
Lambdoidal suture	Maxilla		

Figure 13.9 Identify the bones and features indicated on this inferior view of the skull, using the terms provided.

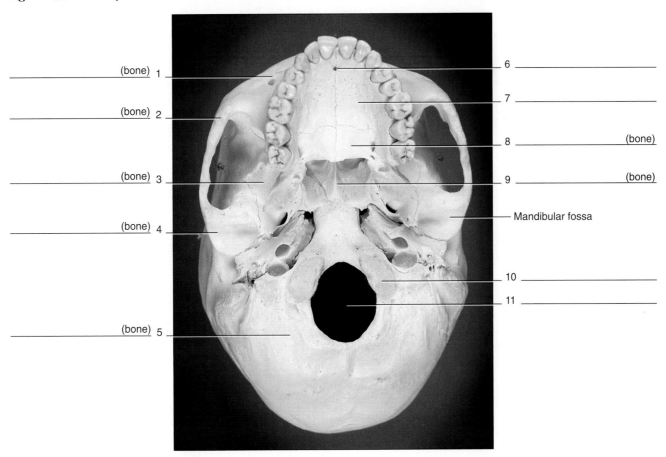

_____ (bone) 1

_____ (bone) 2

_____ (bone) 3

_____ (bone) 4

_____ (bone) 5

6 _____

7 _____

8 _____ (bone)

9 _____ (bone)

Mandibular fossa

10 _____

11 _____

Terms:

Foramen magnum
Incisive foramen
Maxilla
Occipital bone

Occipital condyle
Palatine bone
Palatine process of maxilla
Sphenoid bone

Temporal bone
Vomer bone
Zygomatic bone

Figure 13.10 Identify the bones and features on this floor of the cranial cavity of a skull, using the terms provided.

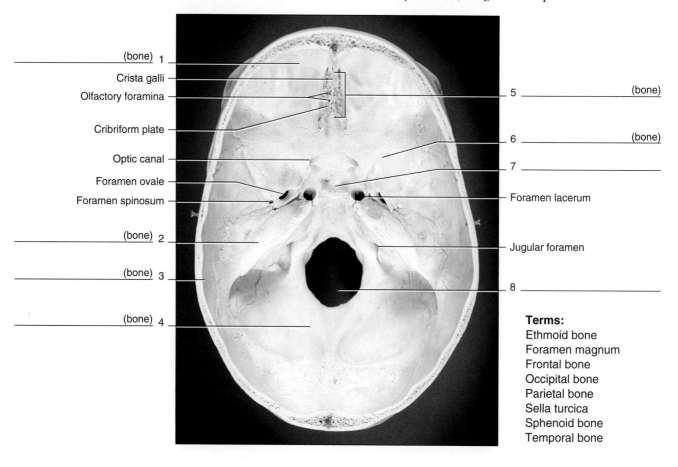

_____ (bone) 1

Crista galli

Olfactory foramina

Cribriform plate

Optic canal

Foramen ovale

Foramen spinosum

_____ (bone) 2

_____ (bone) 3

_____ (bone) 4

5 _____ (bone)

6 _____ (bone)

7 _____

Foramen lacerum

Jugular foramen

8 _____

Terms:
Ethmoid bone
Foramen magnum
Frontal bone
Occipital bone
Parietal bone
Sella turcica
Sphenoid bone
Temporal bone

Figure 13.11 Identify the bones on this disarticulated skull, using the terms provided.

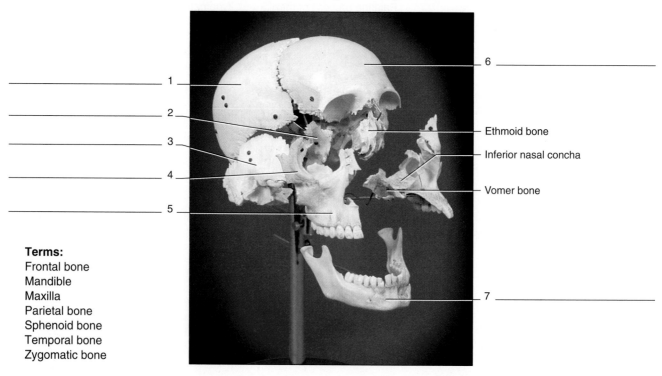

_____ 1

_____ 2

_____ 3

_____ 4

_____ 5

6 _____

Ethmoid bone

Inferior nasal concha

Vomer bone

7 _____

Terms:
Frontal bone
Mandible
Maxilla
Parietal bone
Sphenoid bone
Temporal bone
Zygomatic bone

VERTEBRAL COLUMN AND THORACIC CAGE

MATERIALS NEEDED

Textbook
Human skeleton, articulated
Samples of cervical, thoracic, and lumbar vertebrae
Human skeleton, disarticulated

The vertebral column, consisting of twenty-six bones, extends from the skull to the pelvis and forms the vertical axis of the human skeleton. The column is composed of many vertebrae, which are separated from one another by cartilaginous intervertebral discs and are held together by ligaments.

The thoracic cage surrounds the thoracic and upper abdominal cavities. It includes the ribs, the thoracic vertebrae, the sternum, and the costal cartilages.

PURPOSE OF THE EXERCISE

To examine the vertebral column and the thoracic cage of the human skeleton and to identify the bones and major features of these parts.

LEARNING OBJECTIVES

After completing this exercise, you should be able to

1. Identify the major features of the vertebral column.
2. Name the features of a typical vertebra.
3. Distinguish among a cervical, thoracic, and lumbar vertebra and locate the sacrum and coccyx.
4. Identify the structures of the thoracic cage.
5. Distinguish between true and false ribs.

PROCEDURE A—THE VERTEBRAL COLUMN

1. Review the section entitled "Vertebral Column" in chapter 7 of the textbook.
2. As a review activity, label figures 14.1, 14.2, 14.3, and 14.4.
3. Examine the vertebral column of the human skeleton and locate the following bones and features. At the same time, locate as many of the corresponding bones and features in your own skeleton as possible.

atlas	1
axis	1
cervical vertebrae	7
thoracic vertebrae	12
lumbar vertebrae	5
intervertebral discs	
vertebral canal	
sacrum	1
coccyx	1
cervical curvature	
thoracic curvature	
lumbar curvature	
pelvic curvature	
intervertebral foramina	

Critical Thinking Application

Note the four curvatures of the vertebral column. What functional advantages exist with curvatures for skeletal structure instead of a straight vertebral column?

_____

_____

_____

_____

4. Compare the available samples of cervical, thoracic, and lumbar vertebrae by noting differences in size and shape, and by locating the following features:

body

pedicles

vertebral foramen

laminae

spinous process

vertebral arch

Figure 14.1 Label the bones and features of the vertebral column.

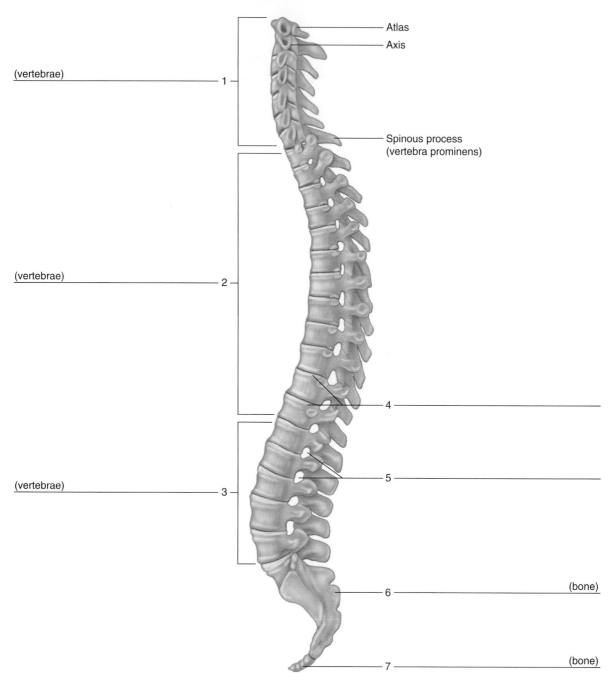

(vertebrae) —— 1

(vertebrae) —— 2

(vertebrae) —— 3

Atlas
Axis
Spinous process
(vertebra prominens)

4 ————————

5 ————————

6 ———————— (bone)

7 ———————— (bone)

transverse processes

superior articular processes

inferior articular processes

transverse foramina

facets

dens of axis

5. Examine the sacrum and coccyx. Locate the following features:

sacrum

superior articular process

dorsal sacral foramen

pelvic sacral foramen

sacral promontory

sacral canal

tubercles

sacral hiatus

coccyx

Figure 14.2 Label the superior features of (*a*) the atlas and (*b*) the axis.

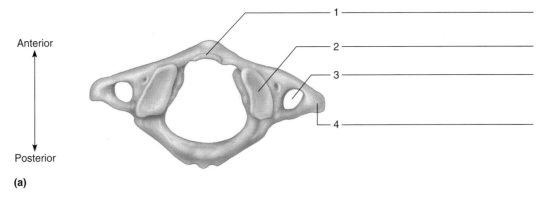

Anterior

Posterior

(a)

1 _____

2 _____

3 _____

4 _____

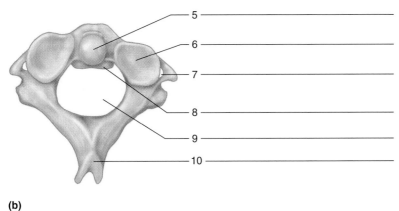

(b)

5 _____

6 _____

7 _____

8 _____

9 _____

10 _____

6. Complete Parts A and B of Laboratory Report 14.

PROCEDURE B—THE THORACIC CAGE

1. Review the section entitled "Thoracic Cage" in chapter 7 of the textbook.
2. As a review activity, label figures 14.5 and 14.6.
3. Examine the thoracic cage of the human skeleton and locate the following bones and features:

   rib

   head

   tubercle

   neck

   shaft

   anterior (sternal) end

   facets

   true ribs

   false ribs

   floating ribs

   costal cartilages

sternum

   sternal notch

   clavicular notch

   manubrium

   body

   xiphoid process

   sternal angle

4. Complete Parts C and D of the laboratory report.

Web Quest

What are the functions of individual bones and features? Search these and review the anatomy of the skeleton at

www.mhhe.com/shier10

Click on Student Edition. Click on Martin Lab Manual, Web Quest.

Figure 14.3 Label the features of the (*a*) cervical, (*b*) thoracic, and (*c*) lumbar vertebrae.

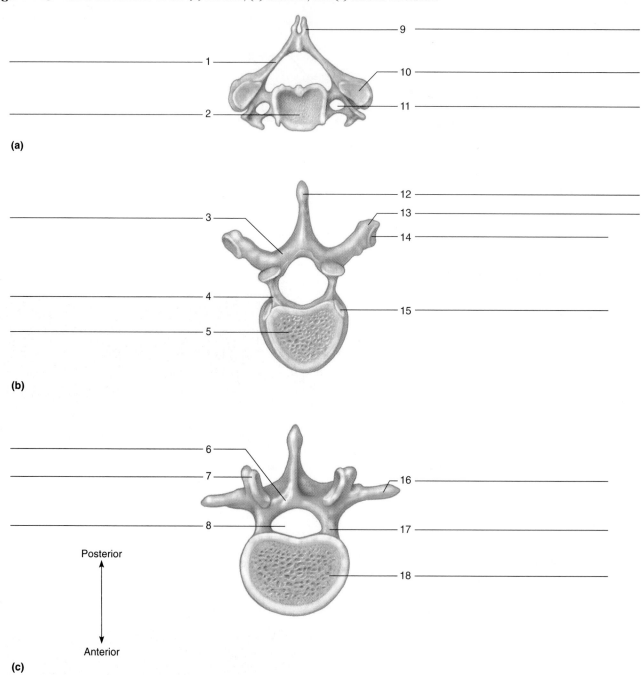

(a)

(b)

Posterior

Anterior

(c)

Figure 14.4 Label the coccyx and the features of the sacrum: (*a*) anterior view; (*b*) posterior view.

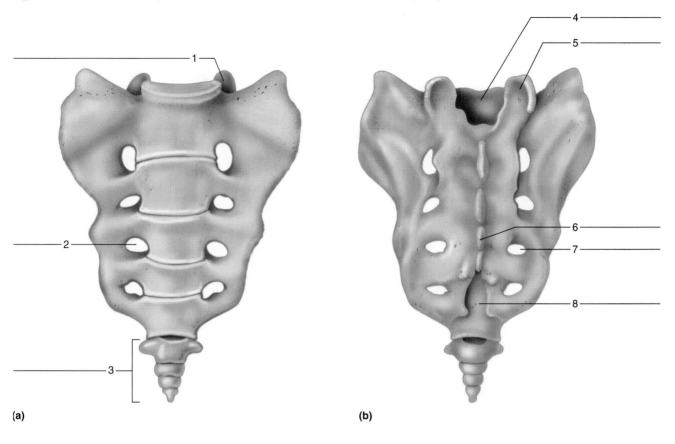

(a)

(b)

Figure 14.5 Label the bones and features of the thoracic cage.

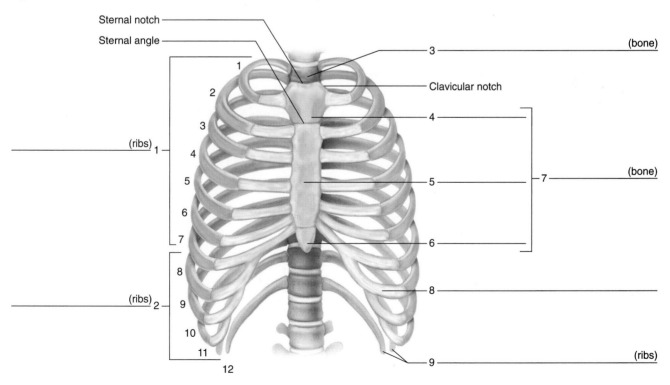

Sternal notch

Sternal angle

Clavicular notch

(ribs) 1

(ribs) 2

(bone)

(bone)

(ribs)

Figure 14.6 Label the features of the ribs: (*a*) posterior view; (*b*) superior view showing articulations with a thoracic vertebra.

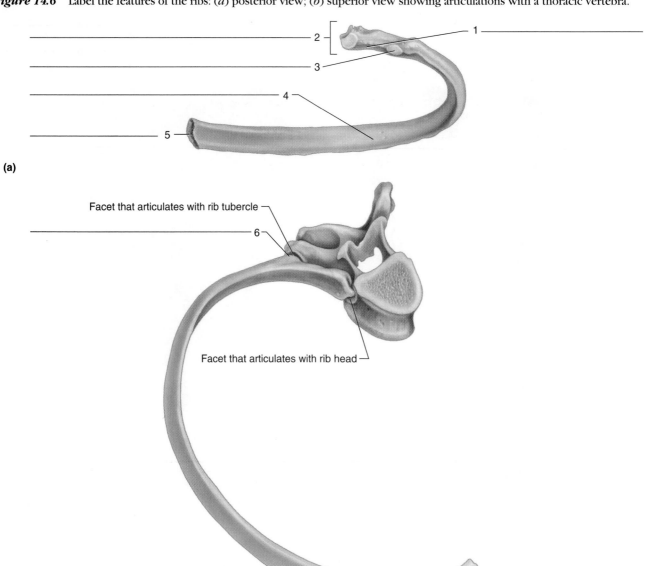

2

1

3

4

5

(a)

Facet that articulates with rib tubercle

6

Facet that articulates with rib head

7

(b)

VERTEBRAL COLUMN AND THORACIC CAGE

PART A

Complete the following statements:

1. The vertebral column encloses and protects the _____ .
2. The number of separate bones in the vertebral column of an infant is _____ .
3. The number of separate bones in the vertebral column of an adult is _____ .
4. The thoracic and pelvic curvatures of the vertebral column are called _____ curves.
5. The _____ of the vertebrae support the weight of the head and trunk.
6. The _____ separate adjacent vertebrae, and they soften the forces created by walking.
7. The pedicles, laminae, and spinous process of a vertebra form the _____ .
8. The intervertebral foramina provide passageways for _____ .
9. Transverse foramina of cervical vertebrae serve as passageways for _____ leading to the brain.
10. The first vertebra is also called the _____ .
11. The second vertebra is also called the _____ .
12. When the head is moved from side to side, the first vertebra pivots around the _____ of the second vertebra.
13. The _____ vertebrae have the largest and strongest bodies.
14. The number of vertebrae that fuse to form the sacrum is _____ .
15. The joint between a coxa of the pelvis and the sacrum is called the _____ joint.
16. The upper, anterior margin of the sacrum that projects forward is called the _____ .
17. An opening called the _____ exists at the tip of the sacral canal.

PART B

Based on your observations, compare typical cervical, thoracic, and lumbar vertebrae in relation to the characteristics indicated in the table. Consider characteristics such as size, shape, presence or absence, and unique features for your responses.

Vertebra	Number	Size	Body	Spinous Process	Transverse Foramina
Cervical					
Thoracic					
Lumbar					

PART C

Complete the following statements:

1. The adult skeleton of most men and women contains a total of _____ (number) bones.
2. The last two pairs of ribs that have no cartilaginous attachments to the sternum are sometimes called _____ ribs.
3. The tubercles of the ribs articulate with facets on the _____ processes of the thoracic vertebrae.
4. Costal cartilages are composed of _____ tissue.
5. The manubrium articulates with the _____ on its superior border.
6. List three general functions of the thoracic cage. _____

_____

PART D

Identify the bones and features indicated in the radiograph (X ray) of the neck in figure 14.7.

Figure 14.7 Identify the bones and features indicated in this radiograph (X ray) of the neck (lateral view), using the terms provided.

Terms:
Atlas
Axis
Body
Intervertebral disc
Spinous process
Transverse process

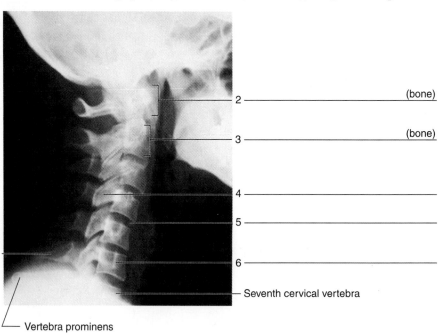

2 _____ (bone)

3 _____ (bone)

4 _____

5 _____

6 _____

— Seventh cervical vertebra

1 _____

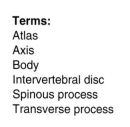

— Vertebra prominens

PECTORAL GIRDLE AND UPPER LIMB

The pectoral girdle consists of two clavicles and two scapulae. These parts support the upper limbs and serve as attachments for various muscles that move these limbs.

Each upper limb includes a humerus, a radius, an ulna, eight carpals, five metacarpals, and fourteen phalanges. These bones form the framework of the arm, forearm, and hand. They also function as parts of levers when the limbs are moved.

PURPOSE OF THE EXERCISE

To examine the bones of the pectoral girdle and upper limb and to identify the major features of these bones.

LEARNING OBJECTIVES

After completing this exercise, you should be able to

1. Locate and identify the bones of the pectoral girdle and their major features.
2. Locate and identify the bones of the upper limb and their major features.

PROCEDURE A—THE PECTORAL GIRDLE

1. Review the section entitled "Pectoral Girdle" in chapter 7 of the textbook.
2. As a review activity, label figures 15.1 and 15.2.

3. Examine the bones of the pectoral girdle and locate the following features. At the same time, locate as many of the corresponding surface bones and features of your own skeleton as possible.

clavicle

   medial (sternal) end

   lateral (acromial) end

scapula

   spine

   lateral (axillary) border

   medial (vertebral) border

   superior border

   acromion process

   coracoid process

   glenoid cavity (fossa)

   supraspinous fossa

   infraspinous fossa

Critical Thinking Application

Why is the clavicle a bone that can easily fracture?

_____

_____

_____

_____

4. Complete Part A of Laboratory Report 15.

Figure 15.1 Label the bones and features of the right shoulder and upper limb (anterior view).

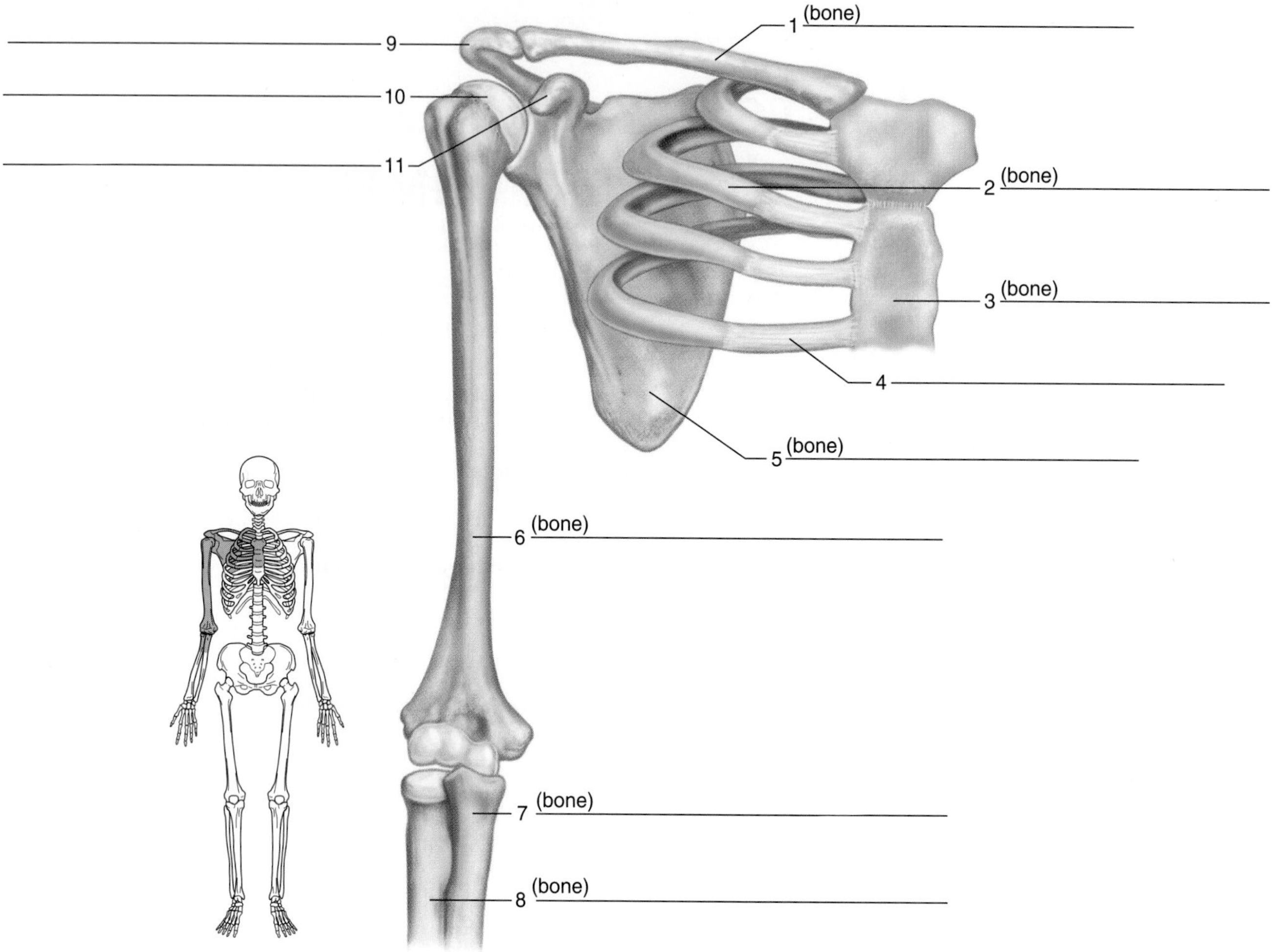

1 (bone) _____

2 (bone) _____

3 (bone) _____

4 _____

5 (bone) _____

6 (bone) _____

7 (bone) _____

8 (bone) _____

9 _____

10 _____

11 _____

Figure 15.2 Label (*a*) the posterior surface of the right scapula and (*b*) the lateral aspect of the right scapula.

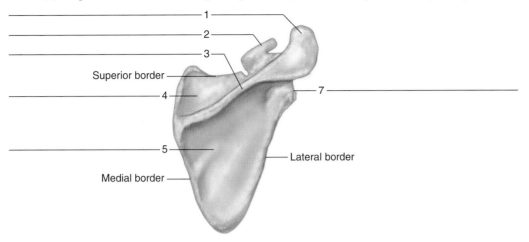

Superior border

Medial border

Lateral border

(a)

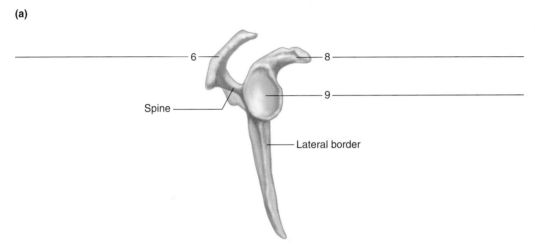

Spine

Lateral border

(b)

PROCEDURE B—THE UPPER LIMB

1. Review the section entitled "Upper Limb" in chapter 7 of the textbook.
2. As a review activity, label figures 15.3, 15.4, 15.5, and 15.6.
3. Examine the following bones and features of the upper limb:

humerus

head

greater tubercle

lesser tubercle

intertubercular groove

anatomical neck

surgical neck

deltoid tuberosity

capitulum

trochlea

medial epicondyle

lateral epicondyle

coronoid fossa

olecranon fossa

radius

head

radial tuberosity

styloid process

ulnar notch

ulna

trochlear notch (semilunar notch)

radial notch

olecranon process

coronoid process

styloid process

head

Figure 15.3 Label the (*a*) anterior surface and (*b*) posterior surface of the right humerus.

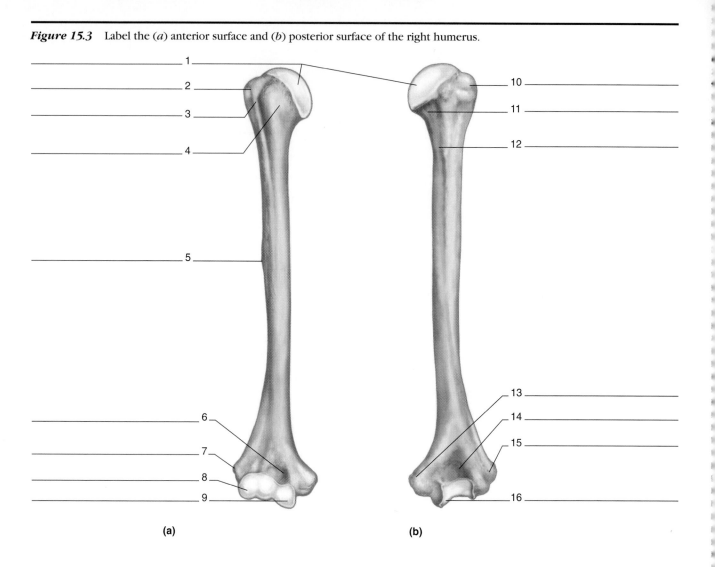

1 _____

2 _____

3 _____

4 _____

5 _____

6 _____

7 _____

8 _____

9 _____

10 _____

11 _____

12 _____

13 _____

14 _____

15 _____

16 _____

(a)

(b)

Figure 15.4 Label the major anterior features of the right radius and ulna.

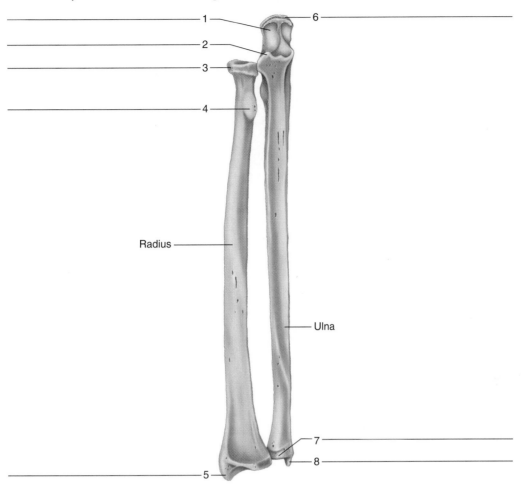

Radius

Ulna

Figure 15.5 Label the bones and features of the right elbow, posterior view.

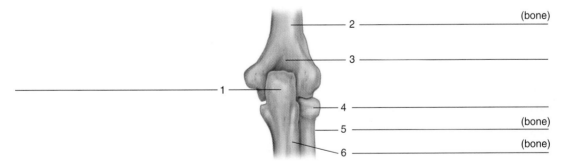

(bone)

(bone)

(bone)

Figure 15.6 Label the bones and groups of bones in this anterior view of the right hand.

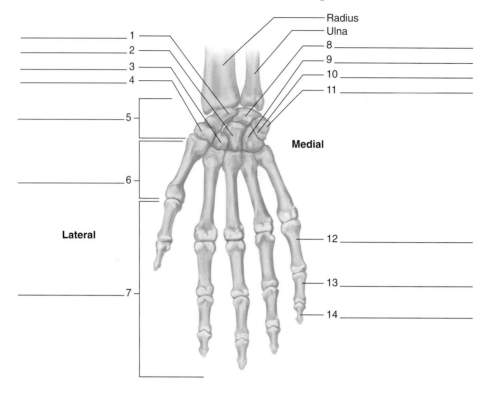

carpal bones

   proximal row (listed lateral to medial)

      scaphoid

      lunate

      triquetrum

      pisiform

   distal row (listed medial to lateral)

      hamate

      capitate

      trapezoid

      trapezium

metacarpal bones

phalanges

   proximal phalanx

   middle phalanx

   distal phalanx

4. Complete Parts B, C, and D of the laboratory report.

The following mnemonic device will help you learn the eight carpals:

So Long Top Part
Here Comes The Thumb

The first letter of each word corresponds to the first letter of a carpal. Notice that this device arranges the carpals in order for the proximal, transverse row of four bones from lateral to medial, followed by the distal, transverse row from medial to lateral, which ends nearest the thumb. This arrangement assumes the hand is in the anatomical position.

OPTIONAL ACTIVITY

Use different colored pencils to distinguish the individual bones in figure 15.6.

Web Quest

What are the functions of individual bones and features? Search these and review the anatomy of the skeleton at
www.mhhe.com/shier10
Click on Student Edition. Click on Martin Lab Manual, Web Quest.

PECTORAL GIRDLE AND UPPER LIMB

PART A

Complete the following statements:

1. The pectoral girdle is an incomplete ring because it is open in the back between the _____ .
2. The medial ends of the clavicles articulate with the _____ of the sternum.
3. The lateral ends of the clavicles articulate with the _____ of the scapulae.
4. The _____ is a bone that serves as a brace between the sternum and the scapula.
5. The _____ divides the scapula into unequal portions.
6. The tip of the shoulder is the _____ of the scapula.
7. At the lateral end of the scapula, the _____ curves anteriorly and inferiorly from the clavicle.
8. The glenoid cavity of the scapula articulates with the _____ of the humerus.

PART B

Match the bones in column A with the bones and features in column B. Place the letter of your choice in the space provided.

Column A	Column B
a. Carpals	_____ 1. Capitate
b. Humerus	_____ 2. Capitulum
c. Metacarpals	_____ 3. Coronoid fossa
d. Phalanges	_____ 4. Five palmar bones
e. Radius	_____ 5. Fourteen bones in digits
f. Ulna	_____ 6. Intertubercular groove
	_____ 7. Lunate
	_____ 8. Olecranon fossa
	_____ 9. Radial notch
	_____ 10. Radial tuberosity
	_____ 11. Trapezium
	_____ 12. Triquetrum
	_____ 13. Trochlea
	_____ 14. Trochlear notch
	_____ 15. Ulnar notch

PART C

Identify the bones and features indicated in the radiographs (X rays) of figures 15.7, 15.8, and 15.9.

Figure 15.7   Identify the bones and features indicated on this radiograph of the elbow, using the terms provided.

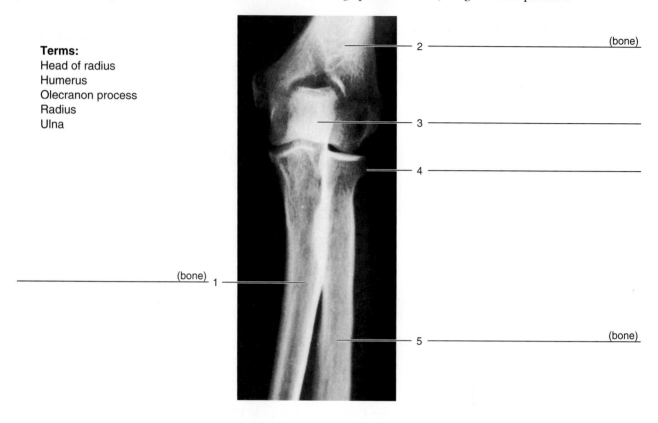

Terms:
Head of radius
Humerus
Olecranon process
Radius
Ulna

2 _____ (bone)

3 _____

4 _____

(bone) 1 _____

5 _____ (bone)

Figure 15.8   Identify the bones and features indicated on this radiograph of the shoulder, using the terms provided.

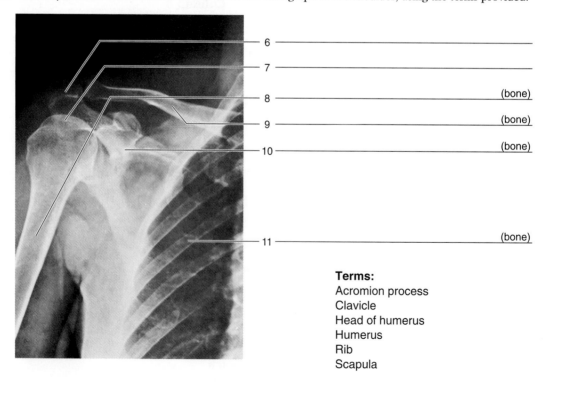

6 _____

7 _____

8 _____ (bone)

9 _____ (bone)

10 _____ (bone)

11 _____ (bone)

Terms:
Acromion process
Clavicle
Head of humerus
Humerus
Rib
Scapula

Figure 15.9 Identify the bones indicated on this radiograph of the left hand, using the terms provided.

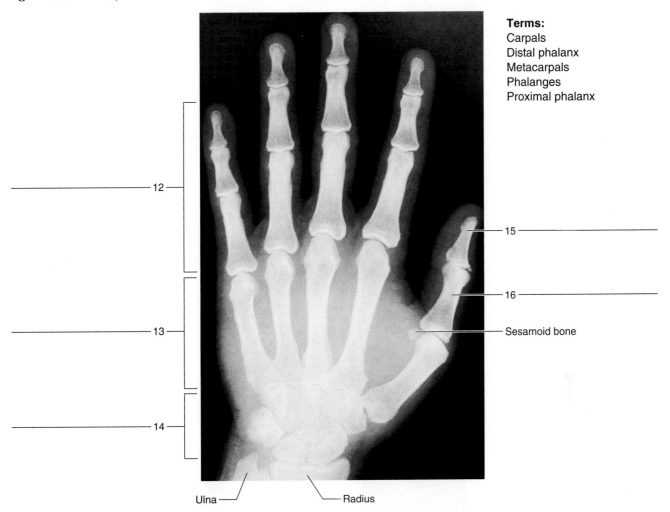

Terms:
Carpals
Distal phalanx
Metacarpals
Phalanges
Proximal phalanx

12

13

14

15

16

Sesamoid bone

Ulna

Radius

PART D

Identify the bones of the hand in figure 15.10.

1. _____
2. _____
3. _____
4. _____
5. _____
6. _____

7. _____
8. _____
9. _____
10. _____
11. _____
12. _____

Figure 15.10 Identify the bones that are numbered on this anterior view of the right hand, using the terms provided.

Terms:
Capitate
Distal phalanges
Hamate
Lunate
Metacarpals
Middle phalanges
Pisiform
Proximal phalanges
Scaphoid
Trapezium
Trapezoid
Triquetrum

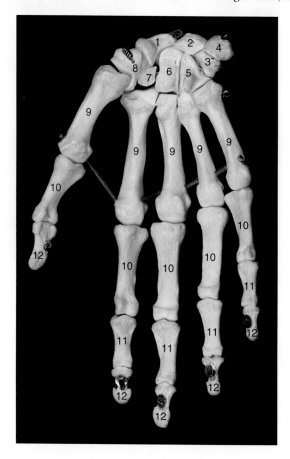

PELVIC GIRDLE AND LOWER LIMB

MATERIALS NEEDED

Textbook
Human skeleton, articulated
Human skeleton, disarticulated
Male and female pelves

For Optional Activity:

Colored pencils

The pelvic girdle includes two coxae (hipbones) that articulate with each other anteriorly at the symphysis pubis and posteriorly with the sacrum. Together, the pelvic girdle, sacrum, and coccyx comprise the pelvis. The pelvis, in turn, provides support for the trunk of the body and provides attachments for the lower limbs.

The bones of the lower limb form the framework of the thigh, leg, and foot. Each limb includes a femur, a patella, a tibia, a fibula, seven tarsals, five metatarsals, and fourteen phalanges.

PURPOSE OF THE EXERCISE

To examine the bones of the pelvic girdle and lower limb and to identify the major features of these bones.

LEARNING OBJECTIVES

After completing this exercise, you should be able to

1. Locate and identify the bones of the pelvic girdle and their major features.
2. Locate and identify the bones of the lower limb and their major features.

PROCEDURE A—THE PELVIC GIRDLE

1. Review the section entitled "Pelvic Girdle" in chapter 7 of the textbook.
2. As a review activity, label figures 16.1 and 16.2.

Figure 16.1   Label the bones of the pelvis (posterior view).

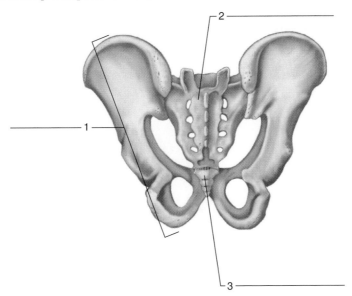

Figure 16.2 Label (*a*) the lateral and (*b*) the medial features of the right coxa.

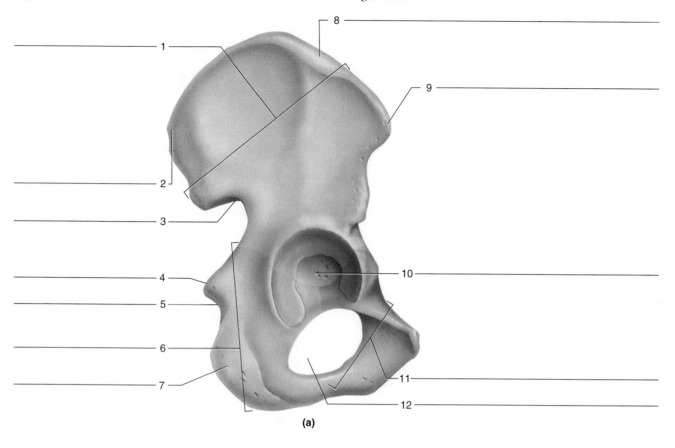

(a)

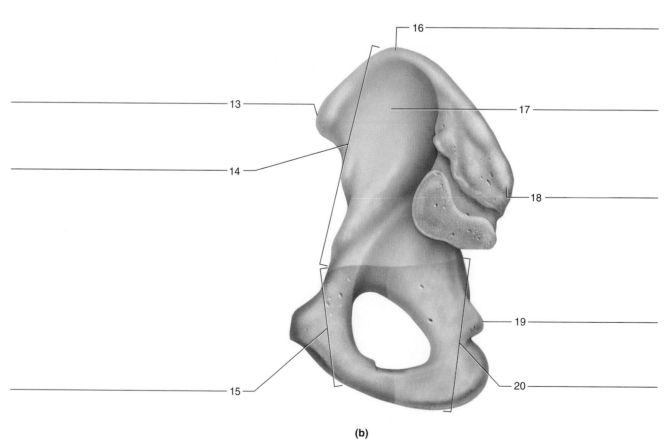

(b)

3. Examine the bones of the pelvic girdle and locate the following:

coxa (hipbone; innominate)

    acetabulum

    ilium

    iliac crest

    iliac fossa

    sacroiliac joint

    anterior superior iliac spine

    posterior superior iliac spine

    greater sciatic notch

    lesser sciatic notch

    ischium

    ischial tuberosity

    ischial spine

    pubis

    symphysis pubis

    pubic arch

    obturator foramen

4. Complete Part A of Laboratory Report 16.

Critical Thinking Application

Examine the male and female pelves. Look for major differences between them. Note especially the flare of the iliac bones, the angle of the pubic arch, the distance between the ischial spines and ischial tuberosities, and the curve and width of the sacrum. In what ways are the differences you observed related to the function of the female pelvis as a birth canal?

_____

_____

_____

_____

PROCEDURE B—THE LOWER LIMB

1. Review the section entitled "Lower Limb" in chapter 7 of the textbook.
2. As a review activity, label figures 16.3, 16.4, 16.5, and 16.6.

3. Examine the bones of the lower limb and locate each of the following:

femur

    head

    fovea capitis

    neck

    greater trochanter

    lesser trochanter

    linea aspera

    lateral condyle

    medial condyle

    lateral epicondyle

    medial epicondyle

patella

tibia

    medial condyle

    lateral condyle

    tibial tuberosity

    anterior crest

    medial malleolus

fibula

    head

    lateral malleolus

tarsal bones

    talus

    calcaneus

    navicular

    cuboid

    lateral cuneiform

    intermediate cuneiform

    medial cuneiform

metatarsal bones

phalanges

    proximal phalanx

    middle phalanx

    distal phalanx

4. Complete Parts B, C, and D of the laboratory report.

Figure 16.3 Label the features of (a) the anterior surface and (b) the posterior surface of the right femur.

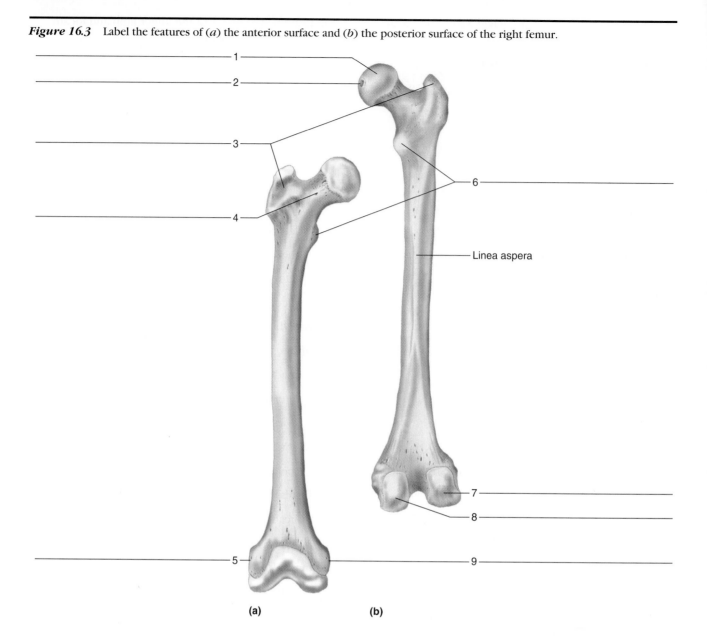

Linea aspera

(a) (b)

OPTIONAL ACTIVITY

Use different colored pencils to distinguish the individual bones in figure 16.6.

Web Quest

What are the functions of individual bones and features? Search these and review the anatomy of the skeleton at

www.mhhe.com/shier10

Click on Student Edition. Click on Martin Lab Manual, Web Quest.

Figure 16.4   Label the bones and features of the right tibia and fibula in this anterior view.

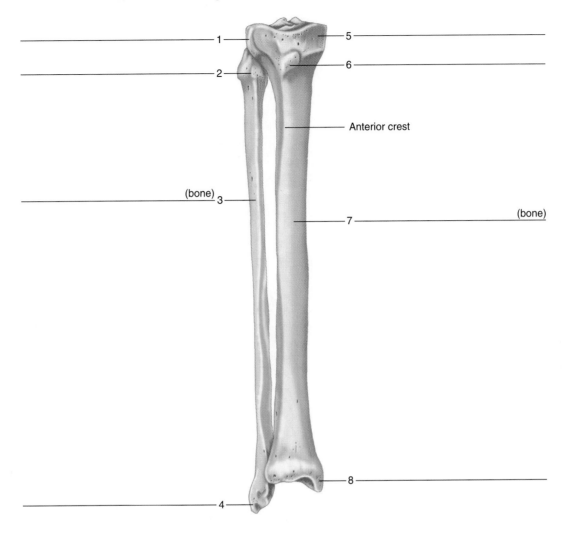

1 —
2 —
(bone) 3 —
4 —
5 —
6 —
Anterior crest
7 — (bone)
8 —

Figure 16.5   Label the bones and features of the right knee, posterior view.

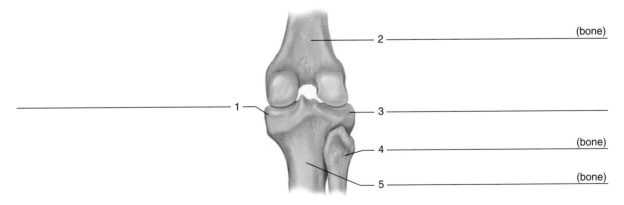

2 — (bone)
1 —
3 —
4 — (bone)
5 — (bone)

Figure 16.6   Label the bones of the superior surface of the right foot.

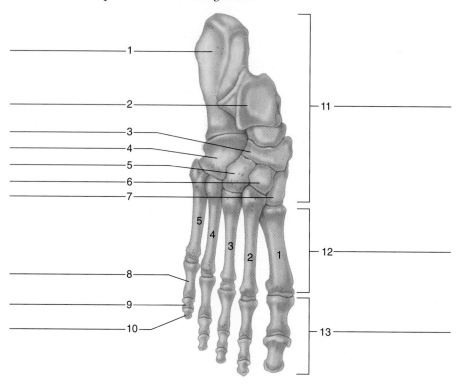

PELVIC GIRDLE AND LOWER LIMB

PART A

Complete the following statements:

1. The pelvic girdle consists of two _____ .
2. The head of the femur articulates with the _____ of the coxa.
3. The _____ is the largest portion of the coxa.
4. The distance between the _____ represents the shortest diameter of the pelvic outlet.
5. The pubic bones come together anteriorly to form the joint called the _____ .
6. The _____ is the portion of the ilium that causes the prominence of the hip.
7. When a person sits, the _____ of the ischium supports the weight of the body.
8. The angle formed by the pubic bones below the symphysis pubis is called the _____ .
9. The _____ is the largest foramen in the skeleton.
10. The ilium joins the sacrum at the _____ joint.

PART B

Match the bones in column A with the features in column B. Place the letter of your choice in the space provided.

Column A	Column B
a. Femur	_____ 1. Middle phalanx
b. Fibula	_____ 2. Lesser trochanter
c. Metatarsals	_____ 3. Medial malleolus
d. Patella	_____ 4. Fovea capitis
e. Phalanges	_____ 5. Calcaneus
f. Tarsals	_____ 6. Lateral cuneiform
g. Tibia	_____ 7. Tibial tuberosity
	_____ 8. Talus
	_____ 9. Linea aspera
	_____ 10. Lateral malleolus
	_____ 11. Sesamoid bone
	_____ 12. Five bones that form the instep

Identify the bones and features indicated in the radiographs (X rays) of figures 16.7, 16.8, and 16.9.

Figure 16.7   Identify the bones and features indicated on this radiograph (X ray) of the pelvic region, using the terms provided.

Terms:
Head of femur
Ilium
Obturator foramen
Pubis
Sacrum
Symphysis pubis

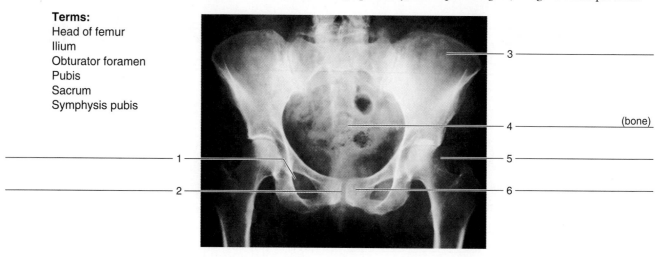

3 _____

4 _____ (bone)

1 _____

5 _____

2 _____

6 _____

Figure 16.8   Identify the bones and features indicated on this radiograph of the knee, using the terms provided.

(bone) 7 _____

Terms:
Femur
Fibula
Head of fibula
Lateral condyle
Lateral epicondyle
Tibia

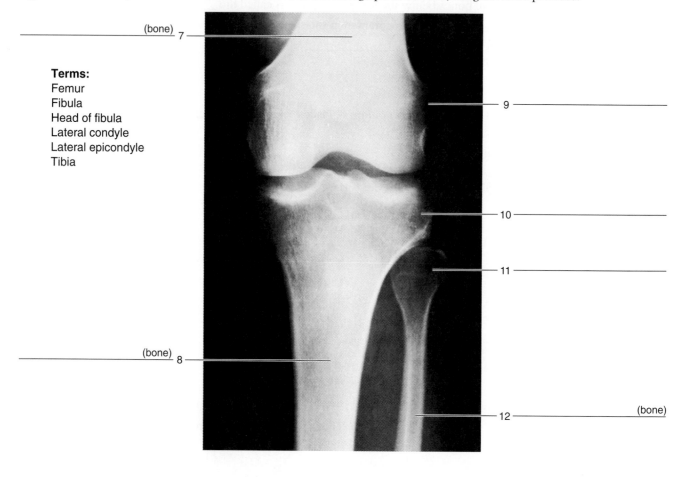

9 _____

10 _____

11 _____

(bone) 8 _____

12 _____ (bone)

Figure 16.9 Identify the bones indicated on this radiograph of the left foot, using the terms provided.

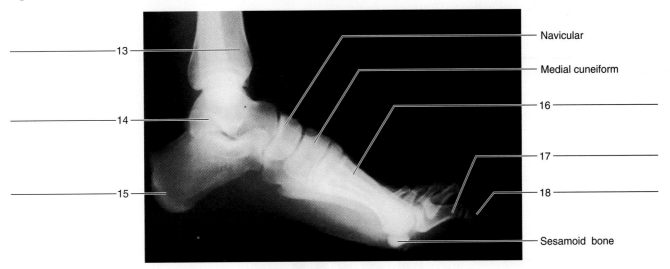

13 _____

14 _____

15 _____

Navicular

Medial cuneiform

16 _____

17 _____

18 _____

Sesamoid bone

Terms:
Calcaneus
Distal phalanx
Metatarsal
Proximal phalanx
Talus
Tibia

PART D

Identify the bones of the foot in figure 16.10.

Figure 16.10   Identify the bones indicated on this superior view of the right foot, using the terms provided.

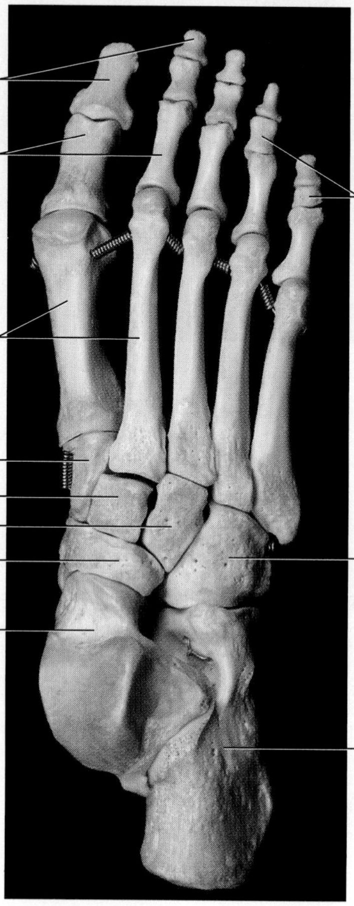

Terms:
Calcaneus
Cuboid
Distal phalanges
Intermediate cuneiform
Lateral cuneiform
Medial cuneiform
Metatarsals
Middle phalanges
Navicular
Proximal phalanges
Talus

LABORATORY EXERCISE 17

THE JOINTS

MATERIALS NEEDED

Textbook
Human skull
Human skeleton, articulated
Models of synovial joints (shoulder, elbow, hip, and knee)

For Demonstration:

Fresh animal joint (knee joint preferred)
Radiographs of major joints

SAFETY

- Wear disposable gloves when handling the fresh animal joint.
- Wash your hands before leaving the laboratory.

Joints are junctions between bones. Although they vary considerably in structure, they can be classified according to the type of tissue that binds the bones together. Thus, the three groups of joints can be identified as (1) fibrous joints, (2) cartilaginous joints, and (3) synovial joints.

Movements occurring at freely movable synovial joints are due to the contractions of skeletal muscles. In each case, the type of movement depends on the kind of joint involved and the way in which the muscles are attached to the bones on either side of the joint.

PURPOSE OF THE EXERCISE

To examine examples of the three types of joints, to identify the major features of these joints, and to review the types of movements produced at synovial joints.

LEARNING OBJECTIVES

After completing this exercise, you should be able to

1. Distinguish among fibrous, cartilaginous, and synovial joints.

2. Identify examples of each type of joint.
3. Identify the major features of each type of joint.
4. Locate and identify examples of each of the six types of synovial joints.
5. Identify the types of movements that occur at synovial joints.
6. Describe the structure of the shoulder, elbow, hip, and knee joints.

PROCEDURE A—TYPES OF JOINTS

1. Review the section entitled "Classification of Joints" in chapter 8 of the textbook.
2. Examine the human skull and articulated skeleton to locate examples of the following types of joints:

   fibrous joints

   syndesmosis

   suture

   gomphosis

   cartilaginous joints

   synchondrosis

   symphysis

   synovial joints

3. Complete Part A of Laboratory Report 17.
4. Locate examples of the following types of synovial joints in the skeleton. At the same time, examine the corresponding joints in the models and in your own body. Experiment with each joint to experience its range of movements.

   ball-and-socket joint

   condyloid joint

   gliding joint

   hinge joint

   pivot joint

   saddle joint

5. Complete Parts B and C of the laboratory report.

PROCEDURE B—JOINT MOVEMENTS

1. Review the section entitled "Types of Joint Movements" in chapter 8 of the textbook.
2. When the body is in anatomical position, most joints are extended and/or adducted. Skeletal muscle action involves the movable end (*insertion*) being pulled toward the stationary end (*origin*). In the limbs, the origin is usually proximal to the insertion; in the trunk, the origin is usually medial to the insertion. Use these concepts as reference points as you move joints. Move various parts of your own body to demonstrate the following joint movements:

> flexion
>
> extension
>
> hyperextension
>
> dorsiflexion
>
> plantar flexion
>
> abduction
>
> adduction
>
> rotation
>
> circumduction
>
> supination
>
> pronation
>
> eversion
>
> inversion
>
> protraction
>
> retraction
>
> elevation
>
> depression

3. Have your laboratory partner do some of the preceding movements and see if you can identify correctly the movements made.

Critical Thinking Application

Describe a body position that can exist when all major body parts are flexed.

_____

_____

_____

4. Complete Part D of the laboratory report.

PROCEDURE C—EXAMPLES OF SYNOVIAL JOINTS

1. Study and compare the shoulder, elbow, hip, and knee joints in figures 17.1, 17.2, 17.3, 17.4, and 17.5.
2. Examine models of the shoulder, elbow, hip, and knee joints. Locate as many features as possible on the models that are illustrated in figures 17.1 through 17.5.
3. Complete Part E of the laboratory report.

Figure 17.1 Shoulder joint structures (coronal section).

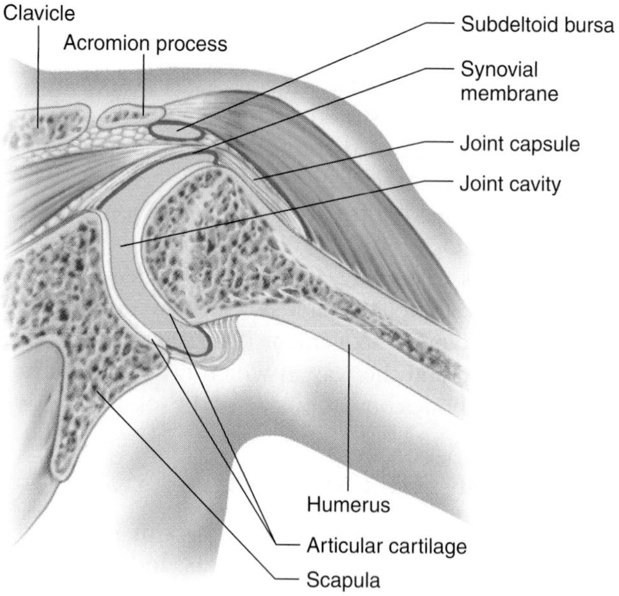

Clavicle
Acromion process
Subdeltoid bursa
Synovial membrane
Joint capsule
Joint cavity
Humerus
Articular cartilage
Scapula

Figure 17.2 Elbow joint structures (sagittal section).

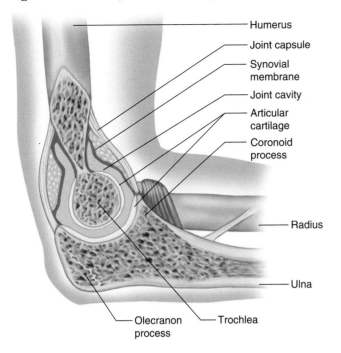

- Humerus
- Joint capsule
- Synovial membrane
- Joint cavity
- Articular cartilage
- Coronoid process
- Radius
- Ulna
- Olecranon process
- Trochlea

Figure 17.4 Knee joint structures (sagittal section).

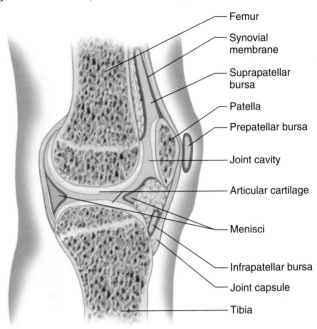

- Femur
- Synovial membrane
- Suprapatellar bursa
- Patella
- Prepatellar bursa
- Joint cavity
- Articular cartilage
- Menisci
- Infrapatellar bursa
- Joint capsule
- Tibia

Figure 17.3 Hip joint structures (coronal section).

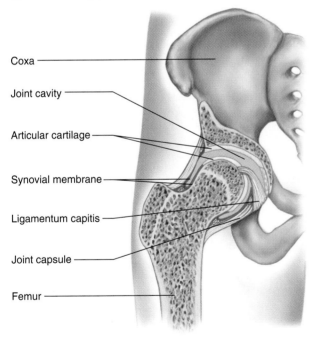

- Coxa
- Joint cavity
- Articular cartilage
- Synovial membrane
- Ligamentum capitis
- Joint capsule
- Femur

Figure 17.5 Anterior view of right knee (patella removed).

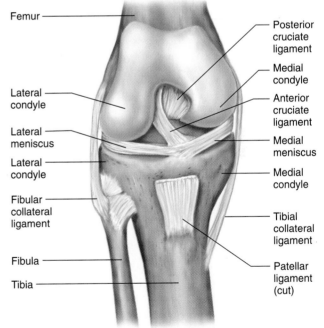

- Femur
- Lateral condyle
- Lateral meniscus
- Lateral condyle
- Fibular collateral ligament
- Fibula
- Tibia
- Posterior cruciate ligament
- Medial condyle
- Anterior cruciate ligament
- Medial meniscus
- Medial condyle
- Tibial collateral ligament
- Patellar ligament (cut)

DEMONSTRATION

Study the available radiographs of joints by holding the films in front of a light source. Identify the type of joint and the bones incorporated in the joint. Also identify other major features that are visible.

Web Quest

Why does the shoulder joint allow extensive movement? What is joint fluoroscopy? Search for these answers and review joints at

www.mhhe.com/shier10

Click on Student Edition. Click on Martin Lab Manual, Web Quest.

Laboratory Report **17**

Name _____

Date _____

Section _____

THE JOINTS

PART A

Match the terms in column A with the descriptions in column B. Place the letter of your choice in the space provided.

Column A

a. Gomphosis
b. Suture
c. Symphysis
d. Synchondrosis
e. Syndesmosis

Column B

_____ 1. Immovable joint between flat bones of the skull united by a thin layer of connective tissue

_____ 2. Articular cartilages at joint attached by pad of fibrocartilage

_____ 3. Temporary joint in which bones are united by bands of hyaline cartilage

_____ 4. Slightly movable joint in which bones are united by interosseous ligament

_____ 5. Joint formed by union of cone-shaped bony process in bony socket

PART B

Identify the types of joints that are numbered in figure 17.6.

1. _____
2. _____
3. _____
4. _____
5. _____
6. _____
7. _____
8. _____
9. _____
10. _____

Figure 17.6 Identify the types of joints that are numbered in these illustrations.

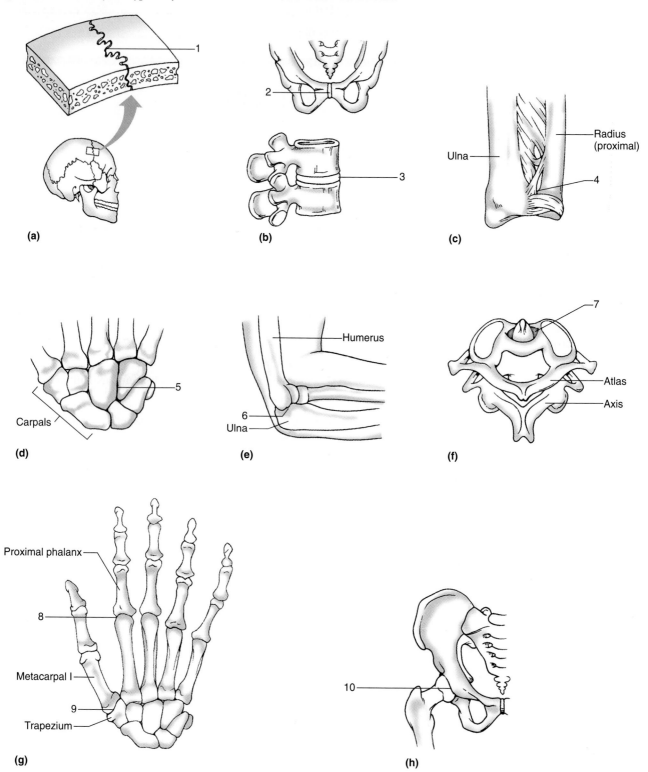

(a)

(b)

(c)

Radius (proximal)

Ulna

Carpals

(d)

Humerus

Ulna

(e)

Atlas

Axis

(f)

Proximal phalanx

Metacarpal I

Trapezium

(g)

(h)

PART C

Match the types of synovial joints in column A with the examples in column B. Place the letter of your choice in the space provided.

Column A		Column B
a.	Ball-and-socket	_____ 1.	Hip joint
b.	Condyloid	_____ 2.	Metacarpal-phalanx
c.	Gliding	_____ 3.	Proximal radius-ulna
d.	Hinge	_____ 4.	Humerus-ulna of the elbow joint
e.	Pivot	_____ 5.	Phalanx-phalanx
f.	Saddle	_____ 6.	Shoulder joint
		_____ 7.	Knee joint
		_____ 8.	Carpal-metacarpal of the thumb
		_____ 9.	Carpal-carpal
		_____ 10.	Tarsal-tarsal

PART D

Identify the types of joint movements that are numbered in figure 17.7.

1. _____ (of head)
2. _____ (of shoulder)
3. _____ (of shoulder)
4. _____ (of hand)
5. _____ (of hand)
6. _____ (of arm at shoulder)
7. _____ (of arm at shoulder)
8. _____ (of hand at wrist)
9. _____ (of hand at wrist)
10. _____ (of thigh at hip)
11. _____ (of thigh at hip)
12. _____ (of lower limb at hip)
13. _____ (of chin/mandible)
14. _____ (of chin/mandible)
15. _____ (of vertebral column/trunk)
16. _____ (of vertebral column/trunk)
17. _____ (of head and neck)
18. _____ (of head and neck)
19. _____ (of arm at shoulder)
20. _____ (of arm at shoulder)
21. _____ (of forearm at elbow)
22. _____ (of forearm at elbow)
23. _____ (of thigh at hip)
24. _____ (of thigh at hip)
25. _____ (of leg at knee)
26. _____ (of leg at knee)
27. _____ (of foot at ankle)
28. _____ (of foot at ankle)

PART E

Complete the following table:

Name of Joint	Type of Joint	Bones Included	Types of Movement Possible
Shoulder joint			
Elbow joint			
Hip joint			
Knee joint			

Figure 17.7 Identify each of the types of movements that are numbered and illustrated: (*a*) anterior view; (*b*) lateral view of head; (*c*) lateral view. (Original illustration drawn by Ross Martin)

(a)

(b)

(c)

SKELETAL MUSCLE STRUCTURE

MATERIALS NEEDED

Textbook
Compound light microscope
Prepared microscope slide of skeletal muscle tissue
Torso with musculature
Model of skeletal muscle fiber

For Demonstration:

Fresh round beefsteak

SAFETY

- Wear disposable gloves when handling the fresh beefsteak.
- Wash your hands before leaving the laboratory.

A skeletal muscle represents an organ of the muscular system and is composed of several kinds of tissues. These tissues include skeletal muscle tissue, nervous tissue, blood, and various connective tissues.

Each skeletal muscle is surrounded by connective tissue. The connective tissue often extends beyond the end of a muscle, providing an attachment to other muscles or to bones. Connective tissue also extends into the structure of a muscle and separates it into compartments.

PURPOSE OF THE EXERCISE

To review the structure of a skeletal muscle.

LEARNING OBJECTIVES

After completing this exercise, you should be able to

1. Describe how connective tissue is associated with muscle tissue within a skeletal muscle.
2. Name and locate the major structures of a skeletal muscle fiber on a model.

3. Distinguish between the origin and insertion of a muscle.
4. Describe the general actions of prime movers, synergists, and antagonists.

PROCEDURE

1. Review the section entitled "Skeletal Muscle Tissue" in chapter 5 of the textbook.
2. Reexamine the microscopic structure of skeletal muscle by observing a prepared microscope slide of this tissue. Use figure 18.1 of skeletal muscle tissue to locate the following features:

   skeletal muscle fiber (cell)

   nuclei

   striations (alternating light and dark)

3. Review the section entitled "Structure of a Skeletal Muscle" in chapter 9 of the textbook.
4. As a review activity, label figures 18.2 and 18.3.
5. Examine the torso and locate examples of fascia, tendons, and aponeuroses. Locate examples of tendons in your own body.
6. Complete Part A of Laboratory Report 18.

DEMONSTRATION

Examine the fresh round beefsteak. It represents a cross section through the beef thigh muscles. Note the white lines of connective tissue that separate the individual skeletal muscles. Also note how the connective tissue extends into the structure of a muscle and separates it into small compartments of muscle tissue. Locate the epimysium and the perimysium of the deep fascia.

Figure 18.1 Structures found in skeletal muscle fibers (cells) (250× micrograph enlarged to 700×).

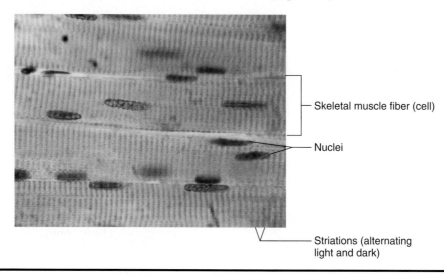

— Skeletal muscle fiber (cell)

— Nuclei

— Striations (alternating light and dark)

Figure 18.2 Label the structures of a skeletal muscle.

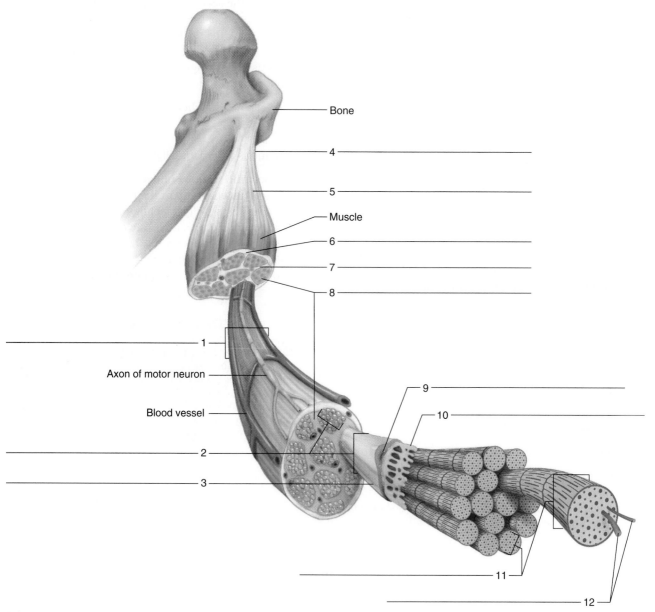

Bone

4

5

Muscle

6

7

8

1

Axon of motor neuron

Blood vessel

2

3

9

10

11

12

Figure 18.3 Label the structures of this skeletal muscle fiber.

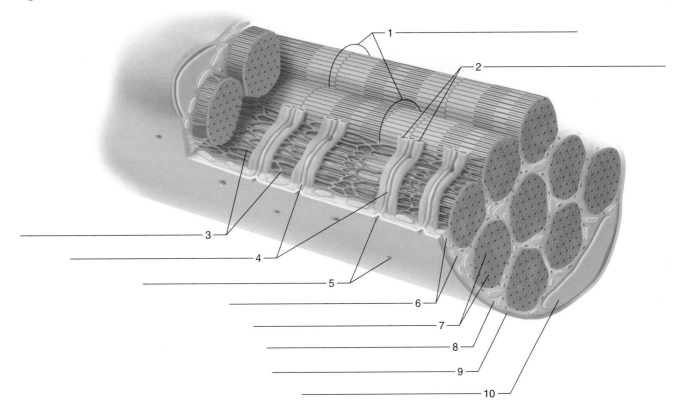

7. Examine the model of the skeletal muscle fiber and locate the following:

sarcolemma

sarcoplasm

myofibril

   myosin filament

   actin filament

sarcomere

   A band

   I band

   H zone

   M line

   Z line

sarcoplasmic reticulum

   cisternae

transverse tubules

8. Complete Part B of the laboratory report.
9. Review the section entitled "Skeletal Muscle Actions" in chapter 9 of the textbook.
10. Provide labels for figure 18.4.
11. Locate the biceps brachii and its origins and insertions in the torso and in your own body.
12. Make various movements with your upper limb at the shoulder and elbow. For each movement, determine the location of the muscles functioning as prime movers and as antagonists. **(Remember, when a prime mover contracts, its antagonist must relax.)**
13. Complete Part C of the laboratory report.

Figure 18.4 Label the major features of the upper limb.

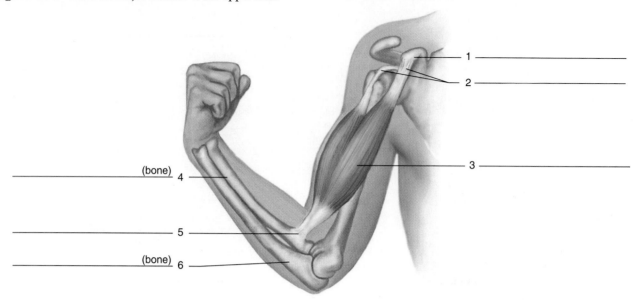

(bone) 4

5

(bone) 6

1

2

3

SKELETAL MUSCLE CONTRACTION

MATERIALS NEEDED

Textbook
Recording system (kymograph, Physiograph, etc.)
Stimulator and connecting wires
Live frog
Dissecting tray
Dissecting instruments
Probe for pithing
Heavy thread
Frog Ringer's solution

For Demonstration A—the Kymograph:

Kymograph recording system
Electronic stimulator (or inductorium)
Frog muscle (from pithed frog)
Probe for pithing
Dissecting instruments
Frog Ringer's solution

For Demonstration B—the Physiograph:

Physiograph
Myograph and stand
Frog muscle (from pithed frog)
Probe for pithing
Dissecting instruments
Frog Ringer's solution

SAFETY

- Wear disposable gloves when handling the frogs.
- Dispose of gloves and frogs as instructed.
- Wash your hands before leaving the laboratory.

To study the characteristics of certain physiological events such as muscle contractions, it often is necessary to use a recording device, such as a *kymograph* or a *Physiograph.* These devices can provide accurate recordings of various physiological changes.

To observe the phenomenon of skeletal muscle contractions, muscles can be removed from anesthetized frogs. These muscles can be attached to recording systems and stimulated by electrical shocks of varying strength, duration, and frequency. Recordings obtained from such procedures can be used to study the basic characteristics of skeletal muscle contractions.

PURPOSE OF THE EXERCISE

To observe and record the responses of an isolated frog muscle to electrical stimulation of varying strength and frequency.

LEARNING OBJECTIVES

After completing this exercise, you should be able to

1. Make use of a recording system and stimulator to record frog muscle responses to electrical stimulation.
2. Determine the threshold level of electrical stimulation in frog muscle.
3. Determine the intensity of stimulation needed for maximal muscle contraction.
4. Record a single muscle twitch and identify its phases.
5. Record the response of a muscle to increasing frequency of stimulation and identify the patterns of tetanic contraction and fatigue.

DEMONSTRATION A—THE KYMOGRAPH

1. Observe the kymograph and, at the same time, study figure 19.1 to learn the names of its major parts.
2. Note that the kymograph consists of a cylindrical *drum* around which a sheet of paper is wrapped. The drum is mounted on a motor-driven *shaft,* and the speed of the motor can be varied. Thus, the drum can be rotated rapidly if rapid physiological events are being recorded or rotated slowly for events that occur more slowly.

   A *stylus* that can mark on the paper is attached to a *movable lever,* and the lever, in turn, is connected to an isolated muscle. The origin of the muscle is fixed in position by a *clamp,* and its insertion is hooked to the muscle lever. The muscle also is connected by wires to an *electronic stimulator* (or inductorium). The stimulator can deliver single or multiple electrical shocks to the muscle, and it can be adjusted so that the intensity (voltage), duration (milliseconds), and frequency (stimuli per second) can be varied. Another stylus, on the *signal marker,* records the time each stimulus is given to the muscle. As the muscle responds, the

Figure 19.1 Kymograph to record frog muscle contractions.

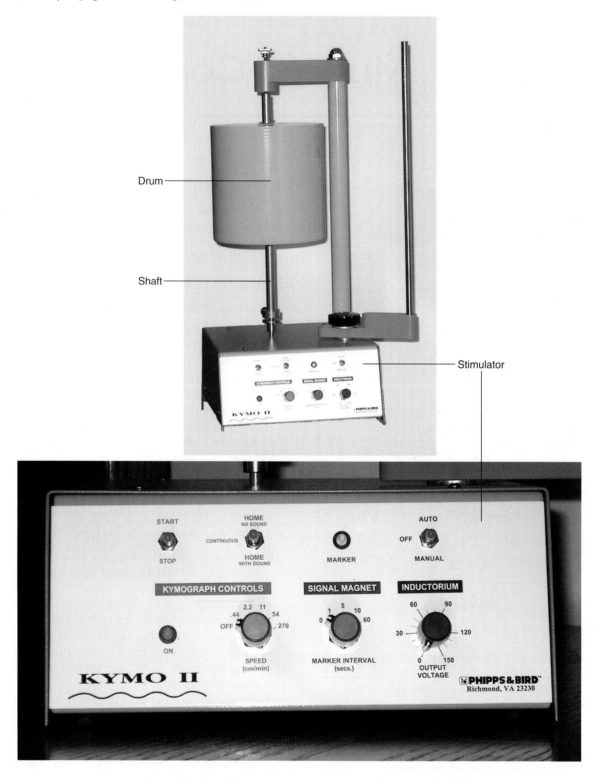

Drum

Shaft

Stimulator

START
STOP

HOME
NO SOUND
CONTINUOUS
HOME
WITH SOUND

MARKER

AUTO
OFF
MANUAL

KYMOGRAPH CONTROLS

.44 OFF
2.2 11
54
0   270

SPEED
(cm/min)

SIGNAL MAGNET

1   5   10
0   60

MARKER INTERVAL
(secs.)

INDUCTORIUM

60   90
30   120
0   150
OUTPUT
VOLTAGE

KYMO II

PHIPPS & BIRD™
Richmond, VA 23230

duration and relative length of its contraction are recorded by the stylus on the muscle lever.

3. Watch carefully while the laboratory instructor demonstrates the operation of the kymograph to record a frog muscle contraction.

DEMONSTRATION B—THE PHYSIOGRAPH

1. Observe the Physiograph and, at the same time, study figures 19.2 and 19.3 to learn the names of its major parts.

Figure 19.2 Physiograph to record frog muscle contractions.

Figure 19.3 Myograph attached to frog muscle.

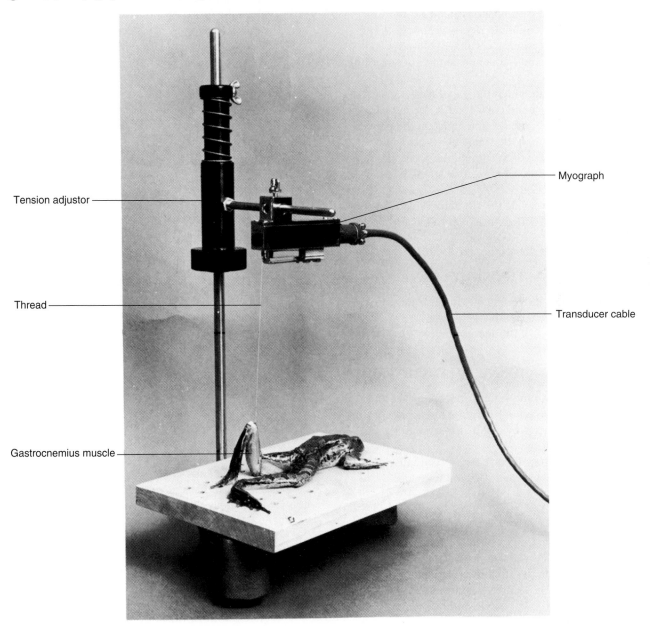

2. Note that the recording system of the Physiograph includes a transducer, an amplifier, and a recording pen. The *transducer* is a sensing device that can respond to some kind of physiological change by sending an electrical signal to the amplifier. The *amplifier* increases the strength of the electrical signal and relays it to an electric motor that moves the *recording pen.* As the pen moves, a line is drawn on paper.

   To record a frog muscle contraction, a transducer called a *myograph* is used (fig. 19.3). The origin of the muscle is held in a fixed position, and its insertion is attached to a small lever in the myograph by a thread. The myograph, in turn, is connected to the amplifier by a transducer cable. The muscle also is connected by wires to the electronic stimulator, which is part of the Physiograph. This stimulator can be adjusted to deliver single or multiple electrical shocks to the muscle, and the intensity (voltage), duration (milliseconds), and frequency (stimuli per second) can be varied.

   The speed at which the paper moves under the recording pen can be controlled. A second pen, driven by a timer, marks time units on the paper and indicates when the stimulator is activated. As the muscle responds to stimuli, the recording pen records the duration and relative length of each muscle contraction.

3. Watch carefully while the laboratory instructor operates the Physiograph to record a frog muscle contraction.

PROCEDURE A—TEXTBOOK REVIEW

1. Review the section entitled "Muscular Responses" in chapter 9 of the textbook.
2. Complete Part A of Laboratory Report 19.

PROCEDURE B—RECORDING SYSTEM

1. Set up the recording system and stimulator to record the contractions of a frog muscle according to the directions provided by the laboratory instructor.
2. Obtain a live frog, and prepare its calf muscle (gastrocnemius) as described in Procedure C.

PROCEDURE C—MUSCLE PREPARATION

1. Prepare the live frog by pithing so that it will have no feelings or movements when its muscle is removed. To do this, follow these steps:
   a. Hold the frog securely in one hand so that its legs are extended downward.
   b. Position the frog's head between your thumb and index finger.

c. Bend the frog's head forward at an angle of about 90° by pressing on its snout with your index finger (fig. 19.4).
   d. Use a sharp probe to locate the foramen magnum between the occipital condyles in the midline between the frog's tympanic membranes.
   e. Insert the probe through the skin and into the foramen magnum, and then quickly move the probe from side to side to separate the brain from the spinal cord.
   f. Slide the probe forward into the braincase, and continue to move the probe from side to side to destroy the brain.
   g. Remove the probe from the braincase, and insert it into the spinal cord through the same opening in the skin.
   h. Move the probe up and down the spinal cord to destroy it. If the frog has been pithed correctly, its legs will be extended and relaxed. Also, the eyes will not respond when touched with a probe.

ALTERNATIVE PROCEDURE

An anesthetizing agent, tricaine methane sulfonate, can be used to prepare frogs for this lab. This procedure eliminates the need to pith frogs.

2. Remove the frog's gastrocnemius muscle by proceeding as follows:
   a. Place the pithed frog in a dissecting tray.
   b. Use scissors to cut through the skin completely around the leg in the thigh.
   c. Pull the skin downward and off the leg.
   d. Locate the gastrocnemius muscle in the calf and the calcaneal tendon (Achilles tendon) at its distal end.
   e. Separate the calcaneal tendon from the underlying tissue, using forceps.
   f. Tie a thread firmly around the tendon (fig. 19.5).
   g. When the thread is secure, free the distal end of the tendon by cutting it with scissors.
   h. Attach the frog muscle to the recording system in the manner suggested by your laboratory instructor (see figs. 19.1 and 19.3).
   i. Insert the ends of the stimulator wires into the muscle so that one wire is located on either side of the belly of the muscle.

      Keep the frog muscle moist at all times by dripping frog Ringer's solution on it. When the muscle is not being used, cover it with some paper toweling that has been saturated with frog Ringer's solution.

      Before you begin operating the recording system and stimulator, have the laboratory instructor inspect your setup.

Figure 19.4 Hold the frog's head between your thumb and index finger to pith (*a*) its brain and (*b*) its spinal cord.

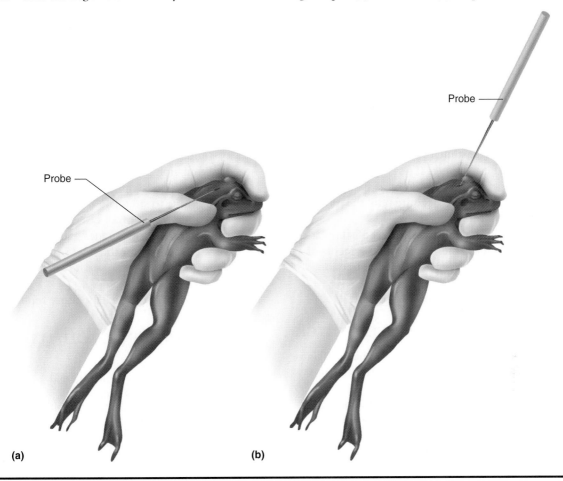

(a) (b)

Figure 19.5 (*a*) Separate the calcaneal (Achilles) tendon from the underlying tissue. (*b*) Tie a thread around the tendon, and cut its distal attachments.

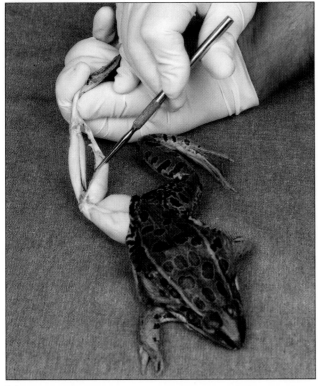 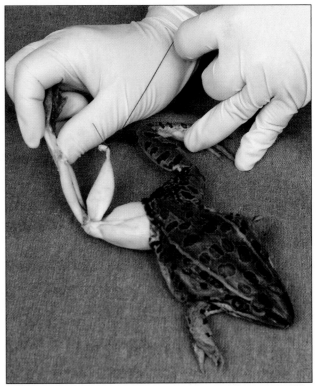

(a) (b) 155

PROCEDURE D— THRESHOLD STIMULATION

1. To determine the threshold or minimal strength of electrical stimulation (voltage) needed to elicit a contraction in the frog muscle, follow these steps:
   a. Set the stimulus duration to a minimum (about 0.1 milliseconds).
   b. Set the voltage to a minimum (about 0.1 volts).
   c. Set the stimulator so that it will administer single stimuli.
2. Administer a single stimulus to the muscle and watch to see if it responds. If no response is observed, increase the voltage to the next higher setting and repeat the procedure until the muscle responds by contracting.
3. After determining the threshold level of stimulation, continue to increase the voltage in increments of 1 or 2 volts until a maximal muscle contraction is obtained.
4. Complete Part B of the laboratory report.

PROCEDURE E—SINGLE MUSCLE TWITCH

1. To record a single muscle twitch, set the voltage for a maximal muscle contraction as determined in Procedure D.
2. Set the paper speed at maximum, and with the paper moving, administer a single electrical stimulus to the frog muscle.
3. Repeat this procedure to obtain several recordings of single muscle twitches.
4. Complete Part C of the laboratory report.

PROCEDURE F— SUSTAINED CONTRACTION

1. To record a sustained muscle contraction, follow these steps:
   a. Set the stimulator for continuous stimulation.
   b. Set the voltage for maximal muscle contraction as determined in Procedure D.
   c. Set the frequency of stimulation at a minimum.
   d. Set the paper speed at about 0.05 cm/sec.
   e. With the paper moving, administer electrical stimulation and slowly increase the frequency of stimulation until the muscle sustains a contraction (tetanic contraction or tetanus).
   f. Continue to stimulate the muscle at the frequency that produces sustained contractions until the muscle fatigues and relaxes.
2. Every 15 seconds for the next several minutes, stimulate the muscle to see how long it takes to recover from the fatigue.
3. Complete Part D of the laboratory report.

OPTIONAL ACTIVITY

To demonstrate the staircase effect (treppe), obtain a fresh frog gastrocnemius muscle and attach it to the recording system as before. Set the paper control for slow speed, and set the stimulator voltage to produce a maximal muscle contraction. Stimulate the muscle once each second for several seconds. How do you explain the differences in the lengths of successive muscle contractions?

_____

_____

_____

_____

SKELETAL MUSCLE CONTRACTION

PART A

Match the terms in column A with the definitions in column B. Place the letter of your choice in the space provided.

Column A	Column B

a. All-or-none
b. Latent period
c. Motor unit
d. Muscle tone
e. Myogram
f. Refractory period
g. Tetanic contraction (tetanus)
h. Threshold stimulus
i. Twitch

_____ 1. Minimal intensity of stimulation necessary to trigger a muscle contraction

_____ 2. Response of a muscle fiber/motor unit complete contraction if stimulated sufficiently

_____ 3. Consists of a single motor neuron and all of the muscle fibers with which the neuron is associated

_____ 4. An action of a muscle contraction and immediate relaxation when exposed to a single stimulus

_____ 5. The time between stimulation and response

_____ 6. The time following a muscle contraction during which the muscle remains unresponsive to stimulation

_____ 7. Forceful, sustained contraction

_____ 8. Some contraction of muscle fibers when a muscle is at rest

_____ 9. The recording of the pattern of a muscle contraction

PART B—THRESHOLD STIMULATION

Complete the following:

1. What was the threshold voltage for stimulation of the frog gastrocnemius muscle? _____

2. What voltage produced maximal contraction of this muscle? _____

Critical Thinking Application

Do you think other frog muscles would respond in an identical way to these voltages of stimulation?

_____

Why or why not?

_____

_____

_____

PART C—SINGLE MUSCLE TWITCH

Complete the following:

1. Fasten a recording of two single muscle twitches in the following space.

2. On a muscle twitch recording, label the *latent period, period of contraction,* and *period of relaxation,* and indicate the time it took for each of these phases to occur.

3. What differences, if any, do you note in the two myograms of a single muscle twitch? How do you explain these differences?_____

   _____

   _____

PART D—SUSTAINED CONTRACTION

Complete the following:

1. Fasten a recording of a sustained contraction in the following space.

2. On the sustained contraction recording, indicate when the muscle twitches began to combine (summate), and label the period of tetanic contraction and the period of fatigue.

3. At what frequency of stimulation did tetanic contraction occur?_____

4. How long did it take for the tetanic muscle to fatigue? _____

5. Is the length of muscle contraction at the beginning of tetanic contraction the same or different from the length of the single muscle contractions before tetanic contraction occurred? _____ How do you explain this?

   _____

   _____

6. How long did it take for the fatigued muscle to become responsive again? _____

MUSCLES OF THE FACE, HEAD, AND NECK

MATERIALS NEEDED

Textbook
Torso with musculature
Human skull
Human skeleton, articulated

The skeletal muscles of the face and head include the muscles of facial expression, which lie just beneath the skin, the muscles of mastication, which are attached to the mandible, and the muscles that move the head, which are located in the neck.

PURPOSE OF THE EXERCISE

To review the locations, actions, origins, and insertions of the muscles of the face, head, and neck.

LEARNING OBJECTIVES

After completing this exercise, you should be able to

1. Locate and identify the muscles of facial expression, the muscles of mastication, and the muscles that move the head.
2. Describe and demonstrate the action of each of these muscles.
3. Locate the origin and insertion of each of these muscles in a human skeleton and the musculature of the torso.

PROCEDURE

1. Review the sections entitled "Muscles of Facial Expression," "Muscles of Mastication," and "Muscles That Move the Head and Vertebral Column" in chapter 9 of the textbook.
2. As a review activity, label figures 20.1, 20.2, and 20.3.

3. Locate the following muscles in the torso and in your own body whenever possible:

    epicranius (frontalis and occipitalis)

    orbicularis oculi

    orbicularis oris

    buccinator

    zygomaticus

    platysma

    masseter

    temporalis

    medial pterygoid

    lateral pterygoid

    sternocleidomastoid

    splenius capitis

    semispinalis capitis

    longissimus capitis (of erector spinae group)

4. Demonstrate the action of these muscles in your own body.
5. Locate the origins and insertions of these muscles in the human skull and skeleton.
6. Complete Parts A, B, and C of Laboratory Report 20.

Web Quest

Determine the origin, insertion, action, nerve innervation, and blood supply of all the major muscles.

Identify muscles and detailed explanations from an interactive site.

Find information about these topics at www.mhhe.com/shier10

Click on Student Edition. Click on Martin Lab Manual, Web Quest.

Figure 20.1 Label the muscles of expression and mastication.

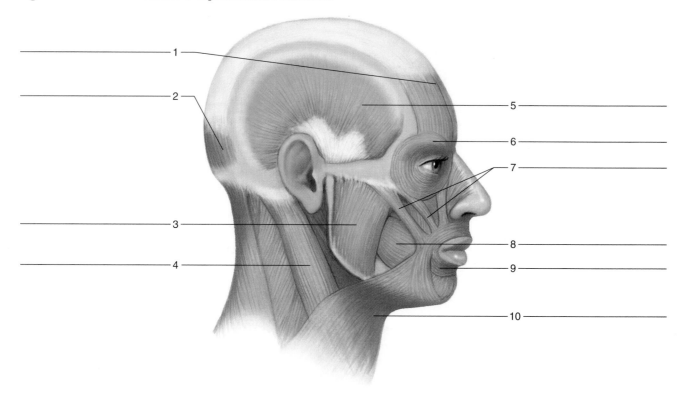

Figure 20.2 Label these muscles of mastication.

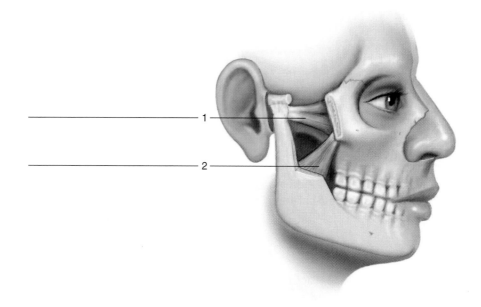

Figure 20.3 Label these deep muscles of the posterior neck (trapezius removed), using the terms provided.

Terms:
Longissimus capitis (erector spinae)
Splenius capitis

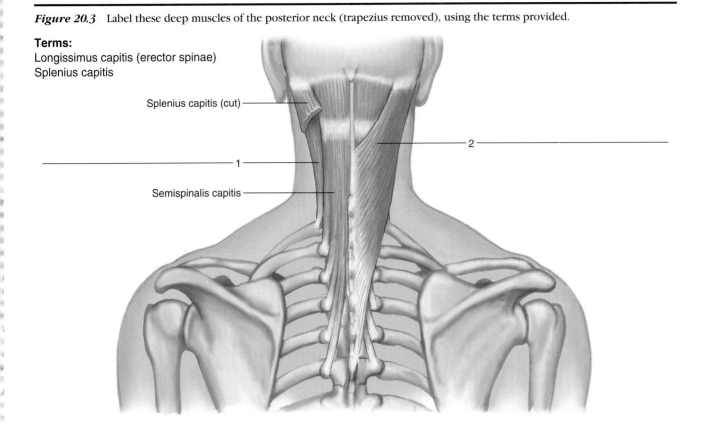

Splenius capitis (cut)

1

2

Semispinalis capitis

MUSCLES OF THE FACE, HEAD, AND NECK

PART A

Complete the following statements:

1. When the _____ contracts, the corner of the mouth is drawn upward.

2. The _____ acts to compress the wall of the cheeks when air is blown out of the mouth.

3. The _____ causes the lips to close and pucker.

4. The _____ and platysma help to lower the mandible.

5. The temporalis acts to _____ .

6. The _____ pterygoid can close the jaw and pull it sideways.

7. The _____ pterygoid can protrude the jaw, pull the jaw sideways, and open the mouth.

8. The _____ can close the eye, as in blinking.

9. The _____ can pull the head toward the chest.

10. The _____ can pull the head to one side, rotate it, or bring it into an upright position (extend head).

11. The muscle used for pouting and to express horror is the _____ .

12. The muscle used to smile and laugh is the _____ .

PART B

Name the muscle indicated by the following combinations of origin and insertion.

Origin	Insertion	Muscle
1. Occipital bone	Skin and muscle around eye	_____
2. Zygomatic bone	Orbicularis oris	_____
3. Zygomatic arch	Lateral surface of mandible	_____
4. Sphenoid bone	Anterior surface of mandibular condyle	_____
5. Anterior surface of sternum and upper clavicle	Mastoid process of temporal bone	_____
6. Outer surfaces of mandible and maxilla	Orbicularis oris	_____
7. Fascia in upper chest	Lower border of mandible and skin around corner of mouth	_____
8. Temporal bone	Coronoid process and anterior ramus of mandible	_____
9. Spinous processes of cervical and thoracic vertebrae	Mastoid process of temporal bone and occipital bone	_____
10. Processes of cervical and thoracic vertebrae	Occipital bone	_____

Critical Thinking Application

Identify the muscles of various facial expressions in the photographs of figure 20.4.

1. _____

2. _____

3. _____

4. _____

5. _____

Figure 20.4   Identify the muscles of expression being contracted in each of these photographs (*a-c*), using the terms provided.

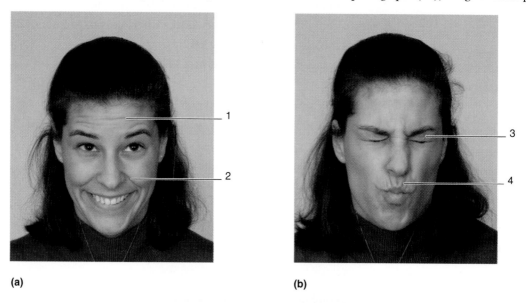

(a)   (b)

Terms:
Epicranius/frontalis
Orbicularis oculi
Orbicularis oris
Platysma
Zygomaticus

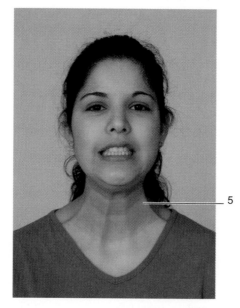

(c)

MUSCLES OF THE CHEST, SHOULDER, AND UPPER LIMB

MATERIALS NEEDED

Textbook
Torso
Human skeleton, articulated
Muscular models of the upper limb

The muscles of the chest and shoulder are responsible for moving the scapula and arm, whereas those within the arm and forearm move joints in the elbow and hand.

PURPOSE OF THE EXERCISE

To review the locations, actions, origins, and insertions of the muscles in the chest, shoulder, and upper limb.

LEARNING OBJECTIVES

After completing this exercise, you should be able to

1. Locate and identify the muscles of the chest, shoulder, and upper limb.
2. Describe and demonstrate the action of each of these muscles.
3. Locate the origin and insertion of each of these muscles.

PROCEDURE

1. Review the sections entitled "Muscles That Move the Pectoral Girdle," "Muscles That Move the Arm," "Muscles That Move the Forearm," and "Muscles That Move the Hand" in chapter 9 of the textbook.
2. As a review activity, label figures 21.1, 21.2, 21.3, and 21.4.
3. Locate the following muscles in the torso and models of the upper limb. Also locate in your own body as many of the muscles as you can.

muscles that move the pectoral girdle

trapezius

rhomboideus major

levator scapulae

serratus anterior

pectoralis minor

muscles that move the arm

coracobrachialis

pectoralis major

teres major

latissimus dorsi

supraspinatus

deltoid

subscapularis

infraspinatus

teres minor

muscles that move the forearm

biceps brachii

brachialis

brachioradialis

triceps brachii

supinator

pronator teres

pronator quadratus

muscles that move the hand

flexor carpi radialis

flexor carpi ulnaris

palmaris longus

flexor digitorum profundus

flexor digitorum superficialis

extensor carpi radialis longus

extensor carpi radialis brevis

extensor carpi ulnaris

extensor digitorum

Figure 21.1 Label the muscles of the posterior shoulder. Superficial muscles are illustrated on the left side and deep muscles on the right side.

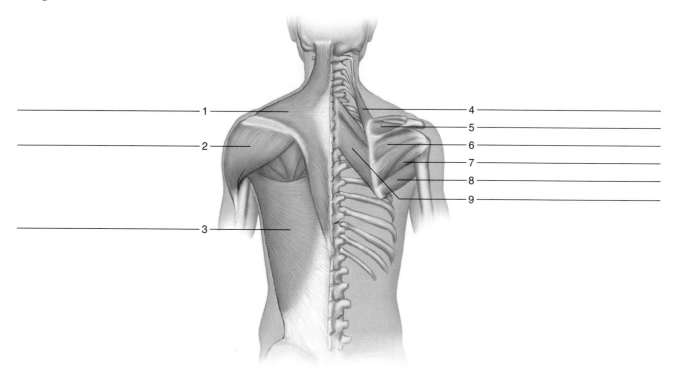

Figure 21.2 Label the muscles of the anterior chest. Superficial muscles are illustrated on the left side and deep muscles on the right side.

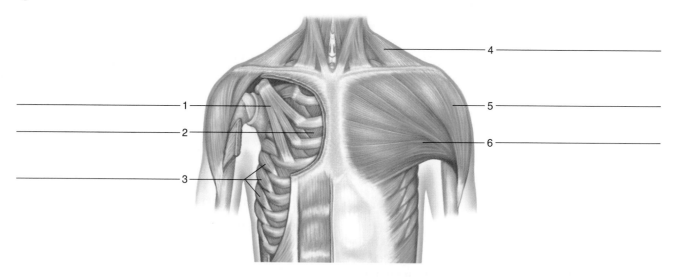

4. Demonstrate the action of these muscles in your own body.
5. Locate the origins and insertions of these muscles in the human skeleton.
6. Complete Parts A, B, C, and D of Laboratory Report 21.

Figure 21.3   Label (*a*) the muscles of the posterior shoulder and arm and (*b*) the muscles of the anterior shoulder and arm.

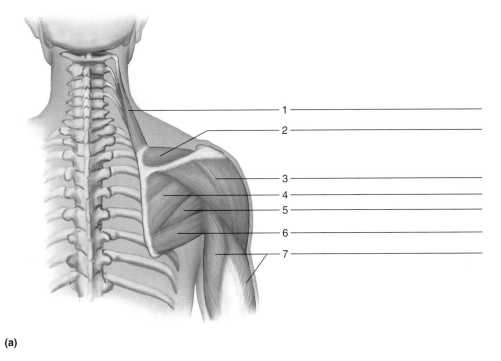

1 _____

2 _____

3 _____
4 _____
5 _____
6 _____
7 _____

(a)

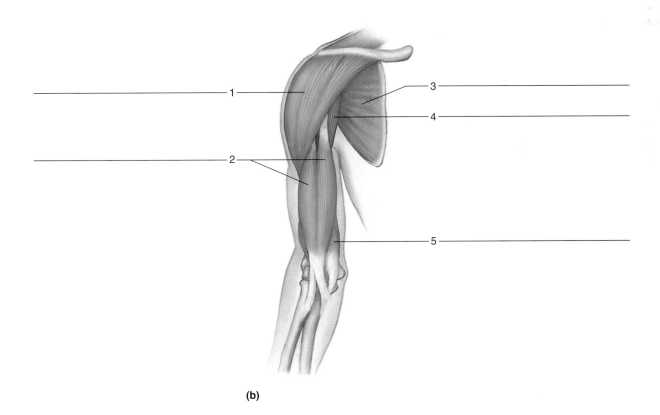

_____ 1

_____ 2

3 _____
4 _____

5 _____

(b)

Figure 21.4 Label (*a*) the muscles of the anterior forearm and (*b*) the muscles of the posterior forearm.

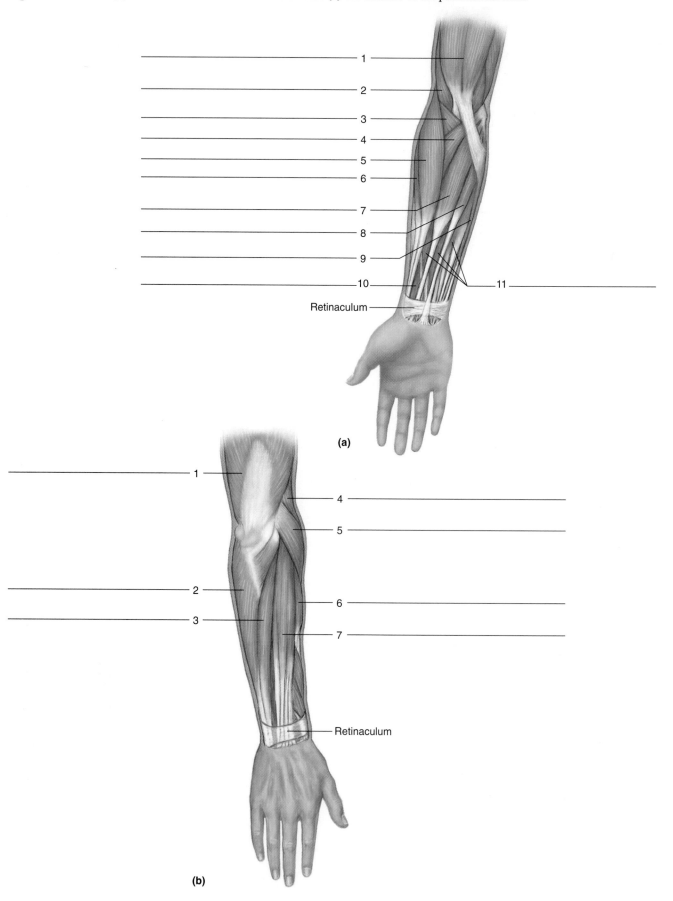

1
2
3
4
5
6
7
8
9
10
Retinaculum
11

(a)

1
2
3
4
5
6
7
Retinaculum

(b)

MUSCLES OF THE CHEST, SHOULDER, AND UPPER LIMB

PART A

Match the muscles in column A with the actions in column B. Place the letter of your choice in the space provided.

<table>
<tr><td colspan="2">Column A</td><td colspan="2">Column B</td></tr>
<tr><td>a.</td><td>Brachialis</td><td>_____ 1.</td><td>Abducts arm</td></tr>
<tr><td>b.</td><td>Coracobrachialis</td><td rowspan="2">_____ 2.</td><td rowspan="2">Pulls arm forward and across chest and rotates arm medially</td></tr>
<tr><td>c.</td><td>Deltoid</td></tr>
<tr><td>d.</td><td>Extensor carpi ulnaris</td><td>_____ 3.</td><td>Flexes and adducts hand at the wrist</td></tr>
<tr><td>e.</td><td>Flexor carpi ulnaris</td><td>_____ 4.</td><td>Raises and adducts scapula</td></tr>
<tr><td>f.</td><td>Flexor digitorum profundus</td><td>_____ 5.</td><td>Rotates forearm medially</td></tr>
<tr><td>g.</td><td>Infraspinatus</td><td rowspan="2">_____ 6.</td><td rowspan="2">Raises ribs in forceful inhalation or pulls scapula forward and downward</td></tr>
<tr><td>h.</td><td>Pectoralis major</td></tr>
<tr><td>i.</td><td>Pectoralis minor</td><td>_____ 7.</td><td>Turns forearm laterally</td></tr>
<tr><td>j.</td><td>Pronator teres</td><td rowspan="2">_____ 8.</td><td rowspan="2">Used to thrust shoulder anteriorly, as when pushing something</td></tr>
<tr><td>k.</td><td>Rhomboideus major</td></tr>
<tr><td>l.</td><td>Serratus anterior</td><td>_____ 9.</td><td>Flexes the forearm at the elbow</td></tr>
<tr><td>m.</td><td>Supinator</td><td rowspan="2">_____ 10.</td><td rowspan="2">Flexes and adducts arm at the shoulder along with pectoralis major</td></tr>
<tr><td>n.</td><td>Teres major</td></tr>
<tr><td>o.</td><td>Triceps brachii</td><td>_____ 11.</td><td>Extends the forearm at the elbow</td></tr>
<tr><td></td><td></td><td>_____ 12.</td><td>Extends, adducts, and rotates arm medially</td></tr>
<tr><td></td><td></td><td>_____ 13.</td><td>Extends and adducts hand at the wrist</td></tr>
<tr><td></td><td></td><td>_____ 14.</td><td>Rotates arm laterally</td></tr>
<tr><td></td><td></td><td>_____ 15.</td><td>Flexes distal joints of fingers 2–5</td></tr>
</table>

Figure 21.5 Identify the muscles indicated in this cross section of the right forearm (superior view).

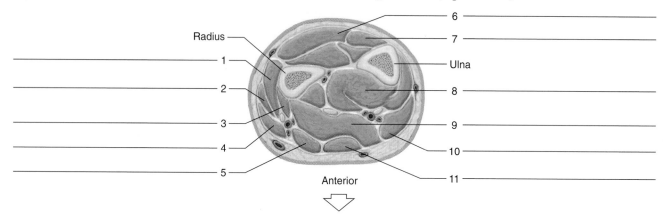

PART B

Identify the muscles indicated in the cross section of the forearm illustrated in figure 21.5.

PART C

Name the muscle indicated by the following combinations of origin and insertion.

Origin	Insertion	Muscle
1. Spines of upper thoracic vertebrae	Medial border of scapula	
2. Outer surfaces of upper ribs	Ventral surface of scapula	
3. Sternal ends of upper ribs	Coracoid process of scapula	
4. Coracoid process of scapula	Shaft of humerus	
5. Lateral border of scapula	Intertubercular groove of humerus	
6. Anterior surface of scapula	Lesser tubercle of humerus	
7. Lateral border of scapula	Greater tubercle of humerus	
8. Anterior shaft of humerus	Coronoid process of ulna	
9. Medial epicondyle of humerus and coronoid process of ulna	Lateral surface of radius	
10. Anterior distal end of ulna	Anterior distal end of radius	
11. Distal lateral end of humerus	Lateral surface of radius above styloid process	
12. Medial epicondyle of humerus	Base of second and third metacarpals	
13. Medial epicondyle of humerus	Fascia of palm	
14. Distal end of humerus	Base of second metacarpal	
15. Lateral epicondyle of humerus	Base of fifth metacarpal	

Critical Thinking Application

Identify the muscles indicated in figure 21.6.

1. _____

2. _____

3. _____

4. _____

5. _____

6. _____

7. _____

8. _____

9. _____

10. _____

11. _____

12. _____

13. _____

14. _____

15. _____

16. _____

17. _____

18. _____

19. _____

20. _____

21. _____

Figure 21.6 Identify the muscles that appear as body surface features in these photographs (*a, b,* and *c*), using the terms provided.

Terms:
Biceps brachii
Deltoid
External oblique
Pectoralis major
Rectus abdominis
Serratus anterior
Sternocleidomastoid
Trapezius

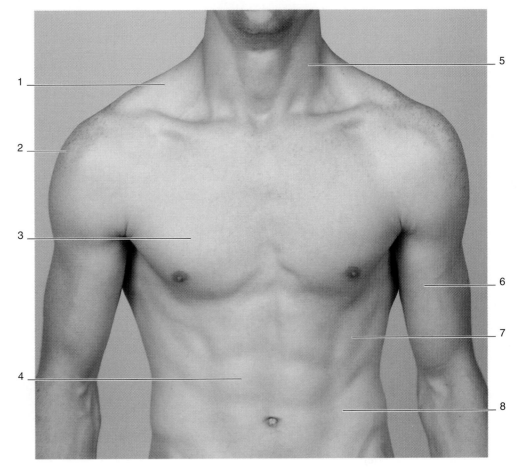

(a)

Figure 21.6 *Continued.*

Terms:
Biceps brachii
Deltoid
Infraspinatus
Latissimus dorsi
Trapezius
Triceps brachii

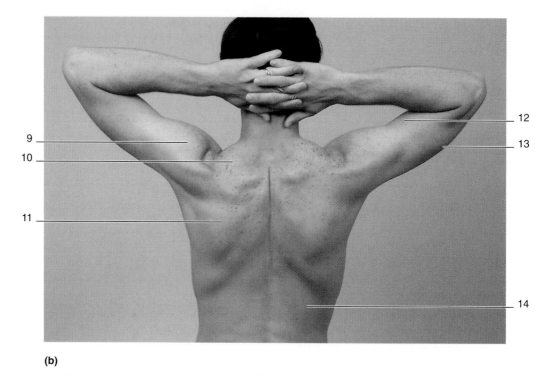

9
10
11
12
13
14

(b)

Terms:
Biceps brachii
Brachioradialis
Deltoid
Pectoralis major
Serratus anterior
Trapezius
Triceps brachii

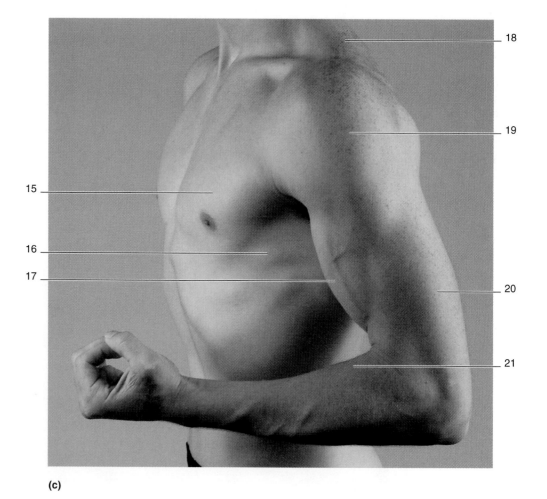

15
16
17
18
19
20
21

(c)

172

MUSCLES OF THE DEEP BACK, ABDOMINAL WALL, AND PELVIC OUTLET

MATERIALS NEEDED

Textbook
Torso with musculature
Human skeleton, articulated
Muscular models of male and female pelves

The deep muscles of the back extend the vertebral column. Because the muscles have numerous origins, insertions, and subgroups, the muscles overlap each other. The deep back muscles can extend the spine when contracting as a group but also help to maintain posture and normal spine curvatures.

The anterior and lateral walls of the abdomen contain broad, flattened muscles arranged in layers. These muscles connect the rib cage and vertebral column to the pelvic girdle.

The muscles of the pelvic outlet are arranged in two muscular sheets: (1) a deeper pelvic diaphragm that forms the floor of the pelvic cavity and (2) a urogenital diaphragm that fills the space within the pubic arch.

PURPOSE OF THE EXERCISE

To review the actions, origins, and insertions of the muscles of the deep back, abdominal wall, and pelvic outlet.

LEARNING OBJECTIVES

After completing this exercise, you should be able to

1. Locate and identify the muscles of the deep back, abdominal wall, and pelvic outlet.
2. Describe the action of each of these muscles.
3. Locate the origin and insertion of each of these muscles.

PROCEDURE

1. Review the sections entitled "Muscles of the Abdominal Wall" and "Muscles of the Pelvic Outlet" in chapter 9 of the textbook.
2. As a review activity, label figures 22.1, 22.2, 22.3, and 22.4.
3. Locate the following muscles in the torso:

   erector spinae group

   iliocostalis (lateral group)

   longissimus (intermediate group)

   spinalis (medial group)

   external oblique

   internal oblique

   transversus abdominis

   rectus abdominis

4. Demonstrate the actions of these muscles in your own body.
5. Locate the origin and insertion of each of these muscles in the human skeleton.
6. Complete Part A of Laboratory Report 22.
7. Locate the following muscles in the models of the male and female pelves:

   levator ani

   coccygeus

   superficial transversus perinei

   bulbospongiosus

   ischiocavernosus

8. Locate the origin and insertion of each of these muscles in the human skeleton.
9. Complete Part B of the laboratory report.

Figure 22.1 Label the three deep back muscle groups of the erector spinae group, using the terms provided. (*Note:* The three erector spinae muscle groups are not illustrated on both sides of the body.)

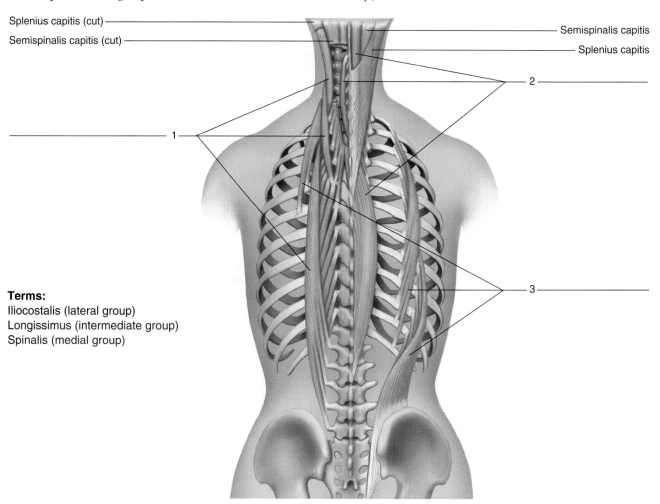

Splenius capitis (cut)

Semispinalis capitis (cut)

Semispinalis capitis

Splenius capitis

1

2

3

Terms:
Iliocostalis (lateral group)
Longissimus (intermediate group)
Spinalis (medial group)

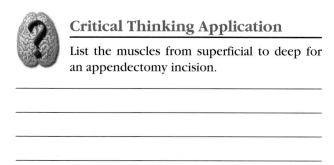

Critical Thinking Application

List the muscles from superficial to deep for an appendectomy incision.

Web Quest

Determine the origin, insertion, action, nerve innervation, and blood supply of all the major muscles.

Identify muscles and detailed explanations from an interactive site.

Find information about these topics at www.mhhe.com/shier10

Click on Student Edition. Click on Martin Lab Manual, Web Quest.

Figure 22.2 Label the muscles of the abdominal wall.

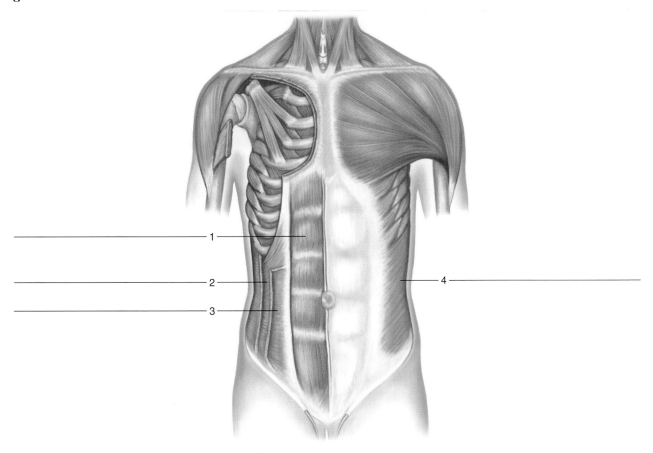

1 ——————————————
2 ——————————————
3 ——————————————
4 ——————————————

Figure 22.3 Label the muscles of the male pelvic outlet.

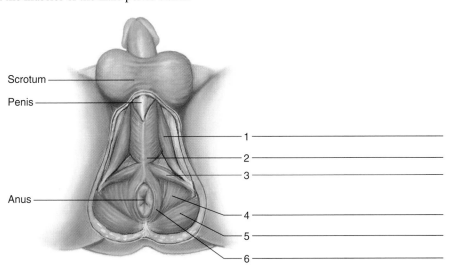

Scrotum ——————
Penis ——————

Anus ——————

1 ——————————————
2 ——————————————
3 ——————————————
4 ——————————————
5 ——————————————
6 ——————————————

Figure 22.4   Label the muscles of the female pelvic outlet.

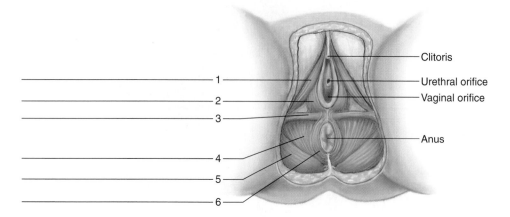

1 _____

2 _____

3 _____

4 _____

5 _____

6 _____

Clitoris

Urethral orifice

Vaginal orifice

Anus

MUSCLES OF THE DEEP BACK, ABDOMINAL WALL, AND PELVIC OUTLET

PART A

Complete the following statements:

1. A band of tough connective tissue in the midline of the anterior abdominal wall called the _____ serves as a muscle attachment.

2. The _____ muscle spans from the ribs and sternum to the pubic bones.

3. The _____ forms the third layer (deepest layer) of the abdominal wall muscles.

4. The action of the external oblique muscle is to_____ .

5. The action of the rectus abdominis is to _____ .

6. The iliocostalis, longissimus, and spinalis muscles together form the _____ .

PART B

Complete the following statements:

1. The levator ani and coccygeus together form the _____ .

2. The levator ani provides a sphincterlike action in the _____ .

3. The action of the coccygeus is to _____

    _____ .

4. The _____ surrounds the base of the penis.

5. In females, the bulbospongiosus acts to _____ .

6. The ischiocavernosus extends from the margin of the pubic arch to the_____ .

7. In the female, the _____ muscles are separated by the vagina, urethra, and anal canal.

8. The action of the superficial transversus perinei is to _____ .

9. The coccygeus extends from the coccyx and sacrum to the _____ .

10. The _____ assists in closing the urethra.

MUSCLES OF THE HIP AND LOWER LIMB

MATERIALS NEEDED

Textbook
Torso with musculature
Human skeleton, articulated
Muscular models of the lower limb

The muscles that move the thigh are attached to the femur and to some part of the pelvic girdle. Those attached anteriorly primarily act to flex the thigh at the hip, whereas those attached posteriorly act to extend, abduct, or rotate the thigh.

The muscles that move the leg connect the tibia or fibula to the femur or to the pelvic girdle. They function to flex or extend the leg at the knee. Other muscles, located in the leg, act to move the foot.

PURPOSE OF THE EXERCISE

To review the actions, origins, and insertions of the muscles that move the thigh, leg, and foot.

LEARNING OBJECTIVES

After completing this exercise, you should be able to

1. Locate and identify the muscles that move the thigh, leg, and foot.
2. Describe and demonstrate the actions of each of these muscles.
3. Locate the origin and insertion of each of these muscles.

PROCEDURE

1. Review the sections entitled "Muscles That Move the Thigh," "Muscles That Move the Leg," and "Muscles That Move the Foot" in chapter 9 of the textbook.
2. As a review activity, label figures 23.1, 23.2, 23.3, 23.4, 23.5, and 23.6.
3. Locate the following muscles in the torso and in the lower limb models. Also locate as many of them as possible in your own body.

muscles that move the thigh

iliopsoas group

psoas major

iliacus

psoas minor

gluteus maximus

gluteus medius

gluteus minimus

tensor fasciae latae

pectineus

adductor longus

adductor magnus

adductor brevis

gracilis

muscles that move the leg

hamstring group

biceps femoris

semitendinosus

semimembranosus

sartorius

quadriceps femoris group

rectus femoris

vastus lateralis

vastus medialis

vastus intermedius

muscles that move the foot

tibialis anterior

fibularis (peroneus) tertius

extensor digitorum longus

gastrocnemius

Figure 23.1 Label the muscles of the anterior right hip and thigh.

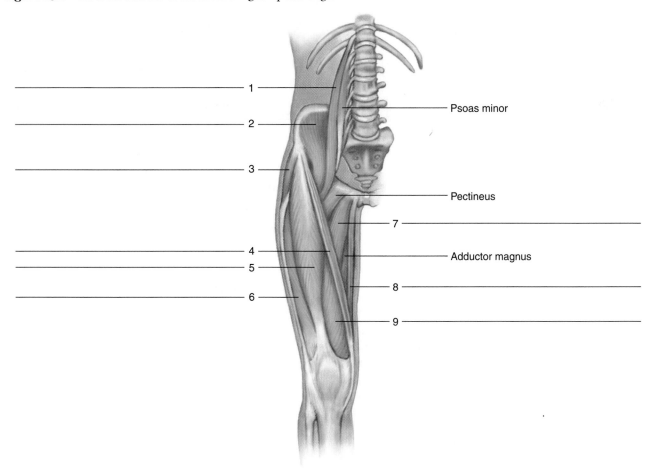

Psoas minor

Pectineus

Adductor magnus

soleus

flexor digitorum longus

tibialis posterior

fibularis (peroneus) longus

fibularis (peroneus) brevis

4. Demonstrate the action of each of these muscles in your own body.
5. Locate the origin and insertion of each of these muscles in the human skeleton.
6. Complete Parts A, B, C, and D of Laboratory Report 23.

Web Quest

Determine the origin, insertion, action, nerve innervation, and blood supply of all the major muscles.

Identify muscles and detailed explanations from an interactive site.

Find information about these topics at www.mhhe.com/shier10

Click on Student Edition. Click on Martin Lab Manual, Web Quest.

Figure 23.2 Label the muscles of the lateral right hip and thigh.

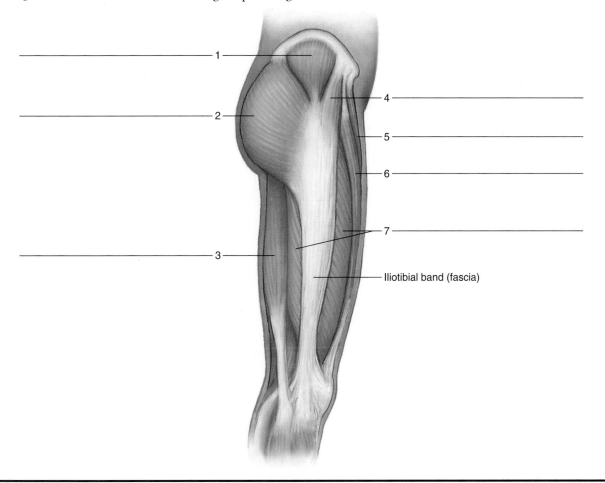

1
2
3
4
5
6
7
Iliotibial band (fascia)

Figure 23.3 Label the muscles of the posterior right hip and thigh.

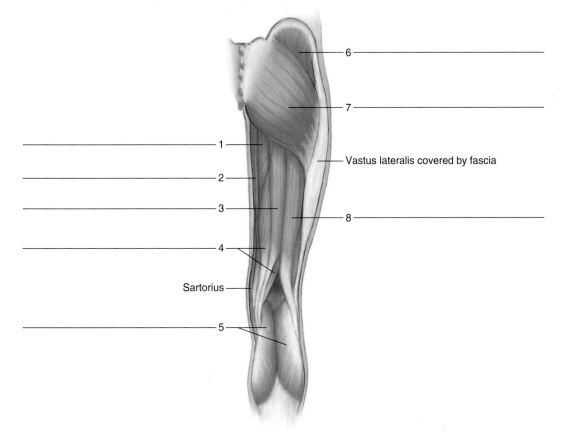

6
7
Vastus lateralis covered by fascia
1
2
3
8
4
Sartorius
5

Figure 23.4   Label the muscles of the anterior right leg.

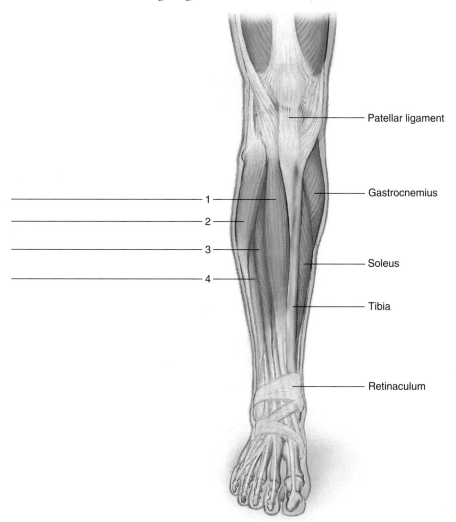

Patellar ligament

Gastrocnemius

Soleus

Tibia

Retinaculum

1

2

3

4

Figure 23.5 Label the muscles of the lateral right leg.

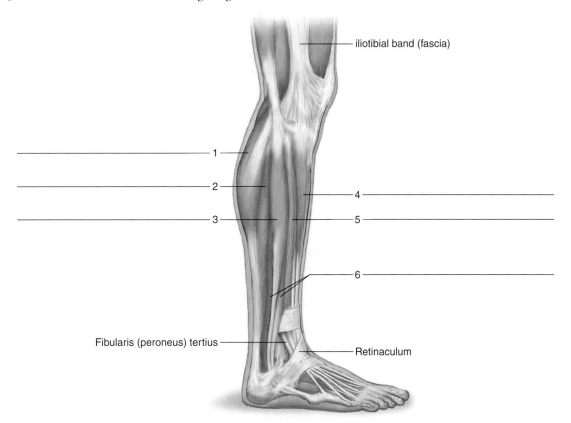

iliotibial band (fascia)

1

2

3

4

5

6

Fibularis (peroneus) tertius

Retinaculum

Figure 23.6 Label the muscles of the posterior right leg.

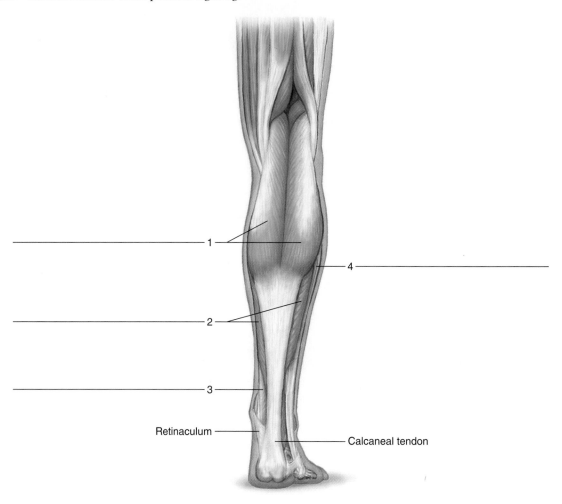

1

2

3

4

Retinaculum

Calcaneal tendon

MUSCLES OF THE HIP AND LOWER LIMB

PART A

Match the muscles in column A with the actions in column B. Place the letter of your choice in the space provided.

Column A		Column B
a. Biceps femoris	_____ 1.	Adducts thigh
b. Fibularis (peroneus) longus	_____ 2.	Plantar flexion and eversion of foot
c. Fibularis (peroneus) tertius	_____ 3.	Flexes thigh at the hip
d. Gluteus medius	_____ 4.	Abducts thigh and rotates it laterally
e. Gracilis	_____ 5.	Dorsiflexion and eversion of foot
f. Psoas major and iliacus	_____ 6.	Abducts thigh and rotates it medially
g. Quadriceps femoris group	_____ 7.	Plantar flexion and inversion of foot
h. Sartorius	_____ 8.	Flexes leg at the knee
i. Tibialis anterior	_____ 9.	Extends leg at the knee
j. Tibialis posterior	_____ 10.	Dorsiflexion and inversion of foot

PART B

Name the muscle indicated by the following combinations of origin and insertion.

Origin	Insertion	Muscle
1. Lateral surface of ilium	Greater trochanter of femur	_____
2. Ischial tuberosity	Posterior surface of femur	_____
3. Anterior superior iliac spine	Medial surface of tibia	_____
4. Lateral and medial condyles of femur	Posterior surface of calcaneus	_____
5. Anterior iliac crest	Fascia (iliotibial band) of the thigh	_____
6. Greater trochanter and posterior surface of femur	Patella to tibial tuberosity	_____
7. Ischial tuberosity	Medial surface of tibia	_____
8. Medial surface of femur	Patella to tibial tuberosity	_____
9. Posterior surface of tibia	Distal phalanges of four lateral toes	_____
10. Lateral condyle and lateral surface of tibia	Cuneiform and first metatarsal	_____

Figure 23.7 Identify the muscles indicated in this cross section of the right leg (superior view).

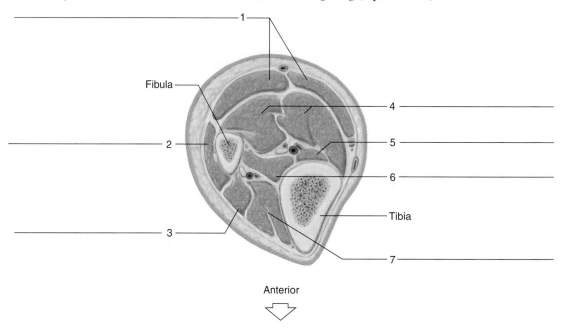

Anterior

PART C

Identify the muscles indicated in the cross section of the leg illustrated in figure 23.7.

PART D

Critical Thinking Application

Identify the muscles indicated in figure 23.8.

1. _____

2. _____

3. _____

4. _____

5. _____

6. _____

7. _____

8. _____

Figure 23.8 Identify the muscles that appear as lower limb surface features in these photographs (*a* and *b*), using the terms provided.

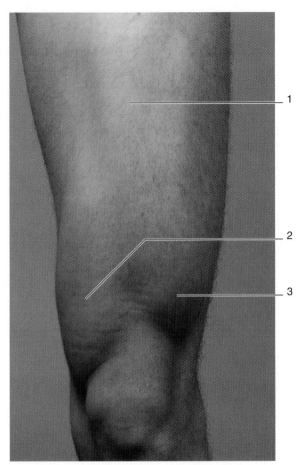

(a) Left thigh, anterior view

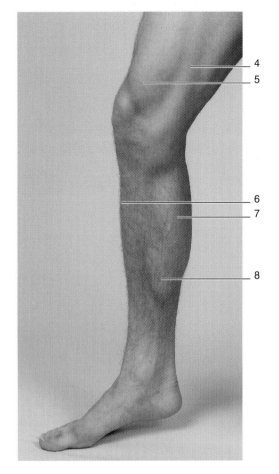

(b) Right lower limb, medial view

Terms:
Gastrocnemius
Rectus femoris
Sartorius
Soleus
Tibialis anterior
Vastus lateralis
Vastus medialis
Vastus medialis

CAT DISSECTION: MUSCULATURE

Although the aim of this exercise is to become more familiar with the human musculature, human cadavers are not always available for dissection. Instead, embalmed cats often are used for dissection because they are relatively small and can be purchased from biological suppliers. Also, as mammals, cats have many features in common with humans, including similar skeletal muscles (with similar names).

On the other hand, cats make use of four limbs for support, whereas humans use only two limbs. Because the musculature of each type of organism is adapted to provide for its special needs, comparisons of the muscles of cats and humans may not be precise.

As you continue your dissection of various systems of the cat, many anatomical similarities between cats and humans will be observed. These fundamental similarities are called homologous structures. Although homologous structures have a similar structure and embryological origin, the functions are sometimes different.

PURPOSE OF THE EXERCISE

To observe the musculature of the cat and to compare it with that of the human.

LEARNING OBJECTIVES

After completing this exercise, you should be able to

1. Name and locate the major skeletal muscles of the cat.
2. Name and locate the corresponding muscles of the human.
3. Name the origins, insertions, and actions of the muscles designated by the laboratory instructor.

MUSCLE DISSECTION TECHNIQUES

Dissect: to expose the entire length of the muscle from origin to insertion. Most of the procedures are accomplished with blunt probes used to separate various connective tissues that hold adjacent structures together. It does not mean to remove or to cut into the muscles or other organs.

Transect: to cut through the muscle near its midpoint. The cut is perpendicular to the muscle fibers.

Reflect: to lift a transected muscle aside to expose deeper muscles or other organs.

PROCEDURE A—SKIN REMOVAL

1. Cats usually are embalmed with a mixture of alcohol, formaldehyde, and glycerol, which prevents microorganisms from causing the tissues to decompose. However, because fumes from this embalming fluid may be annoying and may irritate skin, be sure to work in a well-ventilated room and wear disposable gloves to protect your hands.
2. Obtain an embalmed cat, a dissecting tray, a set of dissecting instruments, a large plastic bag, and an identification tag.
3. If the cat is in a storage bag, dispose of any excess embalming fluid in the bag, as directed by the

laboratory instructor. However, you may want to save some of the embalming fluid to keep the cat moist until you have completed your work later in the year.

4. If the cat is too wet with embalming fluid, blot the cat dry with paper towels.
5. Remove the skin from the cat. To do this, follow these steps:
   a. Place the cat in the dissecting tray with its ventral surface down.
   b. Use a sharp scalpel to make a short, shallow incision through the skin in the dorsal midline at the base of the cat's neck.
   c. Insert the pointed blade of a dissecting scissors through the opening in the skin and cut along the dorsal midline, following the tips of the vertebral spinous processes to the base of the tail (fig. 24.1).
   d. Use the scissors to make an incision encircling the region of the tail, anus, and genital organs. Do not remove the skin from this area.
   e. Use bone shears or a bone saw to sever the tail at its base and discard it (optional procedure).
   f. From the incision at the base of the tail, cut along the lateral surface of each hindlimb to the paw, and encircle the ankle. Also clip off the claws from each paw to prevent them from tearing the plastic storage bag and scratching your skin.
   g. Make an incision from the initial cut at the base of the neck to encircle the neck.

Figure 24.1   Incisions to be made for removing the skin of the cat.

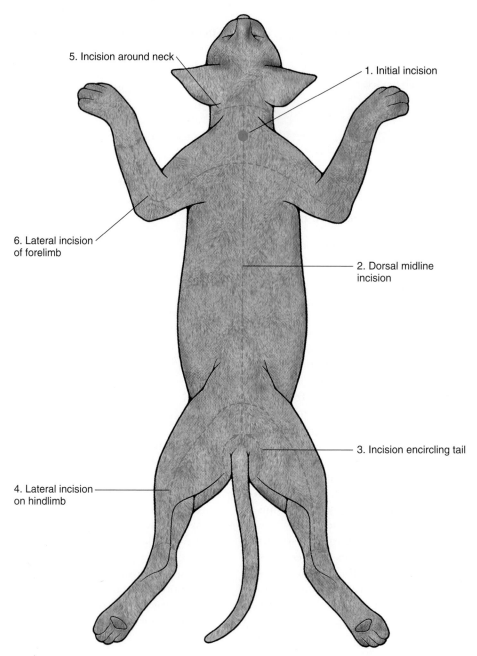

5. Incision around neck

1. Initial incision

6. Lateral incision of forelimb

2. Dorsal midline incision

3. Incision encircling tail

4. Lateral incision on hindlimb

h. From the incision around the neck, cut along the lateral surface of each forelimb to the paw and encircle the wrist (fig. 24.1).

i. Grasp the skin on either side of the dorsal midline incision, and carefully pull the skin laterally away from the body. At the same time, use your fingers or a blunt probe to help remove the skin from the underlying muscles by separating the loose connective tissue (superficial fascia). As you pull the skin away, you may note a thin sheet of muscle tissue attached to it. This muscle is called the *cutaneous maximus,* and it functions to move the cat's skin. Humans lack the cutaneous maximus, but a similar sheet of muscle (platysma) is present in the neck of a human.

j. As you remove the skin, work toward the ventral surface, then work toward the head, and finally work toward the tail. Pull the skin over each limb as if you were removing a glove.

k. If the cat is a female, note the mammary glands on the ventral surface of the thorax and abdomen (fig. 24.2). These glands will adhere to the skin and can be removed from it and discarded.

l. After the skin has been pulled away, carefully remove as much of the remaining connective tissue as possible to expose the underlying skeletal muscles. The muscles should appear light brown and fibrous.

6. After skinning the cat, follow these steps:

a. Discard the tissues you have removed as directed by the laboratory instructor.

Figure 24.2 Mammary glands on the ventral surface of the cat's thorax and abdomen.

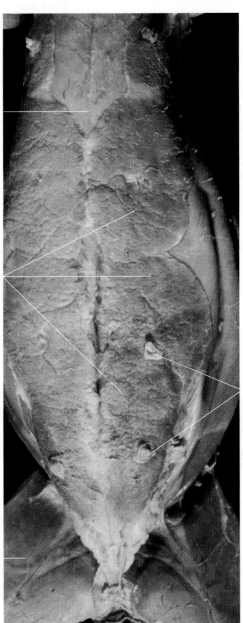

Sternal region

Mammary glands

Mammary papillae (teats or nipples)

Hindlimb

b. Wrap the skin around the cat to help keep its body moist, and place it in a plastic storage bag.
c. Write your name in pencil on an identification tag, and tie the tag to the storage bag so that you can identify your specimen.
7. Observe the recommended safety procedures for the conclusion of a laboratory session.

PROCEDURE B—SKELETAL MUSCLE DISSECTION

1. The purpose of a skeletal muscle dissection is to separate the individual muscles from any surrounding tissues and thus expose the muscles for observation. To do a muscle dissection, follow these steps:
   a. Use the appropriate figure as a guide and locate the muscle to be dissected in the specimen.
   b. Use a blunt probe or a finger to separate the muscle from the surrounding connective tissue along its natural borders. The muscle should separate smoothly. If the border appears ragged, you probably have torn the muscle fibers.
2. If it is necessary to cut through (transect) a superficial muscle to observe a deeper one, use scissors to transect the muscle about halfway between its origin and insertion. Then, lift aside (reflect) the cut ends, leaving their attachments intact.
3. The following procedures will instruct you to dissect some of the larger and more easily identified muscles of the cat. In each case, the procedure will include the names of the muscles to be dissected, figures illustrating their locations, and tables listing the origins, insertions, and actions of the muscles. (More detailed dissection instructions can be obtained by consulting an additional guide or atlas for cat dissection.) As you dissect each muscle, study the figures and tables in chapter 9 of the textbook and identify any corresponding (homologous) muscles of the human body. Also locate these homologous muscles in the torso or models of the human upper and lower limbs.

DEMONSTRATION

Refer to the cat skeleton to help locate origins and insertions of the cat's muscles. Reference to the cat skeleton will also help you visualize the actions of these muscles, if you remember that a muscle's insertion is pulled toward its origin.

PROCEDURE C—MUSCLES OF THE HEAD AND NECK

1. Place the cat in the dissecting tray with its ventral surface up.
2. Remove the skin and underlying connective tissues from one side of the neck, forward to the chin, and up to the ear.
3. Study figures 24.3 and 24.4, and then locate and dissect the following muscles:

sternomastoid	mylohyoid
sternohyoid	masseter
digastric	

Figure 24.3   Muscles on the ventral surface of the cat's head and neck.

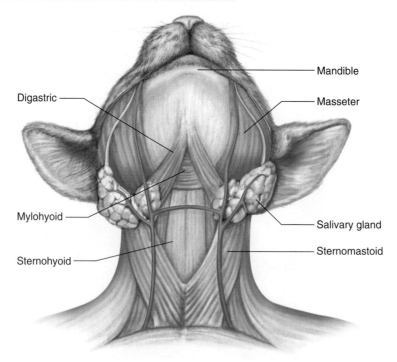

Figure 24.4 Deep muscles on the right ventral surface of the cat's neck.

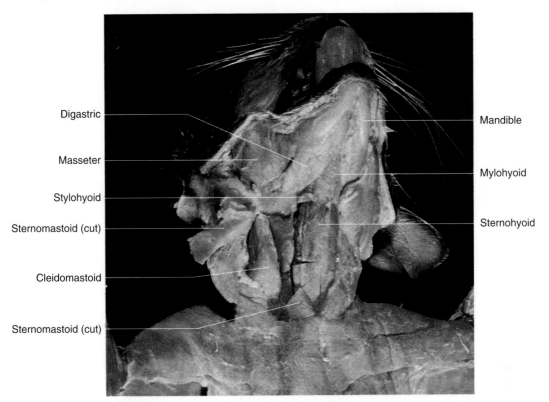

Digastric

Masseter

Stylohyoid

Sternomastoid (cut)

Cleidomastoid

Sternomastoid (cut)

Mandible

Mylohyoid

Sternohyoid

4. To find the deep muscles of the cat's neck and chin, carefully remove the sternomastoid, sternohyoid, and mylohyoid muscles from the right side.

5. Study figure 24.5, and locate the following muscles:

hyoglossus   cleidomastoid

styloglossus   thyrohyoid

geniohyoid   sternothyroid

6. See table 24.1 for the origins, insertions, and actions of these head and neck muscles.

7. Complete Part A of Laboratory Report 24.

PROCEDURE D—MUSCLES OF THE THORAX

1. Place the cat in the dissecting tray with its ventral surface up.

2. Remove any remaining fat and connective tissue to expose the muscles in the walls of the thorax and abdomen.

3. Study figures 24.6 and 24.7. Transect the superficial pectoantebrachialis to expose the entire pectoralis major. Locate and dissect the following pectoral muscles:

pectoantebrachialis   pectoralis minor

pectoralis major   xiphihumeralis

Table 24.1   Muscles of the Head and Neck

Muscle	Origin	Insertion	Action
Sternomastoid	Manubrium of sternum	Mastoid process of temporal bone	Turns and depresses head
Sternohyoid	Costal cartilage	Hyoid bone	Depresses hyoid bone
Digastric	Mastoid process and occipital bone	Ventral border of mandible	Depresses mandible
Mylohyoid	Inner surface of mandible	Hyoid bone	Raises floor of mouth
Masseter	Zygomatic arch	Coronoid fossa of mandible	Elevates mandible
Hyoglossus	Hyoid bone	Tongue	Pulls tongue back
Styloglossus	Styloid process	Tip of tongue	Pulls tongue back
Geniohyoid	Inner surface of mandible	Hyoid bone	Pulls hyoid bone forward
Cleidomastoid	Clavicle	Mastoid process	Turns head to side
Thyrohyoid	Thyroid cartilage	Hyoid bone	Raises larynx
Sternothyroid	Sternum	Thyroid cartilage	Pulls larynx back

Figure 24.5 Deep muscles of the cat's neck and chin regions.

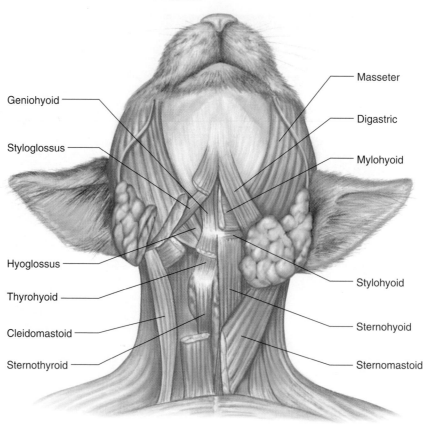

Geniohyoid

Styloglossus

Hyoglossus

Thyrohyoid

Cleidomastoid

Sternothyroid

Masseter

Digastric

Mylohyoid

Stylohyoid

Sternohyoid

Sternomastoid

4. To find the deep thoracic muscles, transect the pectoralis minor, pectoralis major, and pectoantebrachialis muscles, and reflect their cut edges to the sides. Remove the underlying fat and connective tissues to expose the following muscles, as shown in figure 24.8:

scalenus   levator scapulae

serratus ventralis   transversus costarum

5. See table 24.2 for the origins, insertions, and actions of these muscles.

PROCEDURE E—MUSCLES OF THE ABDOMINAL WALL

1. Study figures 24.6 and 24.9
2. Locate the *external oblique muscle* in the abdominal wall.
3. Make a shallow, longitudinal incision through the external oblique. Lift up the cut edge and expose the *internal oblique muscle* beneath (fig. 24.6). Note that the fibers of the internal oblique run at a right angle to those of the external oblique.
4. Make a longitudinal incision through the internal oblique. Lift up the cut edge and expose the *transversus abdominis.*

5. Expose the *rectus abdominis muscle* on one side of the midventral line. This muscle lies beneath an aponeurosis.
6. See table 24.3 for the origins, insertions, and actions of these muscles.
7. Complete Part B of the laboratory report.

PROCEDURE F—MUSCLES OF THE SHOULDER AND BACK

1. Place the cat in the dissecting tray with its dorsal surface up.
2. Remove any remaining fat and connective tissue to expose the muscles of the shoulder and back.
3. Study figures 24.10 and 24.11, and then locate and dissect the following superficial muscles:

clavotrapezius   acromiodeltoid

acromiotrapezius   spinodeltoid

spinotrapezius   latissimus dorsi

clavodeltoid

Note: The clavodeltoid (clavobrachialis) appears as a continuation of the clavotrapezius.

Figure 24.6 Muscles of the cat's thorax and abdomen.

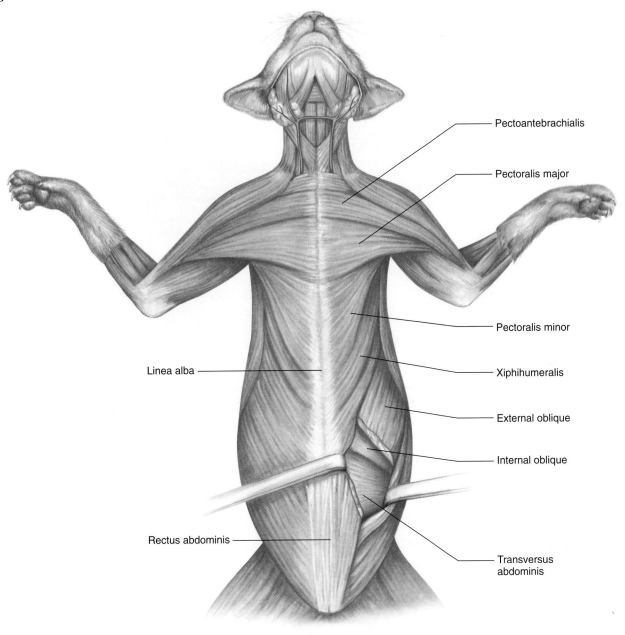

Pectoantebrachialis

Pectoralis major

Pectoralis minor

Linea alba

Xiphihumeralis

External oblique

Internal oblique

Rectus abdominis

Transversus abdominis

Table 24.2 Muscles of the Thorax

Muscle	Origin	Insertion	Action
Pectoantebrachialis	Manubrium of sternum	Fascia of forelimb	Adducts forelimb
Pectoralis major	Manubrium of sternum and costal cartilages	Greater tubercle of humerus	Adducts and rotates forelimb
Pectoralis minor	Sternum	Proximal end of humerus	Adducts and rotates forelimb
Xiphihumeralis	Xiphoid process of sternum	Proximal end of humerus	Adducts forelimb
Scalenus	Ribs	Cervical vertebrae	Flexes neck
Serratus ventralis	Ribs	Scapula	Pulls scapula ventrally and posteriorly
Levator scapulae	Cervical vertebrae	Scapula	Pulls scapula ventrally and anteriorly
Transversus costarum	Sternum	First rib	Pulls sternum toward head

Figure 24.7 Superficial muscles of the cat's thorax.

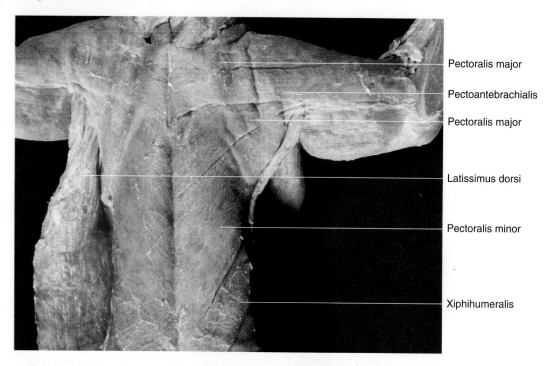

Pectoralis major

Pectoantebrachialis

Pectoralis major

Latissimus dorsi

Pectoralis minor

Xiphihumeralis

Figure 24.8 Deep muscles of the cat's thorax.

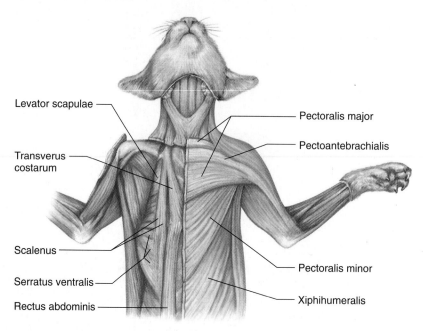

Levator scapulae

Transverus costarum

Scalenus

Serratus ventralis

Rectus abdominis

Pectoralis major

Pectoantebrachialis

Pectoralis minor

Xiphihumeralis

Table 24.3 Muscles of the Abdominal Wall

Muscle	Origin	Insertion	Action
External oblique	Ribs and fascia of back	Linea alba	Compresses abdominal wall
Internal oblique	Fascia of back	Linea alba	Compresses abdominal wall
Transversus abdominis	Lower ribs	Linea alba	Compresses abdominal wall
Rectus abdominis	Symphysis pubis	Sternum and costal cartilages	Compresses abdominal wall and flexes trunk

Figure 24.9 Superficial muscles of the cat's thorax and abdomen.

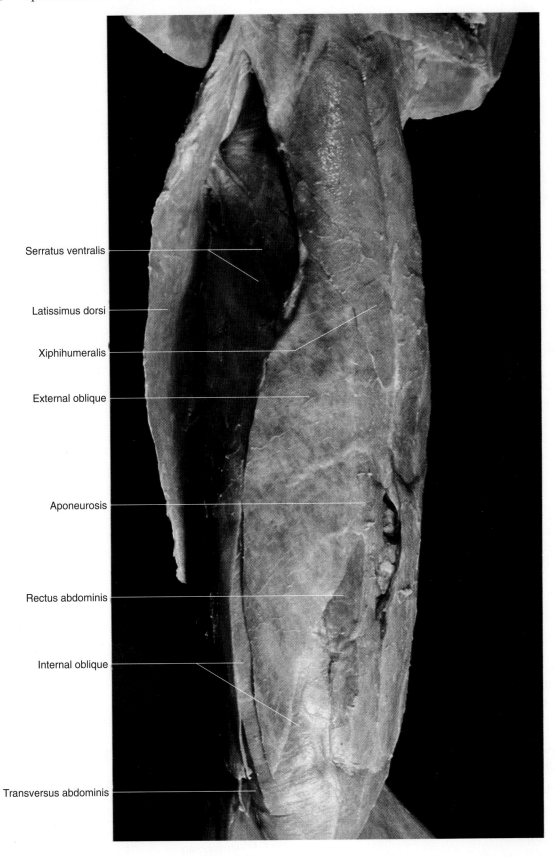

Serratus ventralis

Latissimus dorsi

Xiphihumeralis

External oblique

Aponeurosis

Rectus abdominis

Internal oblique

Transversus abdominis

Figure 24.10 Superficial muscles of the cat's shoulder and back (*right side*); deep muscles of the shoulder and back (*left side*).

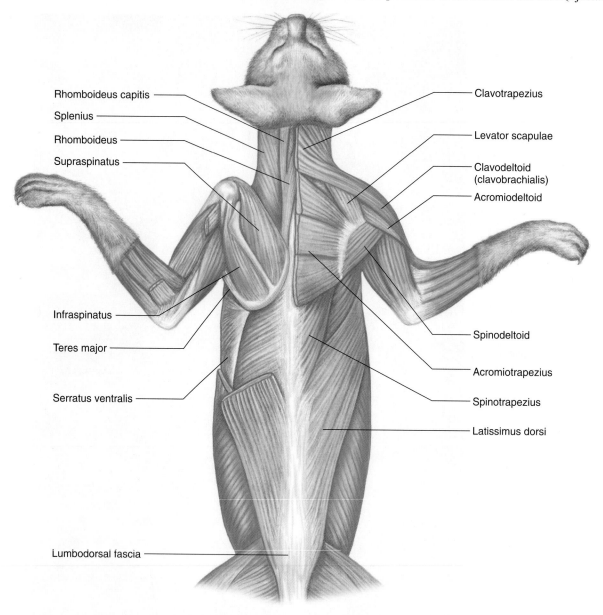

Rhomboideus capitis

Splenius

Rhomboideus

Supraspinatus

Infraspinatus

Teres major

Serratus ventralis

Lumbodorsal fascia

Clavotrapezius

Levator scapulae

Clavodeltoid (clavobrachialis)

Acromiodeltoid

Spinodeltoid

Acromiotrapezius

Spinotrapezius

Latissimus dorsi

4. Using scissors, transect the latissimus dorsi and the group of trapezius muscles. Lift aside their cut edges and remove any underlying fat and connective tissue. Study figures 24.10 and 24.12, and then locate and dissect the following deep muscles of the shoulder and back:

supraspinatus

infraspinatus

teres major

rhomboideus

rhomboideus capitis

splenius

5. See table 24.4 for the origins, insertions, and actions of these muscles in the shoulder and back.
6. Complete Part C of the laboratory report.

PROCEDURE G—MUSCLES OF THE FORELIMB

1. Place the cat in the dissecting tray with its ventral surface up.
2. Remove any remaining fat and connective tissue from a forelimb to expose the muscles.

198

Figure 24.11 Muscles of the cat's shoulder and back.

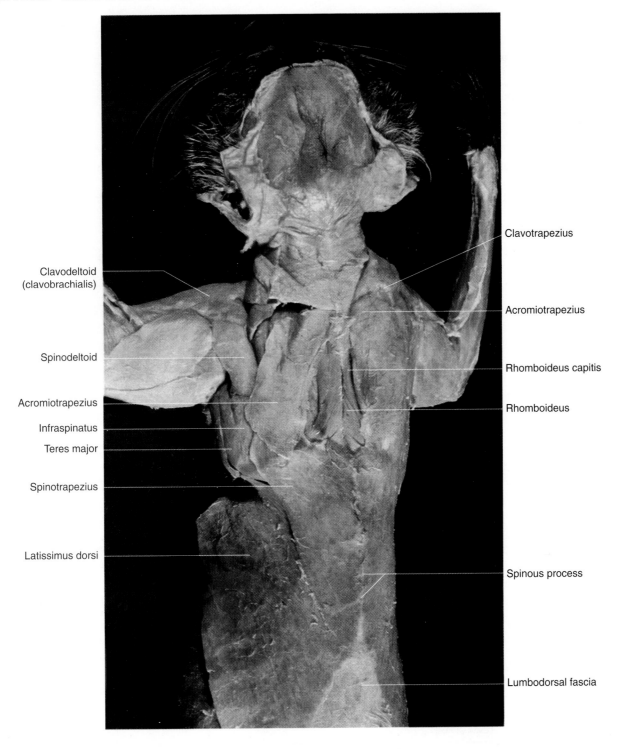

Clavodeltoid (clavobrachialis)

Spinodeltoid

Acromiotrapezius

Infraspinatus

Teres major

Spinotrapezius

Latissimus dorsi

Clavotrapezius

Acromiotrapezius

Rhomboideus capitis

Rhomboideus

Spinous process

Lumbodorsal fascia

3. Study figure 24.13*a*, and then locate and dissect the following muscles from the medial surface of the upper forelimb:

epitrochlearis

triceps brachii

4. Transect the pectoral muscles (pectoantebrachialis, pectoralis major, and pectoralis minor) and lift aside their cut edges.

Figure 24.12 Deep muscles of the cat's back.

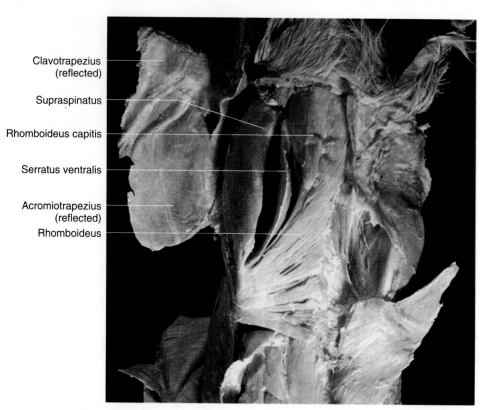

Clavotrapezius (reflected)

Supraspinatus

Rhomboideus capitis

Serratus ventralis

Acromiotrapezius (reflected)

Rhomboideus

Table 24.4 Muscles of the Shoulder and Back

Muscle	Origin	Insertion	Action
Clavotrapezius	Occipital bone of skull and spines of cervical vertebrae	Clavicle	Pulls clavicle upward
Acromiotrapezius	Spines of cervical and thoracic vertebrae	Spine and acromion process of scapula	Pulls scapula upward
Spinotrapezius	Spines of thoracic vertebrae	Spine of scapula	Pulls scapula upward and back
Clavodeltoid	Clavicle	Ulna near semilunar notch	Flexes forelimb
Acromiodeltoid	Acromion process of scapula	Proximal end of humerus	Flexes and rotates forelimb
Spinodeltoid	Spine of scapula	Proximal end of humerus	Flexes and rotates forelimb
Latissimus dorsi	Thoracic and lumbar vertebrae	Proximal end of humerus	Pulls forelimb upward and back
Supraspinatus	Fossa above spine of scapula	Greater tubercle of humerus	Extends forelimb
Infraspinatus	Fossa below spine of scapula	Greater tubercle of humerus	Rotates forelimb
Teres major	Posterior border of scapula	Proximal end of humerus	Rotates forelimb
Rhomboideus	Spines of cervical and thoracic vertebrae	Medial border of scapula	Pulls scapula back
Rhomboideus capitis	Occipital bone of skull	Medial border of scapula	Pulls scapula forward
Splenius	Fascia of neck	Occipital bone	Turns and raises head

5. Study figures 24.13*b* and 24.14, and then locate and dissect the following muscles:

   coracobrachialis

   biceps brachii

   long head of the triceps brachii

   medial head of the triceps brachii

6. Study figures 24.15*a* and 24.16, which show the lateral surface of the upper forelimb.

7. Locate the lateral head of the triceps muscle, transect it, and lift aside its cut edges.

8. Study figure 24.15*b*. Locate and dissect the following deeper muscles:

   anconeus

   brachialis

Figure 24.13 (*a*) Superficial muscles of the cat's medial forelimb. (*b*) Deep muscles of the medial forelimb.

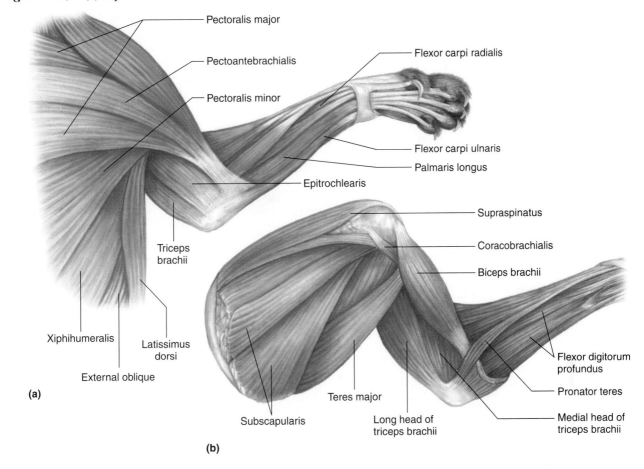

Pectoralis major

Pectoantebrachialis

Pectoralis minor

Flexor carpi radialis

Flexor carpi ulnaris

Palmaris longus

Epitrochlearis

Supraspinatus

Coracobrachialis

Biceps brachii

Triceps brachii

Xiphihumeralis

Latissimus dorsi

External oblique

(a)

Flexor digitorum profundus

Pronator teres

Medial head of triceps brachii

Subscapularis

Teres major

Long head of triceps brachii

(b)

Figure 24.14 Deep muscles of the cat's right medial forelimb.

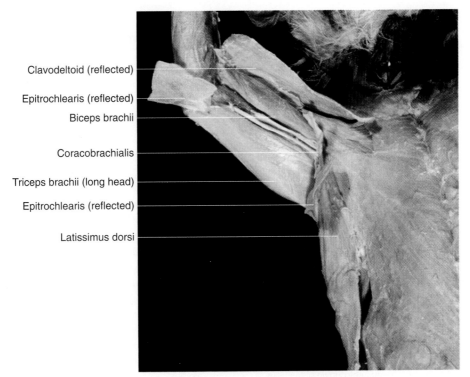

Clavodeltoid (reflected)

Epitrochlearis (reflected)

Biceps brachii

Coracobrachialis

Triceps brachii (long head)

Epitrochlearis (reflected)

Latissimus dorsi

Figure 24.15 (*a*) Superficial muscles of the cat's lateral forelimb. (*b*) Deep muscles of the lateral forelimb.

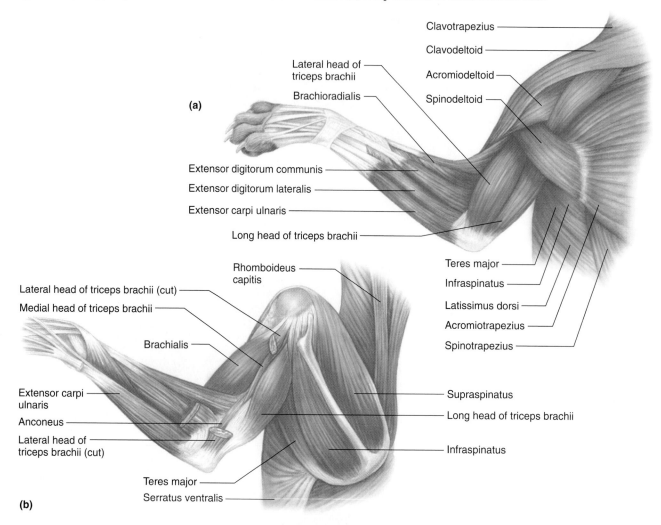

(a)

Clavotrapezius

Clavodeltoid

Acromiodeltoid

Spinodeltoid

Lateral head of triceps brachii

Brachioradialis

Extensor digitorum communis

Extensor digitorum lateralis

Extensor carpi ulnaris

Long head of triceps brachii

Teres major

Infraspinatus

Latissimus dorsi

Acromiotrapezius

Spinotrapezius

Rhomboideus capitis

Lateral head of triceps brachii (cut)

Medial head of triceps brachii

Brachialis

Extensor carpi ulnaris

Anconeus

Lateral head of triceps brachii (cut)

Teres major

Serratus ventralis

Supraspinatus

Long head of triceps brachii

Infraspinatus

(b)

Table 24.5 Muscles of the Forelimb

Muscle	Origin	Insertion	Action
Coracobrachialis	Coracoid process of scapula	Proximal end of humerus	Adducts forelimb
Epitrochlearis	Fascia of latissimus dorsi	Olecranon process of ulna	Extends forelimb
Biceps brachii	Border of glenoid cavity of scapula	Radial tuberosity of radius	Flexes forelimb
Triceps brachii			
Lateral head	Deltoid tuberosity of humerus	Olecranon process of ulna	Extends forelimb
Long head	Border of glenoid cavity of scapula	Olecranon process of ulna	Extends forelimb
Medial head	Shaft of humerus	Olecranon process of ulna	Extends forelimb
Anconeus	Lateral epicondyle of humerus	Olecranon process of ulna	Extends forelimb
Brachialis	Lateral surface of humerus	Proximal end of ulna	Flexes forelimb
Brachioradialis	Shaft of humerus	Distal end of radius	Supinates lower forelimb
Extensor digitorum communis	Lateral surface of humerus	Digits two to five	Extends digits
Extensor digitorum lateralis	Lateral surface of humerus	Digits two to five	Extends digits
Extensor carpi ulnaris	Lateral epicondyle of humerus and ulna	Fifth metacarpal	Extends wrist
Flexor carpi ulnaris	Medial epicondyle of humerus	Carpals	Flexes wrist
Palmaris longus	Medial epicondyle of humerus	Digits	Flexes digits
Flexor carpi radialis	Medial epicondyle of humerus	Second and third metacarpals	Flexes metacarpals
Pronator teres	Medial epicondyle of humerus	Radius	Pronates lower forelimb
Flexor digitorum profundus	Radius, ulna, and medial epicondyle of humerus	Distal phalanges	Flexes digits

Figure 24.16 Superficial muscles of the cat's left lateral forelimb, dorsal view.

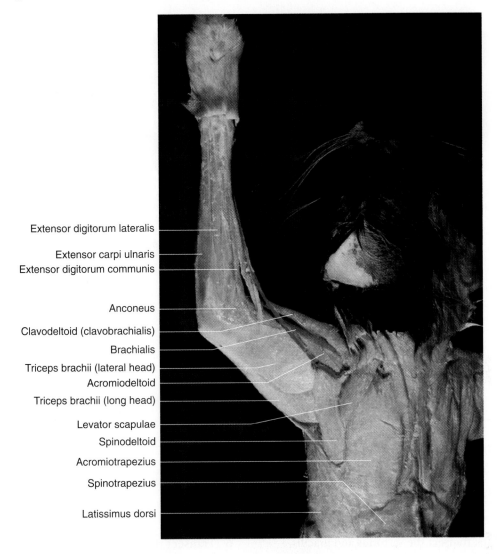

Extensor digitorum lateralis

Extensor carpi ulnaris
Extensor digitorum communis

Anconeus

Clavodeltoid (clavobrachialis)

Brachialis

Triceps brachii (lateral head)
Acromiodeltoid

Triceps brachii (long head)

Levator scapulae

Spinodeltoid

Acromiotrapezius

Spinotrapezius

Latissimus dorsi

9. Separate the muscles in the lower forelimb into an *extensor group* (fig. 24.15*a*) and a *flexor group* (fig. 24.13*a*). Generally, the extensors are located on the posterior surface of the lower forelimb, and the flexors are located on the anterior surface. A thick, tough layer of fascia (antebrachial fascia) surrounds the forelimb. Remove this fascia to expose and dissect the following muscles:

brachioradialis

extensor digitorum communis

extensor digitorum lateralis

extensor carpi ulnaris

flexor carpi ulnaris

palmaris longus

flexor carpi radialis

pronator teres

10. Transect the flexor carpi ulnaris, palmaris longus, and flexor carpi radialis to expose the flexor digitorum profundus.
11. See table 24.5 for the origins, insertions, and actions of these muscles of the forelimb.
12. Complete Part D of the laboratory report.

PROCEDURE H—MUSCLES OF THE HIP AND HINDLIMB

1. Place the cat in the dissecting tray with its ventral surface up.
2. Remove any remaining fat and connective tissue from the hip and hindlimb to expose the muscles.
3. Study figures 24.17*a* and 24.18, and then locate and dissect the following muscles from the medial surface of the thigh:

sartorius

gracilis

203

Figure 24.17 (*a*) Superficial muscles of the cat's medial hindlimb. (*b*) Deep muscles of the medial hindlimb.

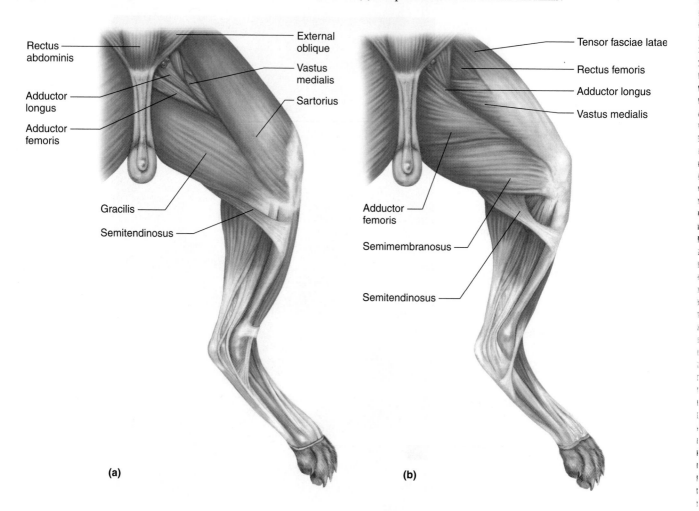

(a)

(b)

4. Using scissors, transect the sartorius and gracilis, and lift aside their cut edges to observe the deeper muscles of the thigh.

5. Study figures 24.17*b* and 24.19, and then locate and dissect the following muscles:

rectus femoris

vastus medialis

adductor longus

adductor femoris

semimembranosus

semitendinosus

tensor fasciae latae

6. Transect the tensor fasciae latae and rectus femoris muscles and turn their ends aside. Locate the *vastus intermedius* and *vastus lateralis* muscles beneath (fig. 24.20). (*Note:* In some specimens, the vastus intermedius, vastus lateralis, and vastus medialis muscles are closely united by connective tissue and are difficult to separate.)

7. Study figures 24.21*a* and 24.22, and then locate and dissect the following muscles from the lateral surface of the hip and thigh:

biceps femoris

caudofemoralis

gluteus maximus

Figure 24.18 Superficial muscles of the cat's hindlimb, medial view.

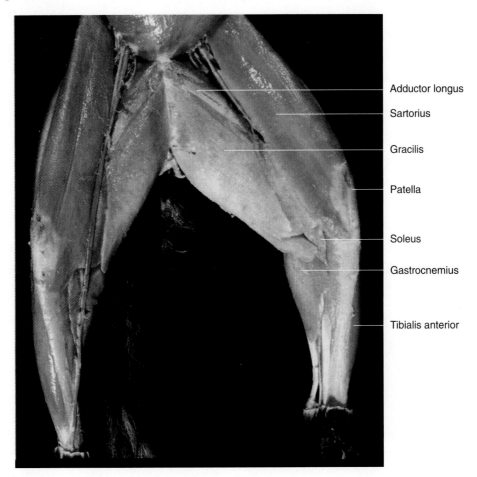

Adductor longus

Sartorius

Gracilis

Patella

Soleus

Gastrocnemius

Tibialis anterior

8. Using scissors, transect the tensor fasciae latae and biceps femoris, and lift aside their cut edges to observe the deeper muscles of the thigh.

9. Locate and dissect the following muscles:

tenuissimus

gluteus medius

(*Note:* The tenuissimus muscle and the sciatic nerve may adhere tightly to the dorsal surface of the biceps femoris muscle. Take care to avoid cutting these structures.)

10. On the lateral surface of the lower hindlimb (figs. 24.21 and 24.23), locate and dissect the following muscles:

gastrocnemius

soleus

tibialis anterior

fibularis (peroneus) group

11. See table 24.6 for the origins, insertions, and actions of these muscles of the hip and hindlimb.

12. Complete Part E of the laboratory report.

Figure 24.19 Deep muscles of the cat's hindlimb, medial view.

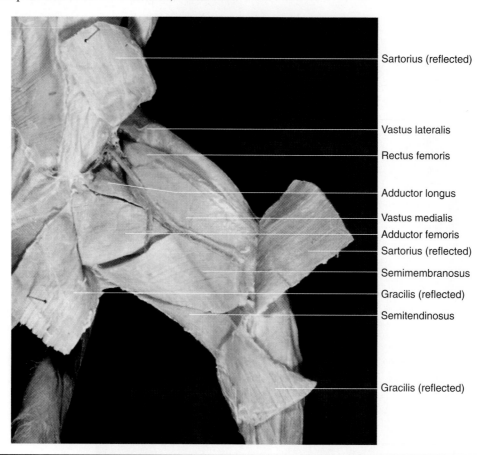

Sartorius (reflected)

Vastus lateralis

Rectus femoris

Adductor longus

Vastus medialis
Adductor femoris
Sartorius (reflected)
Semimembranosus
Gracilis (reflected)
Semitendinosus

Gracilis (reflected)

Figure 24.20 Other deep muscles of the cat's thigh, medial view.

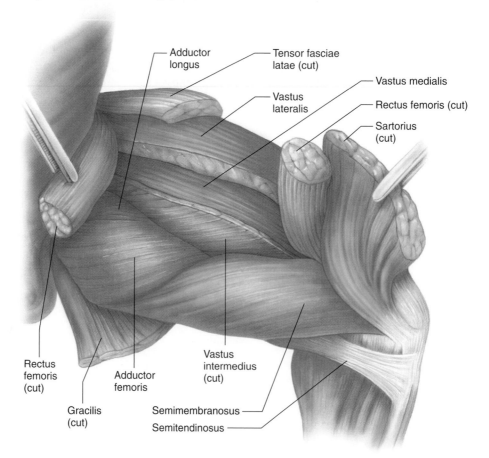

Adductor
longus

Tensor fasciae
latae (cut)

Vastus medialis

Vastus
lateralis

Rectus femoris (cut)

Sartorius
(cut)

Rectus
femoris
(cut)

Adductor
femoris

Gracilis
(cut)

Vastus
intermedius
(cut)

Semimembranosus

Semitendinosus

Figure 24.21 (*a*) Superficial muscles of the cat's lateral hindlimb. (*b*) Deep muscles of the lateral hindlimb.

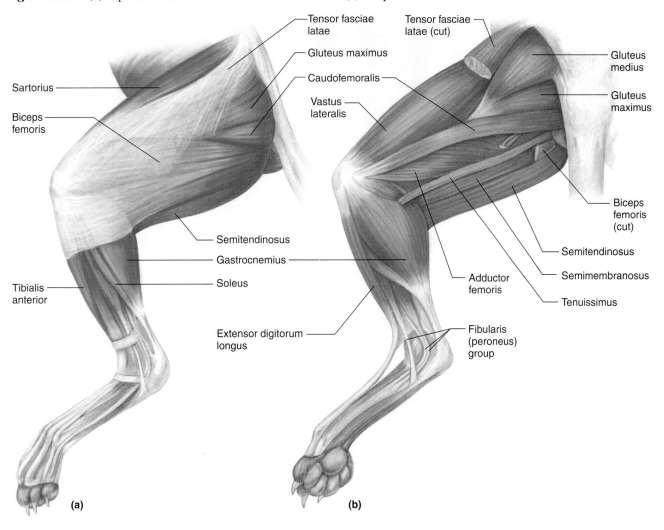

Tensor fasciae latae

Gluteus maximus

Caudofemoralis

Sartorius

Biceps femoris

Vastus lateralis

Semitendinosus

Gastrocnemius

Tibialis anterior

Soleus

Extensor digitorum longus

Tensor fasciae latae (cut)

Gluteus medius

Gluteus maximus

Biceps femoris (cut)

Semitendinosus

Semimembranosus

Tenuissimus

Adductor femoris

Fibularis (peroneus) group

(a)

(b)

Figure 24.22 Muscles of the cat's hindlimb, lateral view.

Sartorius
Gluteus medius
Tensor fasciae latae
Gluteus maximus
Caudofemoralis

Biceps femoris

Semitendinosus

Gastrocnemius

Extensor digitorum longus

Soleus

Fibularis (peroneus)

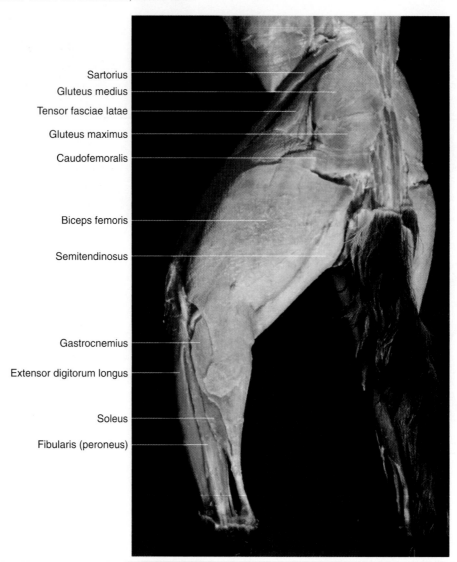

Figure 24.23 Deep muscles of the cat's hindlimb, lateral view.

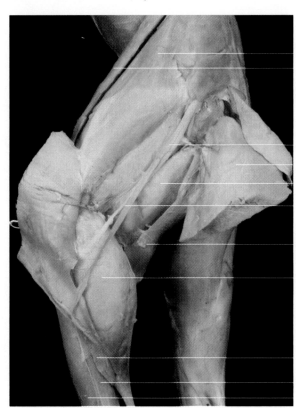

Tensor fasciae latae
Sartorius

Sciatic nerve
Biceps femoris (reflected)
Semimembranosus
Tenuissimus

Semitendinosus

Gastrocnemius

Soleus

Fibularis (peroneus) tertius
Extensor digitorum longus

Table 24.6 Muscles of the Hip and Hindlimb

Muscle	Origin	Insertion	Action
Sartorius	Crest of ilium	Fascia of knee and proximal end of tibia	Rotates and extends hindlimb
Gracilis	Symphysis pubis and ischium	Proximal end of tibia	Adducts hindlimb
Quadriceps femoris			
Vastus lateralis	Shaft of femur and greater trochanter	Patella	Extends hindlimb
Vastus intermedius	Shaft of femur	Patella	Extends hindlimb
Rectus femoris	Ilium	Patella	Extends hindlimb
Vastus medialis	Shaft of femur	Patella	Extends hindlimb
Adductor longus	Pubis	Proximal end of femur	Adducts hindlimb
Adductor femoris	Pubis and ischium	Shaft of femur	Adducts hindlimb
Semimembranosus	Ischial tuberosity	Medial epicondyle of femur	Extends thigh
Tensor fasciae latae	Ilium	Fascia of thigh	Extends thigh
Biceps femoris	Ischial tuberosity	Patella and tibia	Abducts thigh and flexes lower hindlimb
Tenuissimus	Second caudal vertebra	Tibia and fascia of biceps femoris	Abducts thigh and flexes lower hindlimb
Semitendinosus	Ischial tuberosity	Crest of tibia	Flexes lower hindlimb
Caudofemoralis	Caudal vertebrae	Patella	Abducts thigh and extends lower hindlimb
Gluteus maximus	Sacral and caudal vertebrae	Greater trochanter of femur	Abducts thigh
Gluteus medius	Ilium, and sacral and caudal vertebrae	Greater trochanter of femur	Abducts thigh
Gastrocnemius	Lateral and medial epicondyles of femur	Calcaneus	Extends foot
Soleus	Proximal end of fibula	Calcaneus	Extends foot
Tibialis anterior	Proximal end of tibia and fibula	First metatarsal	Flexes foot
Fibularis (peroneus) group	Shaft of fibula	Metatarsals	Flexes foot

CAT DISSECTION: MUSCULATURE

PART A

Complete the following statements:

1. The _____ muscle of the human is homologous to the sternomastoid muscle of the cat.
2. The _____ muscle elevates the mandible in the human and in the cat.
3. Two muscles of the cat that are inserted on the hyoid bone are the _____ and the

   _____ .

PART B

Complete the following:
Name two muscles that are found in the thoracic wall of the cat but are absent in the human.

1. _____
2. _____

Name two muscles that are found in the thoracic wall of the cat and the human.

3. _____
4. _____

Name four muscles that are found in the abdominal wall of the cat and the human.

5. _____
6. _____
7. _____
8. _____

PART C

Complete the following:
Name three muscles of the cat that together correspond to the trapezius muscle in the human.

1. _____
2. _____
3. _____

Name three muscles of the cat that together correspond to the deltoid muscle in the human.

4. _____
5. _____
6. _____

Name the muscle in the cat and in the human that occupies the fossa above the spine of the scapula.

7. _____

Name the muscle in the cat and in the human that occupies the fossa below the spine of the scapula.

8. _____

Name two muscles found in the cat and in the human that can rotate the forelimb.

9. _____

10. _____

PART D

Complete the following:

Name two muscles found in the cat and in the human that can flex the forelimb.

1. _____

2. _____

Name a muscle found in the cat but absent in the human that can extend the forelimb.

3. _____

Name a muscle that has three heads, can extend the forelimb, and is found in the cat and in the human.

4. _____

PART E

Each of the muscles in figure 24.24 is found both in the cat and in the human. Identify each of the numbered muscles in the figure by placing its name in the space next to its number.

Figure 24.24 Identify the numbered muscles that occur in both the cat and the human.

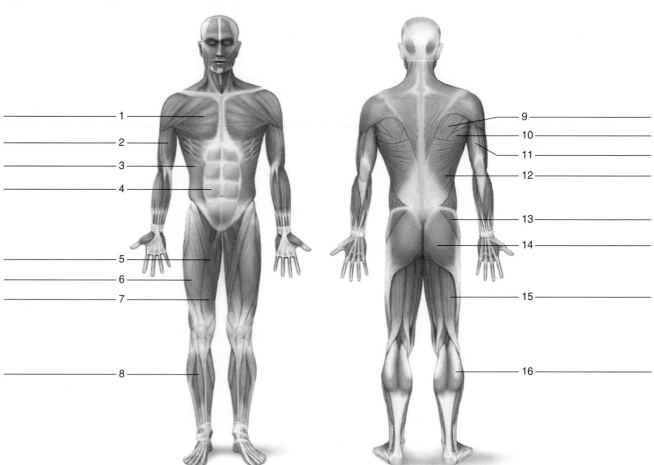

LABORATORY EXERCISE 25

NERVOUS TISSUE AND NERVES

Nervous tissue, which occurs in the brain, spinal cord, and nerves, contains neurons and neuroglial cells. The neurons are the basic structural and functional units of the nervous system involved in decision-making processes, detecting stimuli, and conducting messages. The neuroglial cells perform various supportive and protective functions for neurons.

PURPOSE OF THE EXERCISE

To review the characteristics of nervous tissue and to observe neurons, neuroglial cells, and various features of the nerves.

LEARNING OBJECTIVES

After completing this exercise, you should be able to

1. Describe the general characteristics of nervous tissue.
2. Distinguish between neurons and neuroglial cells.
3. Identify the major structures of a neuron and a nerve.

PROCEDURE

1. Review the sections entitled "General Functions of the Nervous System" and "Classification of Neurons and Neuroglial Cells" in chapter 10 of the textbook.
2. As a review activity, label figures 25.1 and 25.2.

3. Complete Parts A and B of Laboratory Report 25.
4. Obtain a prepared microscope slide of a spinal cord smear. Using low-power magnification, search the slide and locate the relatively large, deeply stained cell bodies of motor neurons (multipolar neurons).
5. Observe a single motor neuron, using high-power magnification, and note the following features:

   cell body

   nucleus

   nucleolus

   chromatophilic substance (Nissl bodies)

   neurofibrils (threadlike structures extending into
      the nerve fibers)

   dendrites

   axon (nerve fiber)

   Compare the slide to the neuron model and to figure 25.3. You also may note small, darkly stained nuclei of neuroglial cells around the motor neuron.
6. Sketch and label a single motor (efferent) neuron in the space provided in Part C of the laboratory report.
7. Obtain a prepared microscope slide of a dorsal root ganglion. Search the slide and locate a cluster of sensory neuron cell bodies. You also may note bundles of nerve fibers passing among groups of neuron cell bodies (fig. 25.4).
8. Sketch and label a single sensory (afferent) neuron cell body in the space provided in Part C of the laboratory report.
9. Obtain a prepared microscope slide of neuroglial cells. Search the slide and locate some darkly stained astrocytes with numerous long, slender processes (fig. 25.5).
10. Sketch a single neuroglial cell in the space provided in Part C of the laboratory report.
11. Obtain a prepared microscope slide of a nerve. Locate the cross section of the nerve and note the many round nerve fibers inside. Also note the dense layer of connective tissue (perineurium) that encircles the nerve fibers and holds them together in a bundle. The individual nerve fibers are surrounded by a layer of more delicate connective tissue (endoneurium) (fig. 25.6).

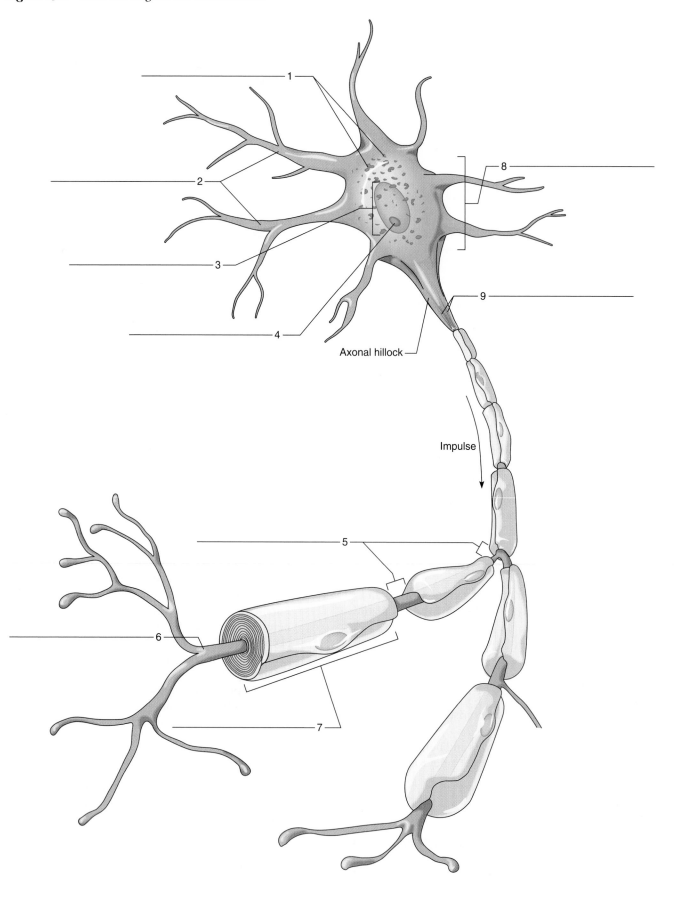

Figure 25.1 Label this diagram of a motor neuron.

Axonal hillock

Impulse

Figure 25.2 Label the features of the myelinated axon (nerve fiber).

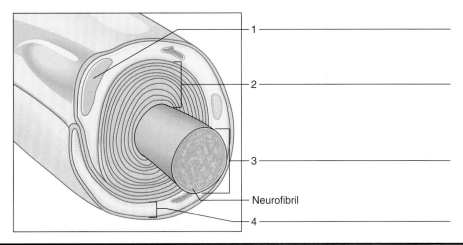

1 _____

2 _____

3 _____

— Neurofibril

4 _____

Figure 25.3 Micrograph of a multipolar neuron and neuroglial cells from a spinal cord smear (100× micrograph enlarged to 600×).

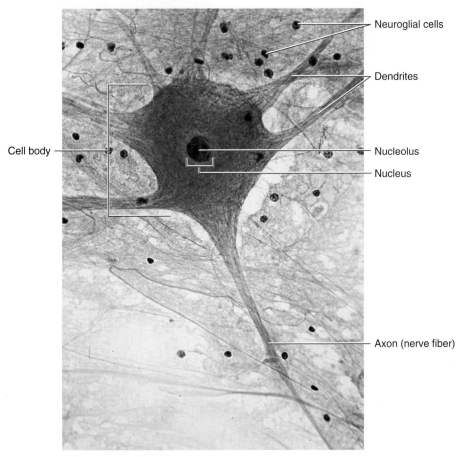

Neuroglial cells

Dendrites

Cell body

Nucleolus

Nucleus

Axon (nerve fiber)

12. Using high-power magnification, observe a single nerve fiber and note the following features:

central axon

myelin around the axon (actually, most of the myelin may have been dissolved and lost during the slide preparation)

neurilemma

13. Sketch and label a single nerve fiber with Schwann cell (cross section) in the space provided in Part D of the laboratory report.

14. Locate the longitudinal section of the nerve on the slide (fig. 25.7). Note the following:

central axons

myelin of Schwann cells

neurilemma of Schwann cells

nodes of Ranvier

15. Sketch and label a single nerve fiber with Schwann cell (longitudinal section) in the space provided in Part D of the laboratory report.

Figure 25.4 Micrograph of a dorsal root ganglion (50× micrograph enlarged to 100×).

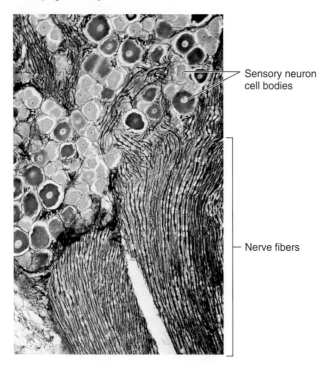

Sensory neuron cell bodies

Nerve fibers

Figure 25.5 Micrograph of astrocytes (250× micrograph enlarged to 1,000×).

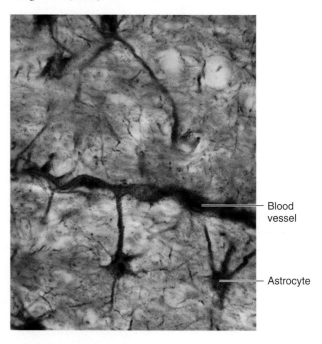

Blood vessel

Astrocyte

Figure 25.6 Cross section of a bundle of neurons within a nerve (400×).

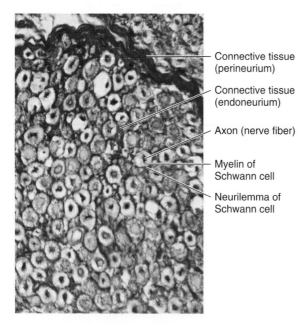

Connective tissue (perineurium)

Connective tissue (endoneurium)

Axon (nerve fiber)

Myelin of Schwann cell

Neurilemma of Schwann cell

Figure 25.7 Longitudinal section of a nerve (250× micrograph enlarged to 2,000×).

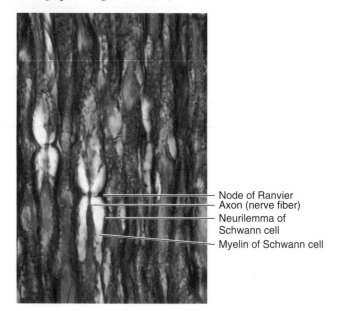

Node of Ranvier
Axon (nerve fiber)
Neurilemma of Schwann cell
Myelin of Schwann cell

OPTIONAL ACTIVITY

Obtain a prepared microscope slide of Purkinje cells. To locate these neurons, search the slide for large, flask-shaped cell bodies. Note that each cell body has one or two large, thick dendrites that give rise to branching networks of fibers. These cells are found in a particular region of the brain (cerebellar cortex).

NERVOUS TISSUE AND NERVES

PART A

Match the terms in column A with the descriptions in column B. Place the letter of your choice in the space provided.

Column A

a. Astrocyte
b. Axon
c. Chromatophilic substance (Nissl body)
d. Collateral
e. Dendrite
f. Myelin
g. Neurilemma
h. Neurofibrils

Column B

_____ 1. Sheath of Schwann cell containing cytoplasm and nucleus that encloses myelin

_____ 2. Corresponds to rough endoplasmic reticulum in other cells

_____ 3. Network of fine threads within nerve fiber

_____ 4. Substance of Schwann cell composed of lipoprotein

_____ 5. Neuron process with many branches that conducts an impulse toward the cell body

_____ 6. Branch of an axon

_____ 7. Star-shaped neuroglial cell between neurons and blood vessels

_____ 8. Nerve fiber arising from a slight elevation of the cell body that conducts an impulse away from the cell body

PART B

Match the terms in column A with the descriptions in column B. Place the letter of your choice in the space provided.

Column A

a. Effector
b. Ependyma
c. Ganglion
d. Interneuron
e. Microglia
f. Motor (efferent) neuron
g. Oligodendrocyte
h. Sensory (afferent) neuron

Column B

_____ 1. Transmits impulse from sensory to motor neuron within central nervous system

_____ 2. Transmits impulse out of the brain or spinal cord to effectors

_____ 3. Transmits impulse into brain or spinal cord from receptors

_____ 4. Myelin-forming neuroglial cell in brain and spinal cord

_____ 5. Phagocytic neuroglial cell

_____ 6. Structure capable of responding to motor impulse

_____ 7. Specialized mass of neuron cell bodies outside the brain or spinal cord

_____ 8. Covers the inside spaces of the ventricles

PART C

1. Sketch and label a single motor neuron.

2. Sketch and label a single sensory neuron cell body.

3. Sketch a single neuroglial cell.

PART D

1. Sketch and label a single nerve fiber with Schwann cell (cross section).

2. Sketch and label a single nerve fiber with Schwann cell (longitudinal section).

NERVE IMPULSE STIMULATION

MATERIALS NEEDED

Textbook
Live frog
Dissecting tray
Dissecting instruments
Frog Ringer's solution
Electronic stimulator
Filter paper
Glass rod
Glass plate
Ring stand and ring
Microscope slides
Bunsen burner
Ice
1% HCl
1% NaCl

For Optional Activity:

2% Novocain solution (procaine hydrochloride)

SAFETY

- Wear disposable gloves when handling the frogs and chemicals.
- Keep loose hair and clothes away from the Bunsen burner.
- Wear heat-resistant gloves when heating the glass rod.
- Dispose of gloves, frogs, and chemicals as instructed.
- Wash your hands before leaving the laboratory.

A nerve cell usually is polarized due to an unequal distribution of ions on either side of its membrane. When such a polarized membrane is stimulated at or above its threshold intensity, a wave of action potentials is triggered to move in all directions away from the site of stimulation. This wave constitutes a nerve impulse, and if it reaches a muscle, the muscle may respond by contracting.

PURPOSE OF THE EXERCISE

To review the characteristics of a nerve impulse and to investigate the effects of certain stimuli on a nerve.

LEARNING OBJECTIVES

After completing this exercise, you should be able to

1. Describe the events that lead to the stimulation of a nerve impulse.
2. List four types of factors that can stimulate a nerve impulse.
3. Test the effects of various factors on a nerve-muscle preparation.

PROCEDURE

1. Review the section entitled "Cell Membrane Potential" in chapter 10 of the textbook.
2. Complete Part A of Laboratory Report 26.
3. Obtain a live frog, and pith its brain and spinal cord as described in Procedure C of Laboratory Exercise 19.

ALTERNATIVE PROCEDURE

An anesthetizing agent, tricaine methane sulfonate, can be used to prepare frogs for this lab. This procedure eliminates the need to pith frogs.

4. Place the pithed frog in a dissecting tray and remove the skin from its hindlimb, beginning at the waist, as described in Procedure C of Laboratory Exercise 19. (As the skin is removed, keep the exposed tissues moist by flooding them with frog Ringer's solution.)
5. Expose the frog's sciatic nerve. To do this, follow these steps:
    a. Use a glass rod to separate the gastrocnemius muscle from the adjacent muscles.
    b. Locate the calcaneal (Achilles) tendon at the distal end of the gastrocnemius, and cut it with scissors.
    c. Place the frog ventral surface down, and separate the muscles of the thigh to locate the sciatic nerve. The nerve will look like a silvery white thread passing through the thigh, dorsal to the femur (fig. 26.1).
    d. Dissect the nerve to its origin in the spinal cord.

Figure 26.1 The sciatic nerve appears as a silvery white thread between the muscles of the thigh.

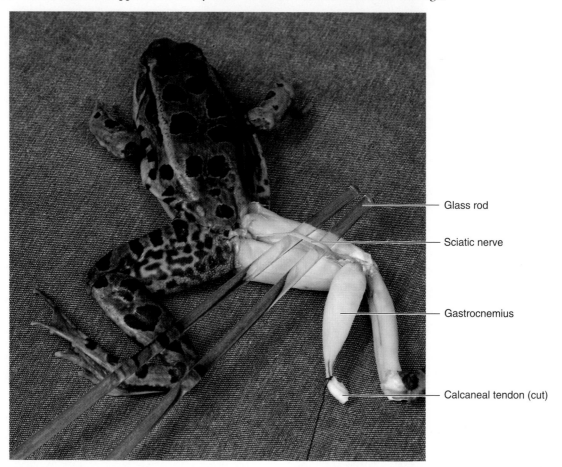

- Glass rod
- Sciatic nerve
- Gastrocnemius
- Calcaneal tendon (cut)

e. Use scissors to cut the nerve at its origin, and carefully snip off all of the branch nerves in the thigh, leaving only its connection to the gastrocnemius muscle.

f. Use a scalpel to free the proximal end of the gastrocnemius.

g. Carefully remove the nerve and attached muscle, and transfer the preparation to a glass plate supported on the ring of a ring stand.

h. Use a glass rod to position the preparation so that the sciatic nerve is hanging over the edge of the glass plate. (Be sure to keep the preparation moistened with frog Ringer's solution at all times.)

6. Determine the threshold voltage and the voltage needed for maximal muscle contraction by using the electronic stimulator, as described in Procedure D of Laboratory Exercise 19.

7. Expose the cut end of the sciatic nerve to each of the following conditions, and observe the response of the gastrocnemius muscle. Add frog Ringer's solution after each of the experiments.

   a. Firmly pinch the end of the nerve between two glass microscope slides or pinch using forceps.

   b. Touch the cut end with a glass rod that is at room temperature.

   c. Touch the cut end with a glass rod that has been cooled in ice water for 5 minutes.

d. Touch the cut end with a glass rod that has been heated in the flame of a Bunsen burner. Wear heat-resistant gloves for this procedure.

e. Dip the cut end in 1% HCl.

f. Dip the cut end in 1% NaCl.

8. Complete Part B of the laboratory report.

OPTIONAL ACTIVITY

Test the effect of Novocain on a frog sciatic nerve. To do this, follow these steps:

1. Place a nerve-muscle preparation on a glass plate supported by the ring of a ring stand, as before.

2. Use the electronic stimulator to determine the voltage needed for maximal muscle contraction.

3. Saturate a small piece of filter paper with 2% Novocain solution, and wrap the paper around the midsection of the sciatic nerve.

4. At 2-minute intervals, stimulate the nerve, using the voltage needed for maximal contraction until the muscle fails to respond.

5. Remove the filter paper, and flood the nerve with frog Ringer's solution.

6. At 2-minute intervals, stimulate the nerve until the muscle responds again. How long did it take for the nerve to recover from the effect of the Novocain?

NERVE IMPULSE STIMULATION

PART A

Complete the following statements:

1. _____ ions tend to pass through cell membranes more easily than sodium ions.

2. When a nerve cell is at rest, there is a relatively greater concentration of _____ ions outside of its membrane.

3. When sodium ions are actively transported outward through a nerve cell membrane, _____ ions are transported inward.

4. The difference in electrical charge between the inside and the outside of a nerve cell membrane is called the _____ .

5. If the resting potential becomes less negative in response to stimulation, the cell membrane is said to be _____ .

6. As a result of an additive phenomenon called _____, the threshold potential of a membrane may be reached.

7. Following depolarization, potassium ions diffuse outward and cause the cell membrane to become _____ .

8. An action potential is a rapid sequence of changes involving depolarization and _____ .

9. The moment following the passage of an action potential during which a threshold stimulus will not trigger another impulse is called the _____ .

10. Muscle fiber contraction and nerve impulse conduction are similar in that both are _____ responses.

11. Myelin contains a high proportion of _____ .

12. Nodes of Ranvier occur between adjacent _____ .

13. The type of conduction in which an impulse seems to jump from node to node is called _____ .

14. The greater the diameter of a nerve fiber, the _____ the impulse travels.

PART B

1. What was the threshold voltage for the frog sciatic nerve? _____

2. What was the voltage needed for maximal contraction of the gastrocnemius muscle? _____

3. Complete the following table:

Factor Tested	Muscle Response	Effect on Nerve
Pinching		
Glass rod (room temperature)		
Glass rod (cooled)		
Glass rod (heated)		
1% HCl		
1% NaCl		

4. Write a statement to summarize the results of these tests.

_____

_____

_____

_____

THE MENINGES AND SPINAL CORD

The meninges consist of layers of membranes located between the bones of the skull and vertebral column and the soft tissues of the central nervous system. They include the dura mater, the arachnoid mater, and the pia mater.

The spinal cord is a column of nerve fibers that extends down through the vertebral canal. Together with the brain, it makes up the central nervous system.

Neurons within the spinal cord provide a two-way communication system between the brain and body parts outside the central nervous system. The cord also contains the processing centers for spinal reflexes.

PURPOSE OF THE EXERCISE

To review the characteristics of the meninges and the spinal cord and to observe the major features of these structures.

LEARNING OBJECTIVES

After completing this exercise, you should be able to

1. Name the layers of the meninges and describe the structure of each.
2. Identify the major features of the spinal cord.
3. Locate the ascending and descending tracts of the spinal cord.

PROCEDURE A—MENINGES

1. Review the section entitled "Meninges" in chapter 11 of the textbook.
2. Complete Part A of Laboratory Report 27.

DEMONSTRATION

Observe the preserved section of spinal cord. Note the heavy covering of dura mater, which is firmly attached to the cord on each side by a set of ligaments (denticulate ligaments) originating in the pia mater. The intermediate layer of meninges, the arachnoid mater, is devoid of blood vessels, but in a live human being, the space beneath this layer contains cerebrospinal fluid. The pia mater, which is closely attached to the surface of the spinal cord, contains many blood vessels. What are the functions of these layers?

_____

_____

_____

_____

PROCEDURE B—STRUCTURE OF THE SPINAL CORD

1. Review the section entitled "Spinal Cord" in chapter 11 of the textbook.
2. As a review activity, label figures 27.1, 27.2, and 27.3.
3. Complete Parts B and C of the laboratory report.
4. Obtain a prepared microscope slide of a spinal cord cross section. Use the low power of the microscope to locate the following features:

posterior median sulcus

anterior median fissure

central canal

gray matter

    gray commissure

    posterior horn

    lateral horn

    anterior horn

white matter

    posterior funiculus

    lateral funiculus

    anterior funiculus

Figure 27.1   Label the features of the spinal cord and the surrounding structures.

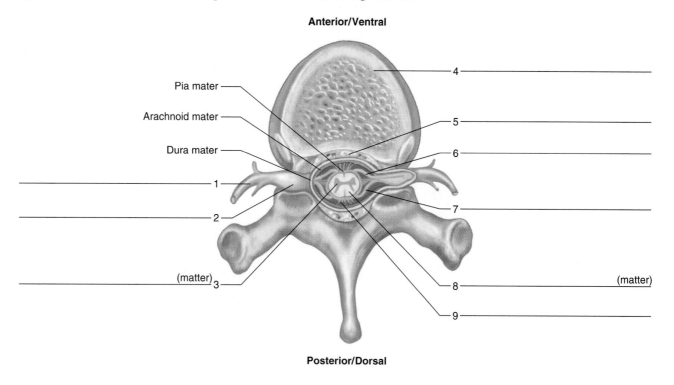

Anterior/Ventral

Pia mater

Arachnoid mater

Dura mater

1

2

(matter) 3

4

5

6

7

8

9

(matter)

Posterior/Dorsal

Figure 27.2   Label this cross section of the spinal cord, including the features of the white and gray matter.

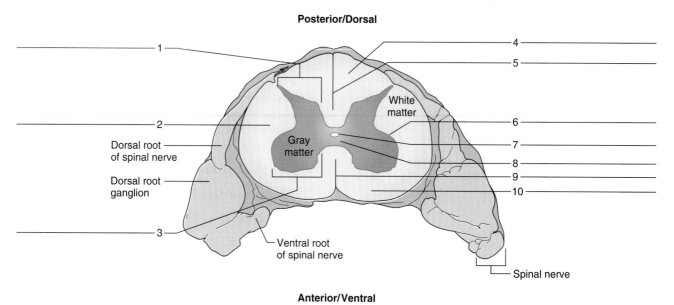

Posterior/Dorsal

1

2

Dorsal root
of spinal nerve

Dorsal root
ganglion

3

White
matter

Gray
matter

4

5

6

7

8

9

10

Ventral root
of spinal nerve

Spinal nerve

Anterior/Ventral

Figure 27.3 Label the ascending and descending tracts of the spinal cord by placing the correct numbers in the spaces provided. (*Note:* These tracts are not visible as individually stained structures on microscope slides.)

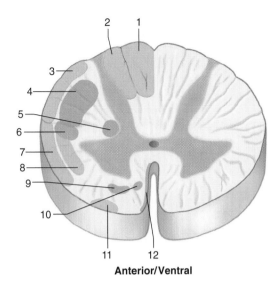

Posterior/Dorsal

Anterior/Ventral

_____ Anterior corticospinal tract

_____ Anterior reticulospinal tract

_____ Anterior spinocerebellar tract

_____ Anterior spinothalamic tract

_____ Fasciculus cuneatus

_____ Fasciculus gracilis

_____ Lateral corticospinal tract

_____ Lateral reticulospinal tract

_____ Lateral spinothalamic tract

_____ Medial reticulospinal tract

_____ Posterior spinocerebellar tract

_____ Rubrospinal tract

roots of spinal nerve

dorsal roots

dorsal root ganglia

ventral roots

5. Observe the model of the spinal cord, and locate the features listed in step 4.
6. Complete Part D of the laboratory report.

Web Quest

Describe the development of the nervous system, and review the brain, cranial nerves, spinal cord, and CSF. Search these at www.mhhe.com/shier10
Click on Student Edition. Click on Martin Lab Manual, Web Quest.

Name _____

Date _____

Section _____

THE MENINGES AND SPINAL CORD

PART A

Match the terms in column A with the descriptions in column B. Place the letter of your choice in the space provided.

Column A

a. Arachnoid mater
b. Denticulate ligament
c. Dural sinus
d. Dura mater
e. Epidural space
f. Pia mater
g. Subarachnoid space

Column B

_____ 1. Band of pia mater that attaches dura mater to cord

_____ 2. Channel through which venous blood flows

_____ 3. Outermost layer of meninges

_____ 4. Follows irregular contours of spinal cord surface

_____ 5. Contains cerebrospinal fluid

_____ 6. Thin, weblike middle membrane

_____ 7. Separates dura mater from bone of vertebra

PART B

Complete the following statements:

1. Each of the thirty-one segments of the spinal cord gives rise to a pair of _____ .

2. The bulge in the spinal cord that gives off nerves to the upper limbs is called the _____ .

3. The bulge in the spinal cord that gives off nerves to the lower limbs is called the _____ .

4. The _____ is a groove that extends the length of the spinal cord posteriorly.

5. In a spinal cord cross section, the posterior _____ of the gray matter appear as the upper wings of a butterfly.

6. The cell bodies of motor neurons are found in the _____ horns of the spinal cord.

7. The _____ connects the gray matter on the left and right sides of the spinal cord.

8. The _____ in the gray commissure of the spinal cord contains cerebrospinal fluid and is continuous with the ventricles of the brain.

9. The white matter of the spinal cord is divided into anterior, lateral, and posterior _____ .

10. The longitudinal bundles of nerve fibers within the spinal cord comprise major nerve pathways called _____ .

11. Collectively, the dura mater, arachnoid mater, and pia mater are called the _____ .

PART C

Match the nerve tracts in column A with the descriptions in column B. Place the letter of your choice in the space provided.

Column A

a. Corticospinal
b. Fasciculus gracilis
c. Lateral spinothalamic
d. Posterior spinocerebellar
e. Reticulospinal

Column B

_____ 1. Ascending tract to the brain to interpret touch, pressure, and body movements

_____ 2. Descending tract whose fibers conduct motor impulses to sweat glands and muscles to control tone

_____ 3. Descending tract whose fibers conduct motor impulses to skeletal muscles

_____ 4. Ascending tract to the cerebellum necessary for coordination of skeletal muscles

_____ 5. Ascending tract to the brain to give rise to sensations of temperature and pain

PART D

Identify the features indicated in the spinal cord cross section of figure 27.4.

1. _____
2. _____
3. _____
4. _____

5. _____
6. _____
7. _____
8. _____

Figure 27.4 Micrograph of a spinal cord cross section with spinal nerve roots (35×). Identify the features, using the terms provided.

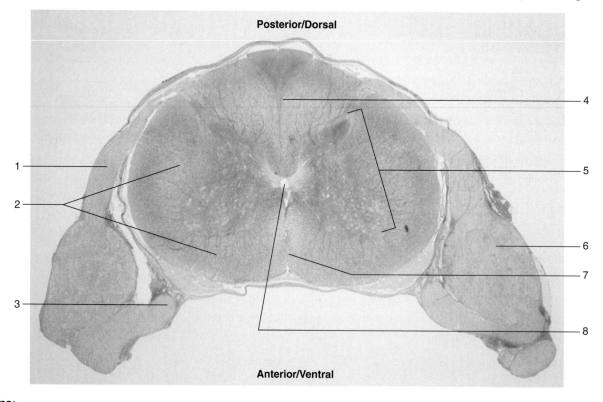

Terms:

Anterior median fissure	Dorsal root ganglion	Gray matter	Ventral root of spinal nerve
Central canal	Dorsal root of spinal nerve	Posterior median sulcus	White matter

228

THE REFLEX ARC AND REFLEXES

MATERIALS NEEDED

Textbook
Rubber percussion hammer

A reflex arc represents the simplest type of nerve pathway found in the nervous system. This pathway begins with a receptor at the dendrite end of a sensory (afferent) neuron. The sensory neuron leads into the central nervous system and may communicate with one or more interneurons. Some of these interneurons, in turn, communicate with motor (efferent) neurons, whose axons (nerve fibers) lead outward to effectors.

Thus, when a sensory receptor is stimulated by some kind of change occurring inside or outside the body, nerve impulses may pass through a reflex arc, and, as a result, effectors may respond. Such an automatic, subconscious response is called a reflex.

Most reflexes demonstrated in this lab are stretch reflexes. When a muscle is stretched by a tap over its tendon, stretch receptors called *muscle spindles* are stretched within the muscle, which initiates an impulse over a reflex arc. A sensory (afferent) neuron conducts an impulse from the muscle spindle into the gray matter of the spinal cord, where it synapses with a motor (efferent) neuron, which conducts the impulse to the effector muscle. The stretched muscle responds by contracting to resist or reverse further stretching. These stretch reflexes are important to maintain proper posture, balance, and movements. Observations of many of these reflexes in clinical tests on patients may indicate damage to a level of the spinal cord or peripheral nerves of the particular reflex arc.

PURPOSE OF THE EXERCISE

To review the characteristics of reflex arcs and reflex behavior and to demonstrate some of the reflexes that occur in the human body.

LEARNING OBJECTIVES

After completing this exercise, you should be able to

1. Describe a reflex arc.
2. Distinguish between a reflex arc and a reflex.
3. Identify and demonstrate several reflex actions that occur in humans.

PROCEDURE

1. Review the section entitled "Reflex Arcs" in chapter 11 of the textbook.
2. As a review activity, label figure 28.1.
3. Complete Part A of Laboratory Report 28.
4. Work with a laboratory partner to demonstrate each of the reflexes listed. (See fig. 28.2 also.) *It is important that muscles involved in the reflexes be totally relaxed in order to observe proper responses.* After each demonstration, record your observations in the table provided in Part B of the laboratory report.
   a. *Knee-jerk reflex (patellar reflex).* Have your laboratory partner sit on a table (or sturdy chair) with legs relaxed and hanging freely over the edge without touching the floor. Gently strike your partner's patellar ligament (just below the patella) with the blunt side of a rubber percussion hammer. The normal response is a moderate extension of the leg at the knee joint.
   b. *Ankle-jerk reflex.* Have your partner kneel on a chair with back toward you and with feet slightly dorsiflexed over the edge and relaxed. Gently strike the calcaneal tendon (just above its insertion on the calcaneus) with the blunt side of the rubber hammer. The normal response is plantar flexion of the foot.
   c. *Biceps-jerk reflex.* Have your partner place a bare arm bent about 90° at the elbow on the table. Press your thumb on the inside of the elbow over the tendon of the biceps brachii, and gently strike your finger with the rubber hammer. Watch the biceps brachii for a response. The response might be a slight twitch of the muscle or flexion of the forearm at the elbow joint.
   d. *Triceps-jerk reflex.* Have your partner lie supine with an upper limb bent about 90° across the abdomen. Gently strike the tendon of the triceps brachii near its insertion just proximal to the olecranon process at the tip of the elbow. Watch the triceps brachii for a response. The response might be a slight twitch of the muscle or extension of the forearm at the elbow joint.

Figure 28.1 Label this diagram of a withdrawal (polysynaptic) reflex arc by placing the correct numbers in the spaces provided. Most reflexes demonstrated in this lab are stretch (monosynaptic) reflex arcs and lack the interneuron.

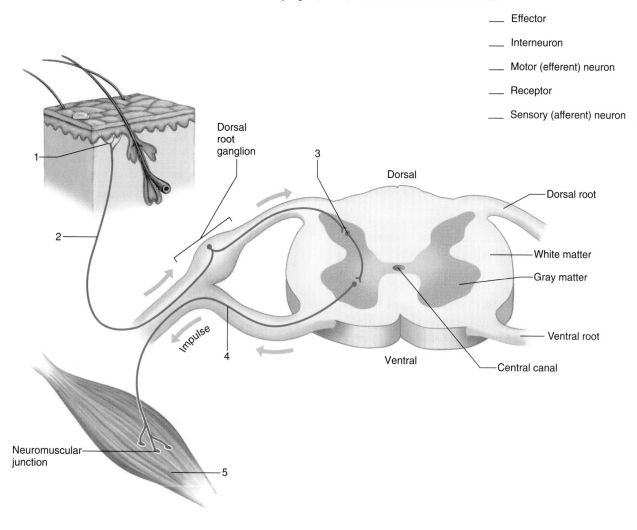

___ Effector

___ Interneuron

___ Motor (efferent) neuron

___ Receptor

___ Sensory (afferent) neuron

e. *Plantar reflex.* Have your partner remove a shoe and sock and lie supine with the lateral surface of the foot resting on the table. Draw the metal tip of the rubber hammer, applying firm pressure, over the sole from the heel to the base of the large toe. The normal response is flexion of the toes and plantar flexion of the foot. If the toes spread apart and dorsiflexion occurs, the reflex is the abnormal *Babinski reflex* response (normal in infants until the nerve fibers have complete myelinization).

5. Complete Part B of the laboratory report.

Figure 28.2 Demonstrate each of the following reflexes: (*a*) knee-jerk reflex; (*b*) ankle-jerk reflex; (*c*) biceps-jerk reflex; (*d*) triceps-jerk reflex; and (*e*) plantar reflex.

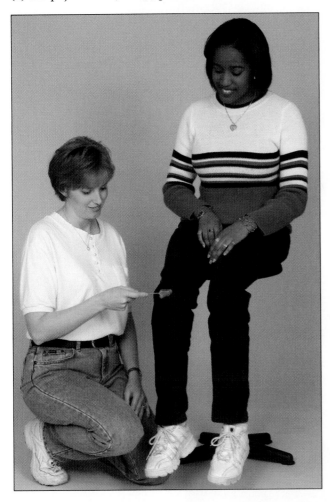

(a) Knee-jerk reflex

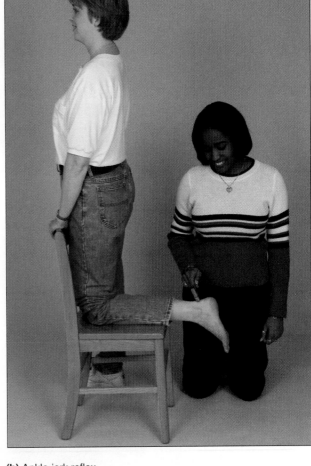

(b) Ankle-jerk reflex

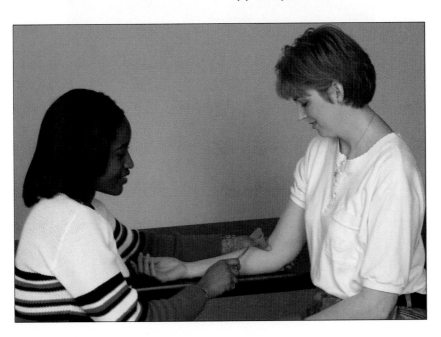

(c) Biceps-jerk reflex

Figure 28.2 *Continued.*

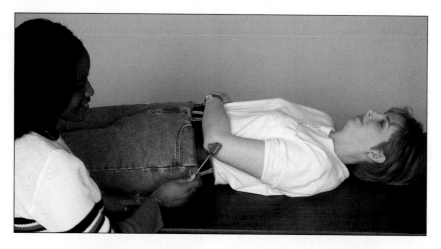

(d) Triceps-jerk reflex

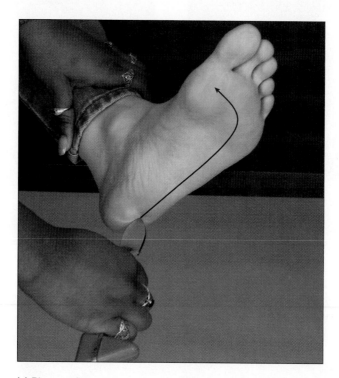

(e) Plantar reflex

Name _____

Date _____

Section _____

THE REFLEX ARC AND REFLEXES

PART A

Complete the following statements:

1. _____ are routes followed by nerve impulses as they pass through the nervous system.

2. Interneurons in a withdrawal reflex are located in the _____ .

3. _____ are automatic subconscious responses to stimuli within or outside the body.

4. Effectors of a reflex arc are glands and _____ .

5. A knee-jerk reflex employs only _____ and motor neurons.

6. The effector of the knee-jerk reflex is the _____ muscle.

7. The sensory stretch receptors of the knee-jerk reflex are located in the _____ .

8. The knee-jerk reflex helps the body to maintain _____ .

9. The sensory receptors of a withdrawal reflex are located in the _____ .

10. _____ muscles in the limbs are the effectors of a withdrawal reflex.

11. The normal plantar reflex results in _____ of toes.

12. Stroking the sole of the foot in infants results in dorsiflexion and toes that spread apart,

    called the _____ reflex.

PART B

1. Complete the following table:

Reflex Tested	Response Observed	Effector Involved
Knee-jerk		
Ankle-jerk		
Biceps-jerk		
Triceps-jerk		
Plantar		

2. List the major events that occur in the knee-jerk reflex, from the striking of the patellar ligament to the resulting response. _____
_____
_____
_____
_____

Critical Thinking Application

What characteristics do the reflexes you demonstrated have in common?

_____
_____
_____

THE BRAIN AND CRANIAL NERVES

MATERIALS NEEDED

Textbook
Dissectible model of the human brain
Preserved human brain
Anatomic charts of the human brain

The brain, the largest and most complex part of the nervous system, contains nerve centers associated with sensory functions and is responsible for sensations and perceptions. It issues motor commands to skeletal muscles and carries on higher mental activities. It also functions to coordinate muscular movements, and it contains centers and nerve pathways necessary for the regulation of internal organs.

Twelve pairs of cranial nerves arise from the ventral surface of the brain and are designated by number and name. Although most of these nerves conduct both sensory and motor impulses, some contain only sensory fibers associated with special sense organs. Others are primarily composed of motor fibers and are involved with the activities of muscles and glands.

PURPOSE OF THE EXERCISE

To review the structural and functional characteristics of the human brain and cranial nerves.

LEARNING OBJECTIVES

After completing this exercise, you should be able to

1. Identify the major structures in the human brain.
2. Locate the major functional regions of the brain.
3. Identify each of the cranial nerves.
4. List the functions of each cranial nerve.

PROCEDURE A—HUMAN BRAIN

1. Review the section entitled "Brain" in chapter 11 of the textbook.
2. As a review activity, label figures 29.1, 29.2, and 29.3.
3. Complete Part A of Laboratory Report 29.
4. Observe the dissectible model and the preserved specimen of the human brain. Locate each of the following features:

cerebrum

cerebral hemispheres

corpus callosum

convolutions (gyri)

sulci

   central sulcus

   lateral sulcus

fissures

   longitudinal fissure

   transverse fissure

lobes

   frontal lobe

   parietal lobe

   temporal lobe

   occipital lobe

   insula

cerebral cortex

basal ganglia

   caudate nucleus

   putamen

   globus pallidus

ventricles

   lateral ventricles

   third ventricle

   fourth ventricle

   choroid plexuses

   cerebral aqueduct

diencephalon

   thalamus

   hypothalamus

   optic chiasma

   mammillary bodies

   pineal gland

Figure 29.1 Label this diagram by placing the correct numbers in the spaces provided.

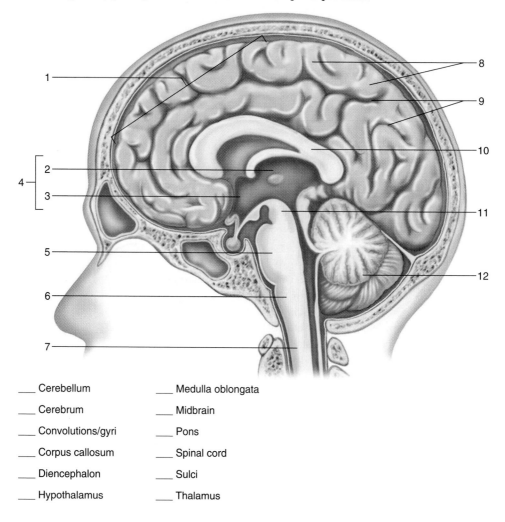

___ Cerebellum		___ Medulla oblongata	
___ Cerebrum		___ Midbrain	
___ Convolutions/gyri		___ Pons	
___ Corpus callosum		___ Spinal cord	
___ Diencephalon		___ Sulci	
___ Hypothalamus		___ Thalamus	

Figure 29.2 Label the lobes of the cerebrum, using the terms provided.

Terms:
Frontal lobe
Occipital lobe
Parietal lobe
Temporal lobe

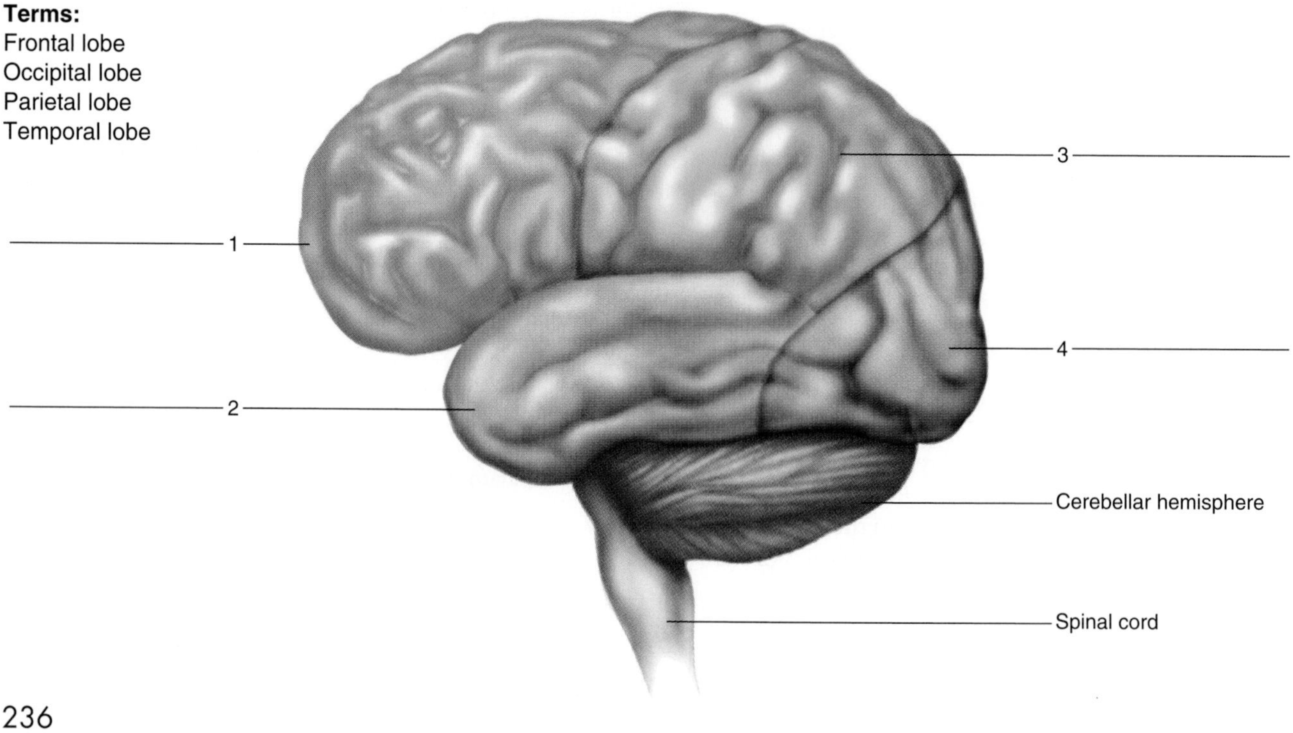

3

4

Cerebellar hemisphere

Spinal cord

1

2

Figure 29.3 Label the functional areas of the cerebrum, using the terms provided. (*Note:* These areas are not visible as distinct parts of the brain.)

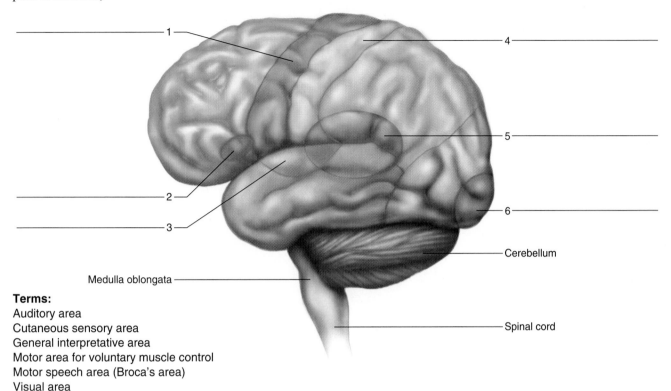

Terms:
Auditory area
Cutaneous sensory area
General interpretative area
Motor area for voluntary muscle control
Motor speech area (Broca's area)
Visual area

cerebellum

    lateral hemispheres

    vermis

    cerebellar cortex

    arbor vitae

    cerebellar peduncles

brainstem

    midbrain

        cerebral aqueduct

        cerebral peduncles

        corpora quadrigemina

    pons

    medulla oblongata

5. Locate the labeled areas in figure 29.3 that represent the following functional regions of the cerebrum:

    motor area for voluntary muscle control

    motor speech area (Broca's area)

    cutaneous sensory area

    auditory area

    visual area

    general interpretative area

6. Complete Parts B and C of the laboratory report.

PROCEDURE B—CRANIAL NERVES

1. Review the section entitled "Cranial Nerves" in chapter 11 of the textbook.
2. As a review activity, label figure 29.4.
3. Observe the model and preserved specimen of the human brain, and locate as many of the following cranial nerves as possible:

    olfactory nerves

    optic nerves

    oculomotor nerves

    trochlear nerves

    trigeminal nerves

    abducens nerves

    facial nerves

    vestibulocochlear nerves

    glossopharyngeal nerves

    vagus nerves

    accessory nerves

    hypoglossal nerves

Figure 29.4 Provide the names of the cranial nerves in this ventral view.

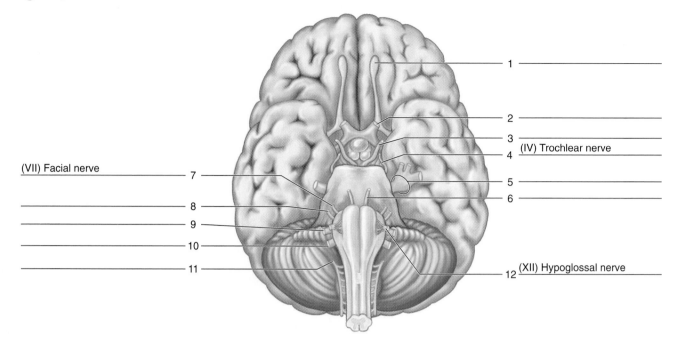

(VII) Facial nerve — 7

1

2

3

4 — (IV) Trochlear nerve

5

6

8

9

10

11

12 — (XII) Hypoglossal nerve

The following mnemonic device will help you learn the twelve pairs of cranial nerves in the proper order:

Old **Op**ie **oc**casionally **tr**ies **trig**onometry, **and** **f**eels **v**ery **glo**omy, **vagu**e, and **hypo**active.[1]

4. Complete Part D of the laboratory report.

Web Quest

Describe the development of the nervous system, and review the brain, cranial nerves, spinal cord, and CSF. Search these at www.mhhe.com/shier10
Click on Student Edition. Click on Martin Lab Manual, Web Quest.

1. From *HAPS-EDucator,* Winter 2002. An official publication of the Human Anatomy & Physiology Society (HAPS).

Name _____

Date _____

Section _____

THE BRAIN AND CRANIAL NERVES

PART A

Match the terms in column A with the descriptions in column B. Place the letter of your choice in the space provided.

Column A

a. Central sulcus
b. Cerebral cortex
c. Convolution (gyrus)
d. Corpus callosum
e. Falx cerebelli
f. Hypothalamus
g. Insula
h. Medulla oblongata
i. Midbrain
j. Optic chiasma
k. Pineal gland
l. Pons
m. Tentorium cerebelli

Column B

_____ 1. Structure formed by the crossing-over of the optic nerves

_____ 2. Part of diencephalon that forms lower walls and floor of third ventricle

_____ 3. Cone-shaped structure attached to upper posterior portion of diencephalon

_____ 4. Connects cerebral hemispheres

_____ 5. Ridge on surface of cerebrum

_____ 6. Separates frontal and parietal lobes

_____ 7. Part of brainstem between diencephalon and pons

_____ 8. Rounded bulge on underside of brainstem

_____ 9. Part of brainstem continuous with the spinal cord

_____ 10. A layer of dura mater that separates cerebellar hemispheres

_____ 11. A layer of dura mater that separates occipital lobe from cerebellum

_____ 12. Cerebral lobe located deep within lateral sulcus

_____ 13. Thin layer of gray matter on surface of cerebrum

PART B

Complete the following table:

Structure	Location	Major Functions
Broca's area		
Cardiac center		
Cerebellar peduncles		
Cerebral peduncles		
Corpora quadrigemina		
Frontal eye fields		
Hypothalamus		
Limbic system		
Respiratory center		
Reticular formation		
Thalamus		
Vasomotor center		

PART C

Identify the features indicated in the midsagittal section of the right half of the human brain in **figure 29.5.**

1. _____

2. _____

3. _____

4. _____

5. _____

6. _____

7. _____

8. _____

9. _____

10. _____

Figure 29.5   Identify the features on this midsagittal section of the right half of the human brain, using the terms provided.

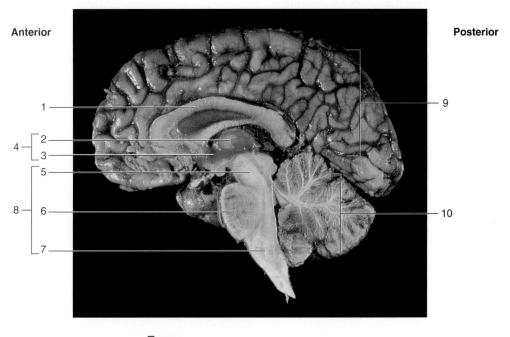

Terms:

Brainstem	Hypothalamus
Cerebellum	Medulla oblongata
Cerebrum	Midbrain
Corpus callosum	Pons
Diencephalon	Thalamus

PART D

Indicate which cranial nerve(s) is (are) most closely associated with each of the following functions:

1. Sense of hearing _____

2. Sense of taste _____

3. Sense of sight _____

4. Sense of smell _____

5. Sense of equilibrium _____

6. Conducting sensory impulses from upper teeth _____

7. Conducting sensory impulses from lower teeth _____

8. Raising eyelids _____

9. Focusing lenses of eyes _____

10. Adjusting amount of light entering eye _____

11. Moving eyes _____

12. Stimulating salivary secretions _____

13. Movement of trapezius and sternocleidomastoid muscles _____

14. Muscular movements associated with speech _____

15. Muscular movements associated with swallowing _____

DISSECTION OF THE SHEEP BRAIN

SAFETY

- Wear disposable gloves when handling the sheep brains.
- Save or dispose of the brains as instructed.
- Wash your hands before leaving the laboratory.

Mammalian brains have many features in common. Because human brains may not be available, sheep brains often are dissected as an aid to understanding mammalian brain structure. However, as in the cat, the adaptations of the sheep differ from the adaptations of the human, so comparisons of their structural features may not be precise.

PURPOSE OF THE EXERCISE

To observe the major features of the sheep brain and to compare these features with those of the human brain.

LEARNING OBJECTIVES

After completing this exercise, you should be able to

1. Identify the major structures of the sheep brain.
2. Locate the larger cranial nerves of the sheep brain.
3. List several differences and similarities between the sheep brain and the human brain.

PROCEDURE

1. Obtain a preserved sheep brain and rinse it thoroughly in water to remove as much of the preserving fluid as possible.

2. Examine the surface of the brain for the presence of meninges. (The outermost layers of these membranes may have been lost during removal of the brain from the cranial cavity.) If meninges are present, locate the following:

   dura mater—the thick, opaque outer layer

   arachnoid mater—the delicate, transparent middle layer that is attached to the undersurface of the dura mater

   pia mater—the thin, vascular layer that adheres to the surface of the brain (should be present)

3. Remove any remaining dura mater by pulling it gently from the surface of the brain.

4. Position the brain with its ventral surface down in the dissecting tray. Study figure 30.1, and locate the following structures on the specimen:

   cerebral hemispheres

   convolutions (gyri)

   sulci

   longitudinal fissure

   frontal lobe

   parietal lobe

   temporal lobe

   occipital lobe

   cerebellum

   medulla oblongata

5. Gently separate the cerebral hemispheres along the longitudinal fissure and expose the transverse band of white fibers within the fissure that connects the hemispheres. This band is the *corpus callosum.*

6. Bend the cerebellum and medulla oblongata slightly downward and away from the cerebrum (fig. 30.2). This will expose the *pineal gland* in the upper midline and the *corpora quadrigemina,* which consists of four rounded structures associated with the midbrain.

Figure 30.1   Dorsal surface of the sheep brain.

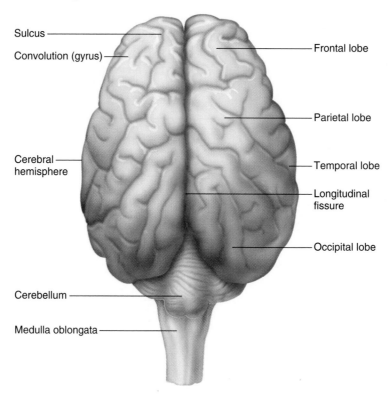

Sulcus

Convolution (gyrus)

Cerebral hemisphere

Cerebellum

Medulla oblongata

Frontal lobe

Parietal lobe

Temporal lobe

Longitudinal fissure

Occipital lobe

Figure 30.2   Gently bend the cerebellum and medulla oblongata away from the cerebrum to expose the pineal gland and the corpora quadrigemina.

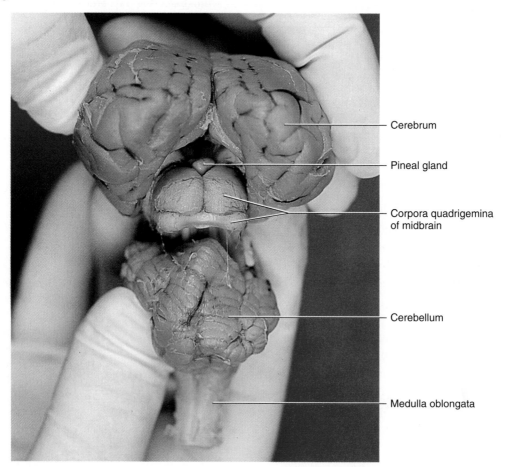

Cerebrum

Pineal gland

Corpora quadrigemina of midbrain

Cerebellum

Medulla oblongata

7. Position the brain with its ventral surface upward. Study figures 30.3 and 30.4, and locate the following structures on the specimen:

longitudinal fissure

olfactory bulbs

optic nerves

optic chiasma

optic tract

mammillary bodies

infundibulum (pituitary stalk)

midbrain

pons

8. Although some of the cranial nerves may be missing or are quite small and difficult to find, locate as many of the following as possible, using figures 30.3 and 30.4 as references:

oculomotor nerves

trochlear nerves

trigeminal nerves

abducens nerves

facial nerves

vestibulocochlear nerves

glossopharyngeal nerves

vagus nerves

accessory nerves

hypoglossal nerves

9. Using a long, sharp knife, cut the sheep brain along the midline to produce a midsagittal section. Study figures 30.5 and 30.6, and locate the following structures on the specimen:

cerebrum

    cerebral hemisphere

    cerebral cortex

    white matter

    gray matter

olfactory bulb

corpus callosum

cerebellum

    white matter

    gray matter

third ventricle

fourth ventricle

diencephalon

    optic chiasma

    infundibulum

    pituitary gland (this structure may be missing)

    mammillary bodies

    thalamus

    hypothalamus

    pineal gland

Figure 30.3 Lateral surface of the sheep brain.

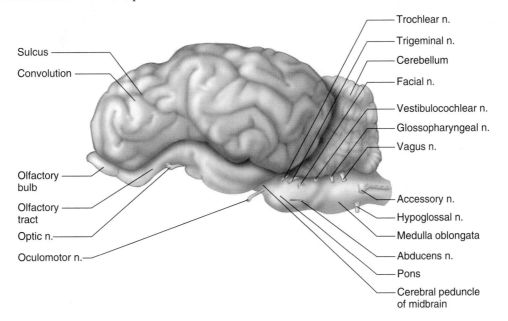

Figure 30.4   Ventral surface of the sheep brain.

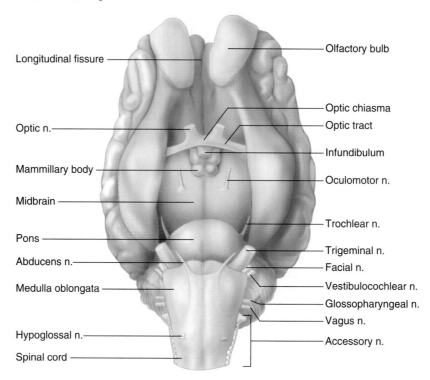

Longitudinal fissure

Optic n.

Mammillary body

Midbrain

Pons

Abducens n.

Medulla oblongata

Hypoglossal n.

Spinal cord

Olfactory bulb

Optic chiasma

Optic tract

Infundibulum

Oculomotor n.

Trochlear n.

Trigeminal n.

Facial n.

Vestibulocochlear n.

Glossopharyngeal n.

Vagus n.

Accessory n.

midbrain

   corpora quadrigemina

   cerebral peduncles

pons

medulla oblongata

DEMONSTRATION

Observe a sheep brain from a coronal section. Note the longitudinal fissure, gray matter, white matter, corpus callosum, lateral ventricles, third ventricle, and thalamus.

10. Dispose of the sheep brain as directed by the laboratory instructor.
11. Complete Parts A and B of Laboratory Report 30.

Figure 30.5   Midsagittal section of the right half of the sheep brain dissection.

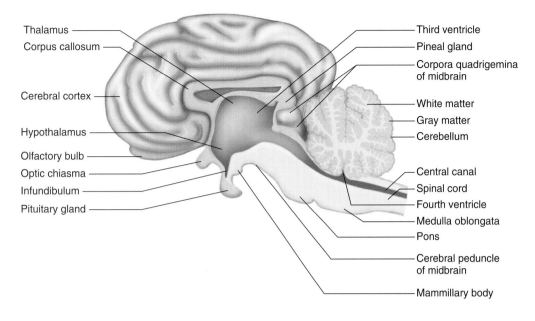

Thalamus
Corpus callosum
Cerebral cortex
Hypothalamus
Olfactory bulb
Optic chiasma
Infundibulum
Pituitary gland

Third ventricle
Pineal gland
Corpora quadrigemina of midbrain
White matter
Gray matter
Cerebellum
Central canal
Spinal cord
Fourth ventricle
Medulla oblongata
Pons
Cerebral peduncle of midbrain
Mammillary body

Figure 30.6   Midsagittal section of the right half of the sheep brain dissection.

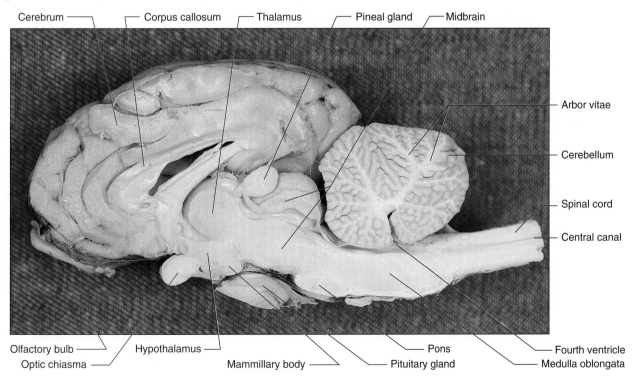

Cerebrum   Corpus callosum   Thalamus   Pineal gland   Midbrain

Arbor vitae
Cerebellum
Spinal cord
Central canal

Olfactory bulb   Hypothalamus   Pons   Fourth ventricle
Optic chiasma   Mammillary body   Pituitary gland   Medulla oblongata

DISSECTION OF THE SHEEP BRAIN

PART A

Answer the following questions:

1. How do the relative sizes of the sheep and human cerebral hemispheres differ? _____

_____

_____

2. How do the convolutions and sulci of the sheep cerebrum compare with the human cerebrum in numbers? _____

_____

_____

3. What is the significance of the differences you noted in your answers for questions 1 and 2? _____

_____

_____

4. What difference did you note in the structures of the sheep cerebellum and the human cerebellum? _____

_____

_____

5. How do the sizes of the olfactory bulbs of the sheep brain compare with those of the human brain? _____

_____

_____

6. Based on their relative sizes, which of the cranial nerves seems to be most highly developed in the sheep brain?

_____

_____

7. What is the significance of the observations you noted in your answers for questions 5 and 6? _____

_____

_____

PART B

Critical Thinking Application

Prepare a list of at least five features to illustrate ways in which the brains of sheep and humans are similar.

1. _____

2. _____

3. _____

4. _____

5. _____

RECEPTORS AND SOMATIC SENSES

MATERIALS NEEDED

Textbook
Marking pen (washable)
Millimeter ruler
Bristle or sharp pencil
Forceps (fine points) or measuring calipers
Blunt metal probes
Three beakers (250 mL)
Hot tap water or 45°C (113°F) water bath
Cold water (ice water)
Thermometer

For Demonstration:

Prepared microscope slides of Meissner's and Pacinian
    corpuscles
Compound light microscope

Sensory receptors are sensitive to changes that occur within the body and its surroundings. When they are stimulated, they initiate nerve impulses that travel into the central nervous system. As a result of the brain interpreting such sensory impulses, the person may experience particular sensations.

The sensory receptors found in skin, muscles, joints, and visceral organs are associated with somatic senses. These senses include touch, pressure, temperature, pain, and the senses of muscle movement and body position.

PURPOSE OF THE EXERCISE

To review the characteristics of sensory receptors and somatic senses and to investigate some of the somatic senses associated with the skin.

LEARNING OBJECTIVES

After completing this exercise, you should be able to

1. Name five general types of receptors.
2. Explain how a sensation results.
3. List the somatic senses.
4. Determine the distribution of touch, heat, and cold receptors in a region of skin.
5. Determine the two-point threshold of a region of skin.

PROCEDURE A—SOMATIC RECEPTORS

1. Review the sections entitled "Receptors and Sensations" and "Somatic Senses" in chapter 12 of the textbook.
2. Complete Part A of Laboratory Report 31.

DEMONSTRATION

Observe the Meissner's corpuscle with the microscope set up by the laboratory instructor. This type of receptor is abundant in the superficial dermis in outer regions of the body, such as in the fingertips, soles, lips, and external genital organs. It is responsible for the sensation of light touch. (See fig. 31.1 and chapter 6 of the textbook.)

Observe the Pacinian corpuscle in the second demonstration microscope. This corpuscle is composed of many layers of connective tissue cells and has a nerve fiber in its central core. Pacinian corpuscles are numerous in the hands, feet, joints, and external genital organs. They are responsible for the sense of deep pressure (fig. 31.2). How are Meissner's and Pacinian corpuscles similar?

_____

_____

How are they different?

_____

_____

PROCEDURE B—SENSE OF TOUCH

1. Investigate the distribution of touch receptors in your laboratory partner's skin. To do this, follow these steps:
    a. Use a marking pen and a millimeter ruler to prepare a square with 2.5 cm on each side on the skin of your partner's inner wrist, near the palm.
    b. Divide the square into smaller squares with 0.5 cm on a side, producing a small grid.
    c. Ask your partner to rest the marked wrist on the tabletop and to keep his or her eyes closed throughout the remainder of the experiment.
    d. Press the end of a bristle on the skin in some part of the grid, using just enough pressure to cause the bristle to bend. A sharp pencil could be used as an alternate device.
    e. Ask your partner to report whenever the touch of the bristle is felt. Record the results in Part B of the laboratory report.

251

Figure 31.1 Meissner's corpuscles, such as this one, are responsible for the sensation of light touch (250×).

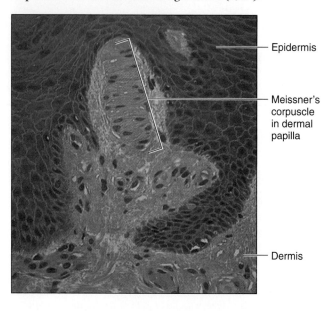

- Epidermis
- Meissner's corpuscle in dermal papilla
- Dermis

Figure 31.2 Pacinian corpuscles, such as this one, are responsible for the sensation of deep pressure (25× micrograph enlarged to 100×).

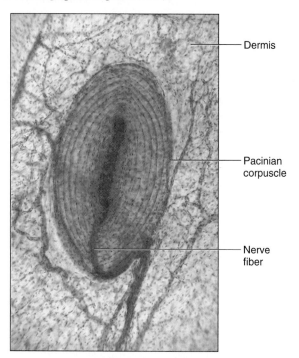

- Dermis
- Pacinian corpuscle
- Nerve fiber

    f. Continue this procedure until you have tested twenty-five different locations on the grid. Move randomly through the grid to help prevent anticipation of the next stimulation site.

2. Test two other areas of exposed skin in the same manner, and record the results in Part B of the laboratory report.

3. Answer the questions in Part B of the laboratory report.

PROCEDURE C—TWO-POINT THRESHOLD

1. Test your partner's ability to recognize the difference between one or two points of skin being stimulated simultaneously. To do this, follow these steps:
   a. Have your partner place a hand with the palm up on the table and close his or her eyes.
   b. Hold the tips of a forceps tightly together and gently touch the skin of your partner's index finger.
   c. Ask your partner to report if it feels like one or two points are touching the finger.
   d. Allow the tips of the forceps or the points of the measuring calipers to spread so they are 1 mm apart, press both points against the skin simultaneously, and ask your partner to report as before.
   e. Repeat this procedure, allowing the tips of the forceps to spread more each time until your partner can feel both tips being pressed against the skin. The minimum distance between the tips of the forceps when both can be felt is called the *two-point threshold.* As soon as you are able to distinguish two points, two separate receptors are being stimulated instead of only one receptor.
   f. Record the two-point threshold for the skin of the index finger in Part C of the laboratory report.
2. Repeat this procedure to determine the two-point threshold of the palm, the back of the hand, the back of the neck, the leg, and the sole. Record the results in Part C of the laboratory report.
3. Answer the questions in Part C of the laboratory report.

PROCEDURE D—SENSE OF TEMPERATURE

1. Investigate the distribution of heat receptors in your partner's skin. To do this, follow these steps:
   a. Mark a square with 2.5-cm sides on your partner's palm.
   b. Prepare a grid by dividing the square into smaller squares, 0.5 cm on a side.
   c. Have your partner rest the marked palm on the table and close his or her eyes.
   d. Heat a blunt metal probe by placing it in a beaker of hot water (about 40–45°C/104–113°F) for a minute or so. *(Be sure the probe does not get so hot that it burns the skin.)* Use a thermometer to monitor the appropriate warm water from the tap or the water bath.
   e. Wipe the probe dry and touch it to the skin on some part of the grid.
   f. Ask your partner to report if the probe feels hot. Then record the results in Part D of the laboratory report.
   g. Keep the probe hot, and repeat the procedure until you have randomly tested twenty-five different locations on the grid.

2. Investigate the distribution of cold receptors by repeating the procedure. Use a blunt metal probe that has been cooled by placing it in ice water for a minute or so. Record the results in Part D of the laboratory report.

3. Answer the questions in Part D of the laboratory report.

OPTIONAL ACTIVITY

Prepare three beakers of water of different temperatures. One beaker should contain warm water (about 40°C/104°F), one should be room temperature (about 22°C/72°F), and one should contain cold water (about 10°C/50°F). Place the index finger of one hand in the warm water and, at the same time, place the index finger of the other hand in the cold water for about 2 minutes. Then, simultaneously move both index fingers into the water at room temperature. What temperature do you sense with each finger? How do you explain the resulting sensations?

_____
_____
_____
_____

RECEPTORS AND SOMATIC SENSES

PART A—SOMATIC RECEPTORS

Complete the following statements:

1. _____ are receptors that are sensitive to changes in the concentrations of chemicals.

2. Whenever tissues are damaged, _____ receptors are likely to be stimulated.

3. Receptors that are sensitive to temperature changes are called _____ .

4. _____ are sensitive to changes in pressure or to movement of fluid.

5. _____ are sensitive to changes in the intensity of light energy.

6. A feeling that occurs when sensory impulses are interpreted by the brain is called a _____ .

7. The process by which the brain can cause a sensation to seem to come from the receptors being stimulated is called _____ .

8. A sensation may seem to fade away when receptors are continuously stimulated as a result of _____ .

9. Meissner's corpuscles are responsible for the sense of light _____ .

10. Pacinian corpuscles are responsible for the sense of deep _____ .

11. Heat receptors are most sensitive to temperatures between _____ .

12. Cold receptors are most sensitive to temperatures between _____ .

13. Pain receptors are widely distributed but are lacking in the _____ .

14. _____ are the only receptors in visceral organs that produce sensations.

15. When visceral pain is felt in some region other than the part being stimulated, the phenomenon is called

_____ .

PART B—SENSE OF TOUCH

1. Record a + to indicate where the bristle was felt and a *0* to indicate where it was not felt.

Skin of wrist

2. Show distribution of touch receptors in two other regions of skin.

Region tested _____   Region tested _____

3. Answer the following questions:

   a. How do you describe the pattern of distribution for touch receptors in the regions of the skin you tested?

     _____

     _____

   b. How does the concentration of touch receptors seem to vary from region to region? _____

     _____

PART C—TWO-POINT THRESHOLD

1. Record the two-point threshold in millimeters for skin in each of the following regions:

   Index finger _____

   Palm _____

   Back of hand _____

   Back of neck _____

   Leg _____

   Sole _____

2. Answer the following questions:

   a. What region of the skin tested has the greatest ability to discriminate two points? _____

   b. What region of the skin has the least sensitivity to this test? _____

   c. What is the significance of these observations in questions *a* and *b*? _____

     _____

     _____

PART D—SENSE OF TEMPERATURE

1. Record a + to indicate where heat was felt and a *0* to indicate where it was not felt.

Skin of palm

2. Record a + to indicate where cold was felt and a *0* to indicate where it was not felt.

Skin of palm

3. Answer the following questions:

   a. How do temperature receptors appear to be distributed in the skin of the palm? _____

     _____

   b. Compare the distribution and concentration of heat and cold receptors in the skin of the palm. _____

     _____

     _____

LABORATORY EXERCISE 32

SENSES OF SMELL AND TASTE

SAFETY

- Prepare fresh solutions for use in Procedure B.
- Wash your hands before starting the taste experiment.
- Use a clean cotton swab for each test. Do not dip a used swab into a test solution.
- Dispose of used cotton swabs and paper towels as directed.
- Wash your hands before leaving the laboratory.

The senses of smell (olfaction) and taste (gustation) are dependent upon chemoreceptors that are stimulated by various chemicals dissolved in liquids. The receptors of smell are found in the olfactory organs, which are located in the upper parts of the nasal cavity and in a portion of the nasal septum. The receptors of taste occur in the taste buds, which are sensory organs primarily found on the surface of the tongue. Chemicals are considered odorless and tasteless if receptor sites for them are absent.

These senses function closely together, because substances that are tasted often are smelled at the same moment, and they play important roles in the selection of foods.

PURPOSE OF THE EXERCISE

To review the structures of the organs of smell and taste and to investigate the abilities of smell and taste receptors to discriminate various chemical substances.

LEARNING OBJECTIVES

After completing this exercise, you should be able to

1. Describe the general characteristics of the smell (olfactory) receptors.
2. Describe the general characteristics of the taste (gustatory) receptors.
3. Explain how the senses of smell and taste function together.
4. Determine the time needed for olfactory sensory adaptation to occur.
5. Determine the distribution of taste receptors on the surface of the tongue and mouth cavity.

PROCEDURE A—SENSE OF SMELL (OLFACTION)

1. Review the section entitled "Sense of Smell" in chapter 12 of the textbook.
2. As a review activity, label figure 32.1.
3. Complete Part A of Laboratory Report 32.

DEMONSTRATION

Observe the olfactory epithelium in the microscope set up by the laboratory instructor. The olfactory receptor cells are spindle-shaped, bipolar neurons with spherical nuclei. They also have six to eight cilia at their distal ends. The supporting cells are pseudostratified columnar epithelial cells. However, in this region the tissue lacks goblet cells. (See fig. 32.2 and chapter 5 of the textbook.)

Figure 32.1 Label this diagram of the olfactory organ by placing the correct numbers in the spaces provided.

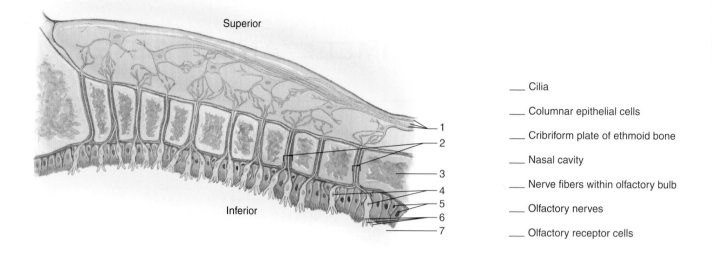

Superior

Inferior

____ Cilia

____ Columnar epithelial cells

____ Cribriform plate of ethmoid bone

____ Nasal cavity

____ Nerve fibers within olfactory bulb

____ Olfactory nerves

____ Olfactory receptor cells

Figure 32.2 Olfactory receptors have cilia at their distal ends (250×).

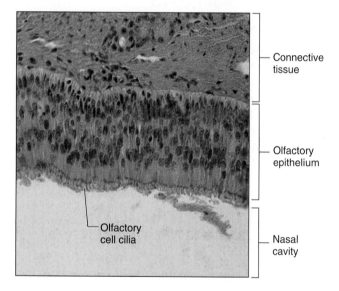

Connective tissue

Olfactory epithelium

Olfactory cell cilia

Nasal cavity

4. Test your laboratory partner's ability to recognize the odors of the bottled substances available in the laboratory. To do this, follow these steps:
   a. Have your partner keep his or her eyes closed.
   b. Remove the stopper from one of the bottles, and hold it about 4 cm under your partner's nostrils for about 2 seconds.
   c. Ask your partner to identify the odor, and then replace the stopper.
   d. Record your partner's response in Part B of the laboratory report.
   e. Repeat steps *b–d* for each of the bottled substances.

5. Repeat the preceding procedure, using the same set of bottled substances, but present them to your partner in a different sequence. Record the results in Part B of the laboratory report.
6. Wait 10 minutes and then determine the time it takes for your partner to experience olfactory sensory adaptation. To do this, follow these steps:
   a. Ask your partner to breathe in through the nostrils and exhale through the mouth.
   b. Remove the stopper from one of the bottles, and hold it about 4 cm under your partner's nostrils.
   c. Keep track of the time that passes until your partner is no longer able to detect the odor of the substance.
   d. Record the result in Part B of the laboratory report.
   e. Wait 5 minutes and repeat this procedure, using a different bottled substance.
   f. Test a third substance in the same manner.
   g. Record the results as before.
7. Complete Part B of the laboratory report.

PROCEDURE B—SENSE OF TASTE (GUSTATION)

1. Review the section entitled "Sense of Taste" in chapter 12 of the textbook.
2. As a review activity, label figure 32.3.
3. Complete Part C of the laboratory report.

DEMONSTRATION

Observe the oval-shaped taste bud in the microscope set up by the laboratory instructor. Note the surrounding epithelial cells. The taste pore, an opening into the taste bud, may be filled with taste hairs (microvilli). Within the taste bud there are supporting cells and thinner taste-receptor cells, which often have lightly stained nuclei (fig. 32.4).

Figure 32.3 Label this diagram by placing the correct numbers in the spaces provided.

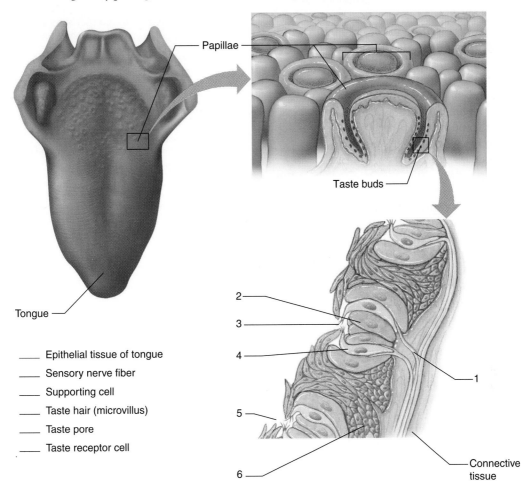

Papillae

Taste buds

Tongue

_____ Epithelial tissue of tongue

_____ Sensory nerve fiber

_____ Supporting cell

_____ Taste hair (microvillus)

_____ Taste pore

_____ Taste receptor cell

2

3

4

5

6

1

Connective tissue

Figure 32.4 Taste receptors are found in taste buds such as this one (250× micrograph enlarged to 1,500×).

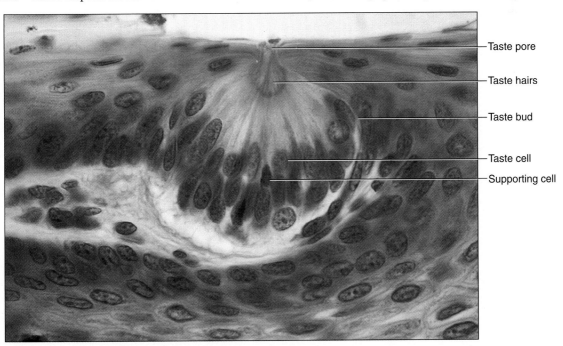

Taste pore

Taste hairs

Taste bud

Taste cell

Supporting cell

4. Map the distribution of the receptors for the primary taste sensations on your partner's tongue. To do this, follow these steps:

   a. Ask your partner to rinse his or her mouth with water and then partially dry the surface of the tongue with a paper towel.

   b. Moisten a clean cotton swab with 5% sucrose solution, and touch several regions of your partner's tongue with the swab.

   c. Each time you touch the tongue, ask your partner to report if a sweet sensation is experienced.

   d. Test the tip, sides, and back of the tongue in this manner.

   e. Test some other representative areas inside the mouth cavity such as the cheek, gums, and roof of the mouth (hard and soft palate) for a sweet sensation.

   f. Record your partner's responses in Part D of the laboratory report.

   g. Have your partner rinse his or her mouth and dry the tongue again, and repeat the preceding procedure, using each of the other three test solutions—NaCl, acetic acid, and quinine or Epsom salt solution. Be sure to use a fresh swab for each test substance and dispose of used swabs and paper towels as directed.

5. Complete Part D of the laboratory report.

OPTIONAL ACTIVITY

Test your laboratory partner's ability to recognize the tastes of apple, potato, carrot, and onion. To do this, follow these steps:

1. Have your partner close his or her eyes and hold the nostrils shut.
2. Place a small piece of one of the test substances on your partner's tongue.
3. Ask your partner to identify the substance without chewing or swallowing it.
4. Repeat the procedure for each of the other substances.

How do you explain the results of this experiment?

_____

_____

_____

SENSES OF SMELL AND TASTE

PART A

Complete the following statements:

1. Olfactory, or smell, receptors are _____ neurons surrounded by columnar epithelial cells.

2. The distal ends of the olfactory neurons are covered with hairlike _____ .

3. Before gaseous substances can stimulate the olfactory receptors, they must be dissolved in _____ that surrounds the cilia.

4. The axons of olfactory receptors pass through small openings in the _____ of the ethmoid bone.

5. Olfactory bulbs lie on either side of the _____ of the ethmoid bone.

6. The sensory impulses pass from the olfactory bulbs through the _____ to the interpreting centers of the brain.

7. The olfactory interpreting centers are located deep within the temporal lobes and at the base of the _____ lobes of the cerebrum.

8. Olfactory sensations usually fade rapidly as a result of _____ .

9. Olfactory receptor neurons are the only parts of the nervous system that come in contact with the _____ directly.

10. A chemical would be considered _____ if a person lacks a particular receptor site on the cilia of the olfactory neurons.

PART B—SENSE OF SMELL

1. Record the results (as +, if recognized; as *0*, if unrecognized) from the tests of odor recognition in the following table:

Substance Tested	Odor Reported	
	First Trial	**Second Trial**

2. Record the results of the olfactory sensory adaptation time in the following table:

Substance Tested	Adaptation Time in Seconds

3. Complete the following:

   a. How do you describe your partner's ability to recognize the odors of the substances you tested? _____

   _____

   _____

   b. Compare your experimental results with others in the class. Did you find any evidence to indicate that

   individuals may vary in their abilities to recognize odors? Explain your answer. _____

   _____

   _____

   _____

Critical Thinking Application

Does the time it takes for sensory adaptation to occur seem to vary with the substances tested? Explain your answer.

_____

_____

_____

PART C

Complete the following statements:

1. Taste receptor cells are modified _____ cells.

2. The opening to a taste bud is called a _____ .

3. The _____ of a taste cell are its sensitive part.

4. Before the taste of a substance can be detected, the substance must be dissolved in _____ .

5. Substances that stimulate taste cells seem to bind with _____ sites on the surfaces

   of taste hairs.

6. Sour receptors are mainly stimulated by _____ .

7. Salt receptors are mainly stimulated by ionized inorganic _____ .

8. Alkaloids usually have a _____ taste.

PART D—SENSE OF TASTE

1. *Taste receptor distribution.* Record a + to indicate where a taste sensation seemed to originate and a *0* if no

   sensation occurred when the spot was stimulated.

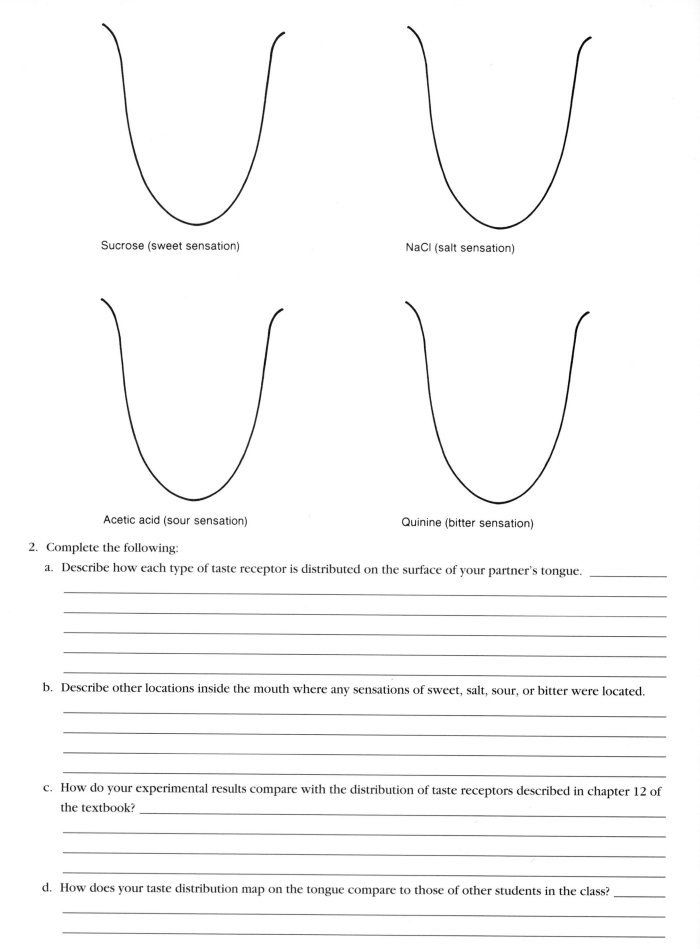

Sucrose (sweet sensation)

NaCl (salt sensation)

Acetic acid (sour sensation)

Quinine (bitter sensation)

2. Complete the following:

   a. Describe how each type of taste receptor is distributed on the surface of your partner's tongue. _____

   _____
   _____
   _____
   _____
   _____

   b. Describe other locations inside the mouth where any sensations of sweet, salt, sour, or bitter were located.

   _____
   _____
   _____
   _____

   c. How do your experimental results compare with the distribution of taste receptors described in chapter 12 of the textbook? _____

   _____
   _____
   _____

   d. How does your taste distribution map on the tongue compare to those of other students in the class? _____

   _____
   _____

THE EAR AND HEARING

The ear is composed of external, middle, and inner parts. The external structures gather sound waves and direct them inward to the tympanic membrane. The parts of the middle ear, in turn, transmit vibrations from the tympanic membrane (eardrum) to the inner ear, where the hearing receptors are located. As they are stimulated, these receptors initiate nerve impulses to pass over the vestibulocochlear nerve into the audi-tory cortex of the brain, where the impulses are interpreted and the sensations of hearing are created.

PURPOSE OF THE EXERCISE

To review the structural and functional characteristics of the ear and to conduct some simple hearing tests.

LEARNING OBJECTIVES

After completing this exercise, you should be able to

1. Identify the major structures of the ear.
2. Describe the functions of the structures of the ear.
3. Trace the pathway of sound vibrations from the tympanic membrane to the hearing receptors.
4. Conduct several simple hearing tests.

PROCEDURE A—STRUCTURE AND FUNCTION OF THE EAR

1. Review the section entitled "Sense of Hearing" in chapter 12 of the textbook.
2. As a review activity, label figures 33.1, 33.2, and 33.3.

Figure 33.1 Label the major structures of the ear.

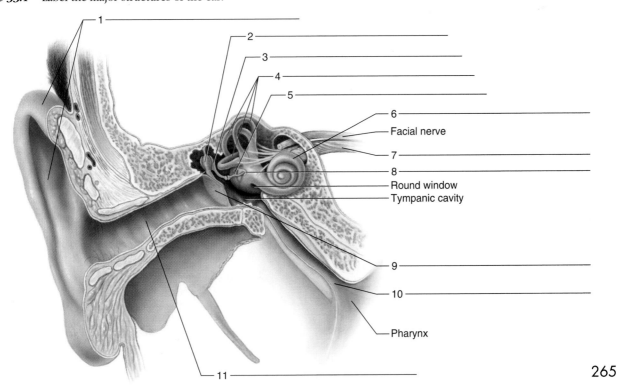

Facial nerve

Round window

Tympanic cavity

Pharynx

Figure 33.2 Label the structures of the inner ear by placing the correct numbers in the spaces provided.

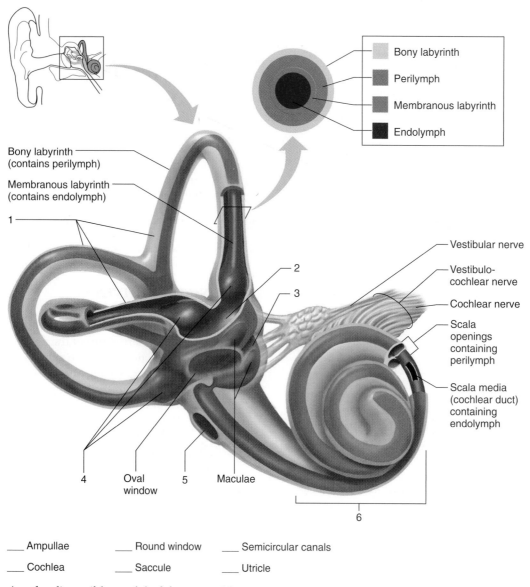

Bony labyrinth
Perilymph
Membranous labyrinth
Endolymph

Bony labyrinth
(contains perilymph)

Membranous labyrinth
(contains endolymph)

1

2

3

Vestibular nerve

Vestibulo-
cochlear nerve

Cochlear nerve

Scala
openings
containing
perilymph

Scala media
(cochlear duct)
containing
endolymph

4   Oval   5   Maculae
         window

6

___ Ampullae   ___ Round window   ___ Semicircular canals

___ Cochlea   ___ Saccule   ___ Utricle

3. Examine the dissectible model of the ear and locate the following features:

external ear
    auricle
    external auditory meatus
    tympanic membrane (eardrum)

middle ear
    tympanic cavity
    auditory ossicles
      malleus
      incus
      stapes
    oval window
    tensor tympani
    stapedius

auditory tube (eustachian tube)

inner ear
    osseous labyrinth
    membranous labyrinth
    cochlea
    round window
    semicircular canals
    vestibule

vestibulocochlear nerve
    vestibular nerve (balance branch)
    cochlear nerve (hearing branch)

DEMONSTRATION

Observe the section of the cochlea in the microscope set up by the laboratory instructor. Locate one of the turns of the cochlea, and using figures 33.3 and 33.4 as a guide, identify the *scala vestibuli, scala media, scala tympani, vestibular membrane, basilar membrane,* and the *organ of Corti.*

Figure 33.3 Label the organ of Corti structures of a cochlea.

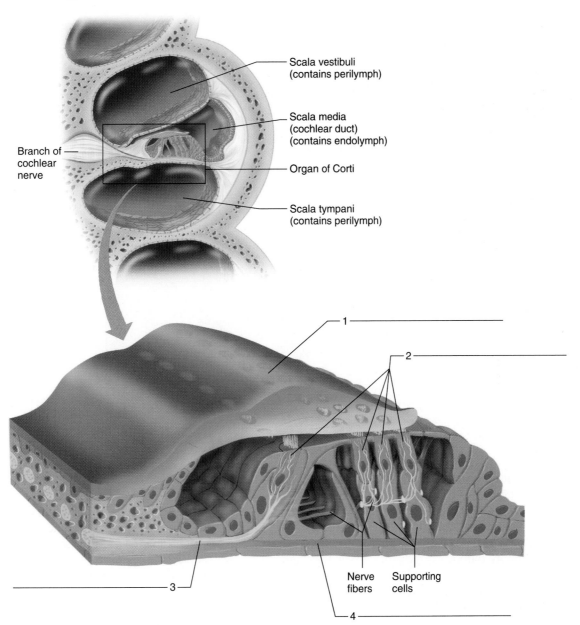

Scala vestibuli
(contains perilymph)

Scala media
(cochlear duct)
(contains endolymph)

Organ of Corti

Scala tympani
(contains perilymph)

Branch of
cochlear
nerve

1

2

3

Nerve
fibers

Supporting
cells

4

4. Complete Parts A and B of Laboratory Report 33.

PROCEDURE B—HEARING TESTS

Perform the following tests in a quiet room, using your laboratory partner as the test subject.

1. *Auditory acuity test.* To conduct this test, follow these steps:
   a. Have the test subject sit with eyes closed.
   b. Pack one of the subject's ears with cotton.
   c. Hold a ticking watch close to the open ear, and slowly move it straight out and away from the ear.
   d. Have the subject indicate when the sound of the ticking can no longer be heard.

   e. Use a meterstick to measure the distance in centimeters from the ear to the position of the watch.
   f. Repeat this procedure to test the acuity of the other ear.
   g. Record the test results in Part C of the laboratory report.
2. *Sound localization test.* To conduct this test, follow these steps:
   a. Have the subject sit with eyes closed.
   b. Hold the ticking watch somewhere within the audible range of the subject's ears, and ask the subject to point to the watch.
   c. Move the watch to another position and repeat the request. In this manner, determine how accurately the subject can locate the watch

Figure 33.4 A section through the cochlea (22×).

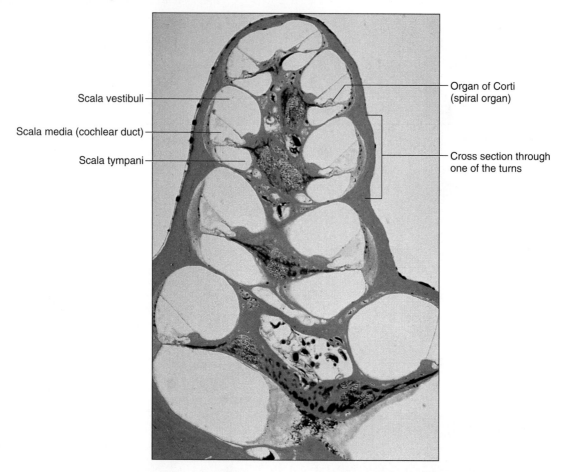

Scala vestibuli

Scala media (cochlear duct)

Scala tympani

Organ of Corti
(spiral organ)

Cross section through
one of the turns

when it is in each of the following positions: in front of the head, behind the head, above the head, on the right side of the head, and on the left side of the head.

d. Record the test results in Part C of the laboratory report.

3. *Rinne test.* This test is done to assess possible conduction deafness by comparing bone and air conduction. To conduct this test, follow these steps:

a. Obtain a tuning fork and strike it with a rubber hammer, or on the heel of your hand, causing it to vibrate.

b. Place the end of the fork's handle against the subject's mastoid process behind one ear. Have the prongs of the fork pointed downward and away from the ear, and be sure nothing is touching them (fig. 33.5*a*). The sound sensation is that of bone conduction. If no sound is experienced, nerve deafness exists.

c. Ask the subject to indicate when the sound is no longer heard.

d. Then quickly remove the fork from the mastoid process and position it in the air close to the opening of the nearby external auditory meatus (fig. 33.5*b*).

If hearing is normal, the sound (from air conduction) will be heard again; if there is

conductive impairment, the sound will not be heard. Conductive impairment involves outer or middle ear defects. Hearing aids can improve hearing for conductive deafness because bone conduction transmits the sound into the inner ear. Surgery could possibly correct this type of defect.

e. Record the test results in Part C of the laboratory report.

4. *Weber test.* This test is used to distinguish possible conduction or sensory deafness. To conduct this test, follow these steps:

a. Strike the tuning fork with the rubber hammer.

b. Place the handle of the fork against the subject's forehead in the midline (fig. 33.6).

c. Ask the subject to indicate if the sound is louder in one ear than in the other or if it is equally loud in both ears.

If hearing is normal, the sound will be equally loud in both ears. If there is conductive impairment, the sound will appear louder in the affected ear. If some degree of sensory (nerve) deafness exists, the sound will be louder in the normal ear. The impairment involves the organ of Corti or the cochlear nerve. Hearing aids will not improve sensory deafness.

d. Have the subject experience the effects of conductive impairment by packing one ear with

Figure 33.5 Rinne test: (*a*) first placement of vibrating tuning fork until sound is no longer heard; (*b*) second placement of tuning fork to assess air conduction.

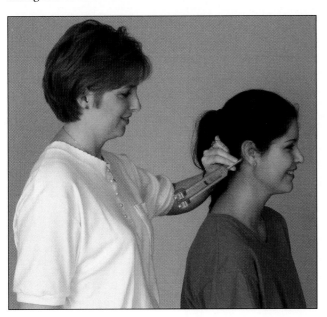

(a)

(b)

Figure 33.6 Weber test.

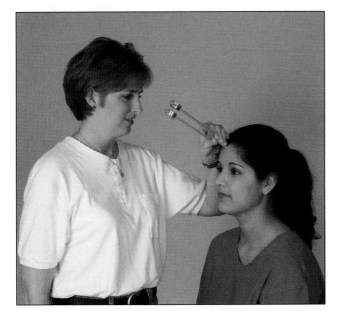

cotton and repeating the Weber test. Usually the sound appears louder in the plugged (or impaired) ear because extraneous sounds from the room are blocked out.

    e. Record the test results in Part C of the laboratory report.

5. Complete Part C of the laboratory report.

 ## Critical Thinking Application

Ear structures from the outer ear into the inner ear are progressively smaller. Using results obtained from the hearing tests, explain this advantage.

_____

_____

_____

_____

DEMONSTRATION

Ask the laboratory instructor to demonstrate the use of the audiometer. This instrument produces sound vibrations of known frequencies that are transmitted to one or both ears of a test subject through earphones. The audiometer can be used to determine the threshold of hearing for different sound frequencies, and, in the case of hearing impairment, it can be used to determine the percentage of hearing loss for each frequency.

Web Quest

What are the causes of hearing impairment? Search this site and review the anatomy and physiology of the ear at
www.mhhe.com/shier10
Click on Student Edition. Click on Martin Lab Manual, Web Quest.

THE EAR AND HEARING

PART A

Match the terms in column A with the descriptions in column B. Place the letter of your choice in the space provided.

Column A		Column B
a.	Auditory tube	_____ 1. Muscle attached to stapes
b.	Ceruminous gland	_____ 2. Muscle attached to malleus
c.	External auditory meatus	_____ 3. Auditory ossicle attached to tympanic membrane
d.	Malleus	_____ 4. Air-filled space containing auditory ossicles
e.	Membranous labyrinth	_____ 5. Contacts hairs of hearing receptors
f.	Osseous labyrinth	_____ 6. Leads from oval window to apex of cochlea
g.	Scala tympani	_____ 7. S-shaped tube leading to tympanic membrane
h.	Scala vestibuli	_____ 8. Wax-secreting structure
i.	Stapedius	_____ 9. Cone-shaped, semitransparent membrane attached to malleus
j.	Stapes	_____ 10. Auditory ossicle attached to oval window
k.	Tectorial membrane	_____ 11. Bony chamber between the cochlea and semicircular canals
l.	Tensor tympani	_____ 12. Contains endolymph
m.	Tympanic cavity	_____ 13. Bony canal of inner ear in temporal bone
n.	Tympanic membrane	_____ 14. Connects middle ear and pharynx
o.	Vestibule	_____ 15. Extends from apex of cochlea to round window

PART B

Label the structures indicated in the micrograph of the organ of Corti (spiral organ) in figure 33.7.

Figure 33.7 Label the structures of this organ of Corti (spiral organ) region of a cochlea, using the terms provided (75× micrograph enlarged to 300×).

Terms:
Basilar membrane
Hair cells
Scala media (cochlear duct)
Scala tympani
Tectorial membrane

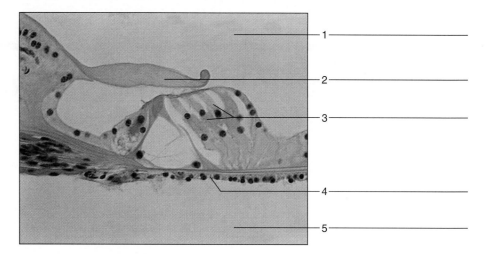

1 _____
2 _____
3 _____
4 _____
5 _____

PART C

1. Results of auditory acuity test:

Ear Tested	Audible Distance (cm)
Right	
Left	

2. Results of sound localization test:

Actual Location	Reported Location
Front of the head	
Behind the head	
Above the head	
Right side of the head	
Left side of the head	

3. Results of experiments using tuning forks:

Test	Left Ear (normal or impaired)	Right Ear (normal or impaired)
Rinne		
Weber		

4. Summarize the results of the hearing tests you conducted on your laboratory partner.

_____
_____
_____
_____
_____
_____
_____
_____

SENSE OF EQUILIBRIUM

SAFETY

- Do not pick subjects that have frequent motion sickness.
- Have four people surround the subject in the swivel chair in case the person falls from vertigo or loss of balance.
- Stop your experiment if the subject becomes nauseated.

The sense of equilibrium involves two sets of sensory organs. One set helps to maintain the stability of the head and body when they are motionless and produces a sense of static equilibrium. The other set is concerned with balancing the head and body when they are moved suddenly and produces a sense of dynamic equilibrium.

The organs associated with the sense of static equilibrium are located within the vestibules of the inner ears, whereas those associated with the sense of dynamic equilibrium are found within the ampullae of the semicircular canals of the inner ear.

PURPOSE OF THE EXERCISE

To review the structure and function of the organs of equilibrium and to conduct some simple tests of equilibrium.

LEARNING OBJECTIVES

After completing this exercise, you should be able to

1. Distinguish between static and dynamic equilibrium.

2. Identify the organs of equilibrium and describe their functions.
3. Explain the role of vision in the maintenance of equilibrium.
4. Conduct the Romberg and Bárány tests of equilibrium.

PROCEDURE A—STRUCTURE AND FUNCTION OF ORGANS OF EQUILIBRIUM

1. Review the section entitled "Sense of Equilibrium" in chapter 12 of the textbook.
2. Complete Part A of Laboratory Report 34.

DEMONSTRATION

Observe the cross section of the semicircular canal through the ampulla in the microscope set up by the laboratory instructor. Note the crista projecting into the lumen of the membranous labyrinth, which in a living person is filled with endolymph (fig. 34.1). The space between the membranous and osseous labyrinths is normally filled with perilymph.

PROCEDURE B—TESTS OF EQUILIBRIUM

Perform the following tests, using a person as a test subject who is not easily disturbed by dizziness or rotational movement. Also have some other students standing close by to help prevent the test subject from falling during the tests. *The tests should be stopped immediately if the test subject begins to feel uncomfortable or nauseated.*

1. *Vision and equilibrium.* To demonstrate the importance of vision in the maintenance of equilibrium, follow these steps:
   a. Have the test subject stand erect on one foot for 1 minute with his or her eyes open.
   b. Observe the subject's degree of unsteadiness.
   c. Repeat the procedure with the subject's eyes closed. *Be prepared to prevent the subject from falling.*

Figure 34.1 A micrograph of a crista ampullaris (1,400×).

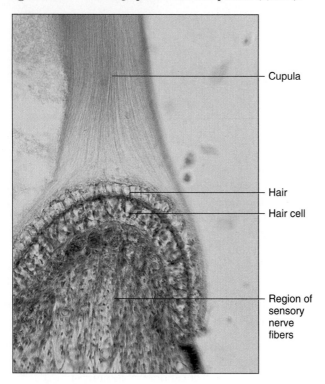

Cupula

Hair

Hair cell

Region of sensory nerve fibers

2. *Romberg test.* To conduct this test, follow these steps:

   a. Position the test subject close to a chalkboard with the back toward the board.

   b. Place a bright light in front of the subject so that a shadow of the body is cast on the board.

   c. Have the subject stand erect with feet close together and eyes staring straight ahead for a period of 3 minutes.

   d. During the test, make marks on the chalkboard along the edge of the shadow of the subject's shoulders to indicate the range of side-to-side swaying.

   e. Measure the maximum sway in centimeters and record the results in Part B of the laboratory report.

   f. Repeat the procedure with the subject's eyes closed.

   g. Position the subject so one side is toward the chalkboard.

   h. Repeat the procedure with the eyes open.

   i. Repeat the procedure with the eyes closed.

The Romberg test is used to evaluate a person's ability to integrate sensory information from proprioceptors and receptors within the organs of equilibrium and to relay appropriate motor impulses to postural muscles. A person who shows little unsteadiness when standing with feet together and eyes open, but who becomes unsteady when the eyes are closed, has a positive Romberg test.

3. *Bárány test.* To conduct this test, follow these steps:

   a. Have the test subject sit on a swivel chair with his or her eyes closed, the head tilted forward about 30°, and the hands gripped firmly to the seat. Position four people around the chair for safety. *Be prepared to prevent the subject and the chair from tipping over.*

   b. Rotate the chair ten rotations within 20 seconds.

   c. Abruptly stop the movement of the chair. The subject will still have the sensation of continuous movement and might experience some dizziness (vertigo).

   d. Have the subject open the eyes, and note the nature of the eye movements and their direction. (Such reflex eye movements are called *nystagmus*.) Also note the time it takes for the nystagmus to cease. Nystagmus will continue until the cupula is returned to an original position.

   e. Record your observations in Part B of the laboratory report.

   f. Allow the subject several minutes of rest, then repeat the procedure with the subject's head tilted nearly 90° onto one shoulder.

   g. After another rest period, repeat the procedure with the subject's head bent forward so that the chin is resting on the chest.

   In this test, when the head is tilted about 30°, the lateral semicircular canals receive maximal stimulation, and the nystagmus is normally from side to side. When the head is tilted at 90°, the superior canals are stimulated, and the nystagmus is up and down. When the head is bent forward with the chin on the chest, the posterior canals are stimulated, and the nystagmus is rotary.

4. Complete Part B of the laboratory report.

THE EYE

MATERIALS NEEDED

Textbook
Dissectible eye model
Compound light microscope
Prepared microscope slide of a mammalian eye (sagittal section)
Sheep or beef eye (fresh or preserved)
Dissecting tray
Dissecting instruments—forceps, sharp scissors, and dissecting needle

For Optional Activity:

Ophthalmoscope

SAFETY

- Wear disposable gloves when working on the eye dissection.
- Dispose of tissue remnants and gloves as instructed.
- Wash the dissecting tray and instruments as instructed.
- Wash your laboratory table.
- Wash your hands before leaving the laboratory.

The eye contains photoreceptors, which are modified neurons located on its inner wall. Other parts of the eye provide protective functions or make it possible to move the eyeball. Still other structures serve to focus light entering the eye so that a sharp image is projected onto the receptor cells. Nerve impulses generated when the receptors are stimulated travel along the optic nerves to the brain, which interprets the impulses and creates the sensation of sight.

PURPOSE OF THE EXERCISE

To review the structure and function of the eye and to dissect a mammalian eye.

LEARNING OBJECTIVES

After completing this exercise, you should be able to

1. Identify the major structures of an eye.
2. Describe the functions of the structures of an eye.

3. List the structures through which light passes as it travels from the cornea to the retina.
4. Dissect a mammalian eye and identify its major features.

PROCEDURE A—STRUCTURE AND FUNCTION OF THE EYE

1. Review the sections entitled "Visual Accessory Organs," "Structure of the Eye," and "Visual Receptors" in chapter 12 of the textbook.
2. As a review activity, label figures 35.1, 35.2, and 35.3.
3. Complete Part A of Laboratory Report 35.
4. Examine the dissectible model of the eye and locate the following features:

eyelid

conjunctiva

orbicularis oculi

levator palpebrae superioris

lacrimal apparatus

    lacrimal gland

    canaliculi

    lacrimal sac

    nasolacrimal duct

extrinsic muscles

    superior rectus

    inferior rectus

    medial rectus

    lateral rectus

    superior oblique

    inferior oblique

trochlea (pulley)

cornea

sclera

optic nerve

Figure 35.1 Label the structures of the lacrimal apparatus.

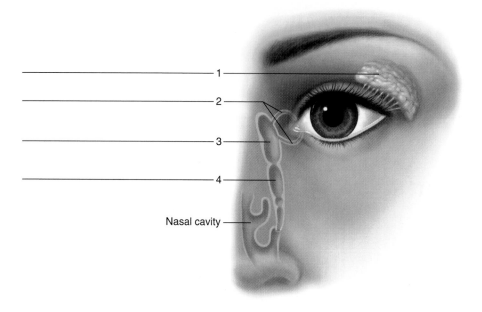

1

2

3

4

Nasal cavity

Figure 35.2 Label the extrinsic muscles of the right eye (lateral view).

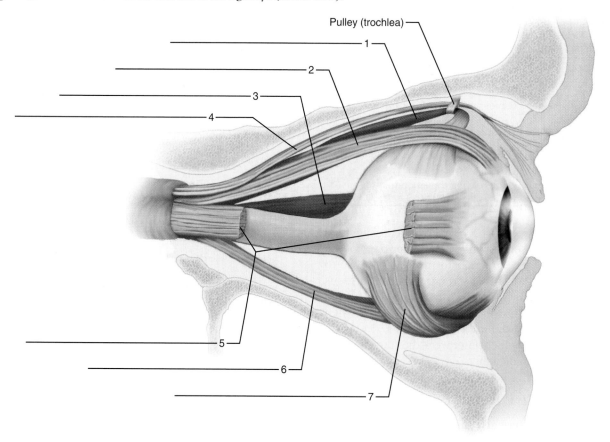

Pulley (trochlea)

1

2

3

4

5

6

7

choroid coat

ciliary body

    ciliary processes

    ciliary muscles

lens

suspensory ligaments

iris

Figure 35.3 Label the structures indicated in this transverse section of the right eye (superior view).

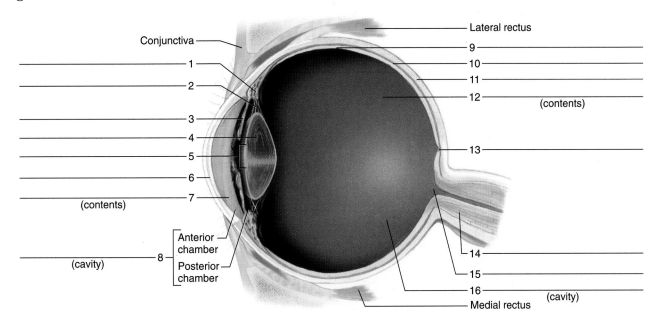

Conjunctiva

Lateral rectus
9
10
11
12
(contents)

1
2
3
4
5
6
7
(contents)

13

(cavity)
8

Anterior chamber
Posterior chamber

14
15
16
(cavity)

Medial rectus

anterior cavity

> anterior chamber
>
> posterior chamber
>
> aqueous humor

pupil

retina

> macula lutea
>
> fovea centralis
>
> optic disc

posterior cavity

> vitreous humor

5. Obtain a microscope slide of a mammalian eye section, and locate as many of the features in step 4 as possible.
6. Using high-power magnification, observe the posterior portion of the eye wall, and locate the sclera, choroid coat, and retina.
7. Using high-power magnification, examine the retina, and note its layered structure (fig. 35.4). Locate the following:

> nerve fibers leading to the optic nerve (innermost layer of the retina)
>
> layer of ganglion cells
>
> layer of bipolar neurons
>
> nuclei of rods and cones
>
> receptor ends of rods and cones
>
> pigmented epithelium (outermost layer of the retina)

OPTIONAL ACTIVITY

Use an ophthalmoscope to examine the interior of your laboratory partner's eye. This instrument consists of a set of lenses held in a rotating disc, a light source, and some mirrors that reflect the light into the test subject's eye.

The examination should be conducted in a dimly lighted room. Have your partner seated and staring straight ahead at eye level. Move the rotating disc of the ophthalmoscope so that the *O* appears in the lens selection window. Hold the instrument in your right hand with the end of your index finger on the rotating disc (fig. 35.5). Direct the light at a slight angle from a distance of about 15 cm into the subject's right eye. The light beam should pass along the inner edge of the pupil. Look through the instrument, and you should see a reddish, circular area—the interior of the eye. Rotate the disc of lenses to higher values until sharp focus is achieved.

Move the ophthalmoscope to within about 5 cm of the eye being examined *being very careful that the instrument does not touch the eye,* and again rotate the lenses to sharpen the focus (fig. 35.6). Locate the optic disc and the blood vessels that pass through it. Also locate the yellowish macula lutea by having your partner stare directly into the light of the instrument (fig. 35.7).

Examine the subject's iris by viewing it from the side and by using a lens with a +15 or +20 value.

PROCEDURE B—EYE DISSECTION

1. Obtain a mammalian eye, place it in a dissecting tray, and dissect it as follows:
   a. Trim away the fat and other connective tissues but leave the stubs of the *extrinsic muscles* and of the *optic nerve.* This nerve projects outward from the posterior region of the eyeball.

Figure 35.4 The cells of the retina are arranged in distinct layers (75×).

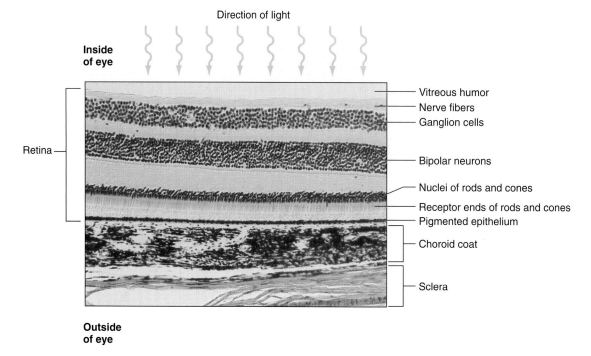

Direction of light

Inside of eye

Retina

Vitreous humor
Nerve fibers
Ganglion cells

Bipolar neurons

Nuclei of rods and cones
Receptor ends of rods and cones
Pigmented epithelium

Choroid coat

Sclera

Outside of eye

b. Note the *conjunctiva,* which lines the eyelid and is reflected over the anterior surface (except cornea) of the eye. Lift some of this thin membrane away from the eye with forceps and examine it.

c. Locate and observe the *cornea, sclera,* and *iris.* Also note the *pupil* and its shape. The cornea from a fresh eye will be transparent; when preserved, it becomes opaque.

d. Use sharp scissors to make a coronal section of the eye. To do this, cut through the wall about 1 cm from the margin of the cornea and continue all the way around the eyeball. Try not to damage the internal structures of the eye (fig. 35.8).

e. Gently separate the eyeball into anterior and posterior portions. Usually the jellylike vitreous humor will remain in the posterior portion, and the lens may adhere to it. Place the parts in the dissecting tray with their contents facing upward.

f. Examine the anterior portion of the eye, and locate the *ciliary body,* which appears as a dark, circular structure. Also note the *iris* and the *lens* if it remained in the anterior portion. The lens is normally attached to the ciliary body by many *suspensory ligaments,* which appear as delicate, transparent threads.

g. Use a dissecting needle to gently remove the lens, and examine it. If the lens is still transparent, hold it up and look through it at something in the distance and note that the lens inverts the image. The lens of a preserved eye is usually too opaque for this experience.

Figure 35.5 The ophthalmoscope is used to examine the interior of the eye.

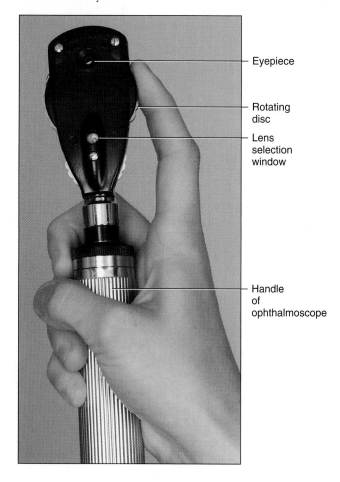

Eyepiece

Rotating disc

Lens selection window

Handle of ophthalmoscope

Figure 35.6 (*a*) Rotate the disc of lenses until sharp focus is achieved. (*b*) Move the ophthalmoscope to within 5 cm of the eye to examine the optic disc.

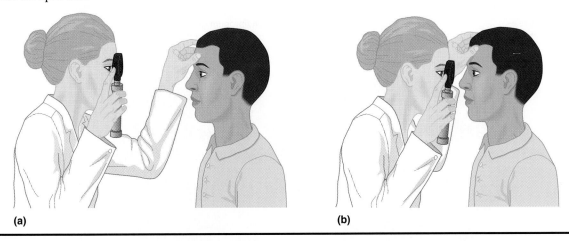

(a)   (b)

Figure 35.7 The interior of the eye as seen using an ophthalmoscope: (*a*) photograph with an arrow indicating the optic disc; (*b*) diagram.

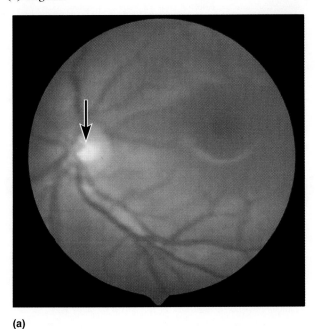

(a)

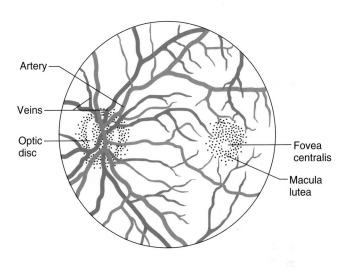

Artery

Veins

Optic disc

Fovea centralis

Macula lutea

(b)

h. Examine the posterior portion of the eye. Note the *vitreous humor*. This jellylike mass helps to hold the lens in place anteriorly and helps to hold the *retina* against the choroid coat.

i. Carefully remove the vitreous humor and examine the retina. This layer will appear as a thin, nearly colorless to cream-colored membrane that detaches easily from the choroid coat. Compare the structures identified to figure 35.9.

j. Locate the *optic disc*—the point where the retina is attached to the posterior wall of the eyeball and where the optic nerve originates. Because there are no receptor cells in the optic disc, this region is also called the "blind spot."

k. Note the iridescent area of the choroid coat beneath the retina. This colored surface in ungulates (mammals having hoofs) is called the *tapetum fibrosum*. It serves to reflect light back

Figure 35.8 Prepare a coronal section of the eye.

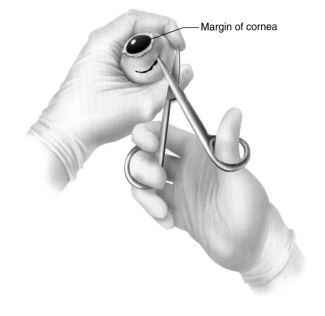

Margin of cornea

Figure 35.9 Internal structures of the beef eye dissection.

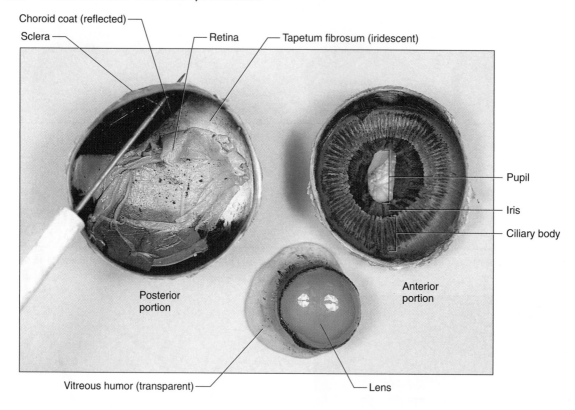

Choroid coat (reflected)
Sclera
Retina
Tapetum fibrosum (iridescent)
Pupil
Iris
Ciliary body
Posterior portion
Anterior portion
Vitreous humor (transparent)
Lens

through the retina, an action that is thought to aid the night vision of some animals. The tapetum fibrosum is lacking in the human eye.

1. Discard the tissues of the eye as directed by the laboratory instructor.
2. Complete Parts B and C of the laboratory report.

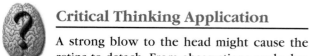

Critical Thinking Application

A strong blow to the head might cause the retina to detach. From observations made during the eye dissection, explain why this could happen.

_____

_____

_____

_____

Web Quest

How do we see? Search this and review the anatomy and physiology of the eye at www.mhhe.com/shier10

Click on Student Edition. Click on Martin Lab Manual, Web Quest.

Name _____

Date _____

Section _____

THE EYE

PART A

Match the terms in column A with the descriptions in column B. Place the letter of your choice in the space provided.

Column A	**Column B**
a. Aqueous humor	_____ 1. Posterior five-sixths of middle or vascular tunic
b. Choroid coat	_____ 2. White part of outer tunic
c. Ciliary muscle	_____ 3. Transparent anterior portion of outer tunic
d. Conjunctiva	_____ 4. Inner lining of eyelid
e. Cornea	_____ 5. Secretes tears
f. Iris	_____ 6. Empties into nasal cavity
g. Lacrimal gland	_____ 7. Fills posterior cavity of eye
h. Lysozyme	_____ 8. Area where optic nerve originates
i. Nasolacrimal duct	_____ 9. Smooth muscle that controls light intensity
j. Optic disc	_____ 10. Fills anterior and posterior chambers of the anterior cavity of the eye
k. Retina	_____ 11. Contains visual receptors called rods and cones
l. Sclera	_____ 12. Connects lens to ciliary body
m. Suspensory ligament	_____ 13. Causes lens to change shape
n. Vitreous humor	_____ 14. Antibacterial agent in tears

Complete the following:

15. List the structures and fluids through which light passes as it travels from the cornea to the retina. _____

_____

_____

_____

16. List three ways in which rods and cones differ in structure or function. _____

_____

_____

_____

_____

PART B

Complete the following:

1. Which tunic/layer of the eye was the most difficult to cut? _____

2. What kind of tissue do you think is responsible for this quality of toughness? _____

3. How do you compare the shape of the pupil in the dissected eye with your own pupil? _____

   _____

   _____

4. Where do you find aqueous humor in the dissected eye? _____

   _____

   _____

5. What is the function of the dark pigment in the choroid coat? _____

   _____

   _____

6. Describe the lens of the dissected eye. _____

   _____

   _____

   _____

7. Describe the vitreous humor of the dissected eye. _____

   _____

   _____

   _____

PART C

Identify the features of the eye indicated in figure 35.10.

a. Anterior portion of eye:

   1. _____

   2. _____

   3. _____

   4. _____

   5. _____

b. Posterior portion of eye:

   6. _____

   7. _____

   8. _____

   9. _____

   10. _____

   11. _____

Figure 35.10 Sections of the eye: (*a*) anterior portion (10×); (*b*) posterior portion (53×). Make your selection of the numbered features, using the terms provided.

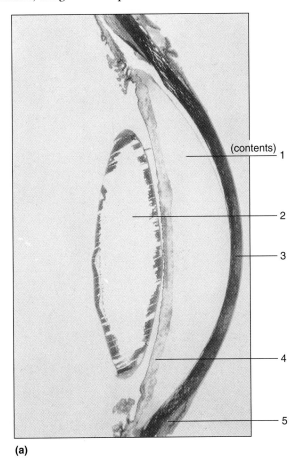

(contents) 1

2

3

4

5

(a)

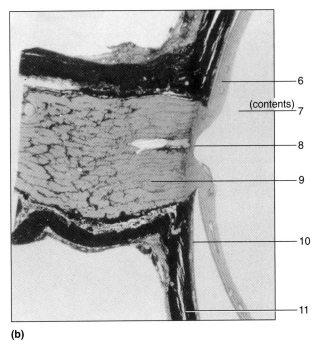

6

(contents) 7

8

9

10

11

(b)

Terms:
Aqueous humor
Choroid coat/middle tunic
Conjunctiva
Cornea
Iris
Lens
Optic disc
Optic nerve
Retina/inner tunic
Sclera/outer tunic
Vitreous humor

VISUAL TESTS AND DEMONSTRATIONS

Normal vision (emmetropia) results when light rays from objects in the external environment are refracted by the cornea and lens of the eye and focused onto the photoreceptors of the retina. Irregular curvatures in the surface of the cornea or lens, inability to change the shape of the lens, or defects in the shape of the eyeball can result in defective vision.

PURPOSE OF THE EXERCISE

To conduct tests for visual acuity, astigmatism, accommodation, color vision, the blind spot, and certain reflexes of the eye.

LEARNING OBJECTIVES

After completing this exercise, you should be able to

1. Describe four conditions that can lead to defective vision.
2. Conduct the tests used to evaluate visual acuity, astigmatism, the ability to accommodate for close vision, and color vision.
3. Demonstrate the blind spot, photopupillary reflex, accommodation pupillary reflex, and convergence reflex.

PROCEDURE A—VISUAL TESTS

Perform the following visual tests, using your laboratory partner as a test subject. If your partner usually wears glasses, test each eye with and without the glasses.

1. *Visual acuity test.* Visual acuity (sharpness of vision) can be measured by using a Snellen eye chart (fig. 36.1). This chart consists of several sets of letters in different sizes printed on a white card.

The letters near the top of the chart are relatively large, and those in each lower set become smaller. At one end of each set of letters is an acuity value in the form of a fraction. One of the sets near the bottom of the chart, for example, is marked 20/20. The normal eye can clearly see these letters from the standard distance of 20 feet and thus is said to have 20/20 vision. The letter at the top of the chart is marked 20/200. The normal eye can read letters of this size from a distance of 200 feet. Thus, an eye that is able to read only the top letter of the chart from a distance of 20 feet is said to have 20/200 vision. This person has less than normal vision. A line of letters near the bottom of the chart is marked 20/15. The normal eye can read letters of this size from a distance of 15 feet, but a person might be able to read it from 20 feet. This person has better than normal vision.

To conduct the visual acuity test, follow these steps:
a. Hang the Snellen eye chart on a well-illuminated wall at eye level.
b. Have your partner stand 20 feet in front of the chart, gently cover the left eye with a 3″ × 5″ card, and read the smallest set of letters possible.
c. Record the visual acuity value for that set of letters in Part A of Laboratory Report 36.
d. Repeat the procedure, using the left eye.
2. *Astigmatism test.* Astigmatism is a condition that results from a defect in the curvature of the cornea or lens. As a consequence, some portions of the image projected on the retina are sharply focused, and other portions are blurred. Astigmatism can be evaluated by using an astigmatism chart (fig. 36.2). This chart consists of sets of black lines radiating from a central spot like the spokes of a wheel. To a normal eye, these lines appear sharply focused and equally dark; however, if the eye has an astigmatism, some sets of lines appear sharply focused and dark, while others are blurred and less dark.

To conduct the astigmatism test, follow these steps:
a. Hang the astigmatism chart on a well-illuminated wall at eye level.

Figure 36.1 The Snellen eye chart looks similar to this but is somewhat larger.

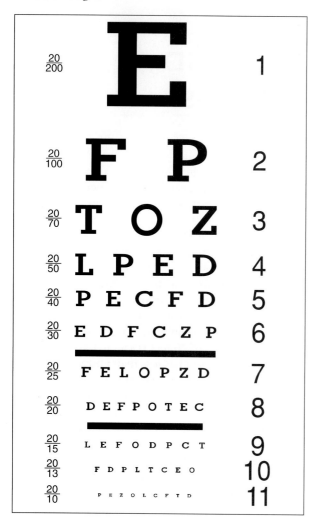

Figure 36.2 Astigmatism is evaluated using a chart such as this one.

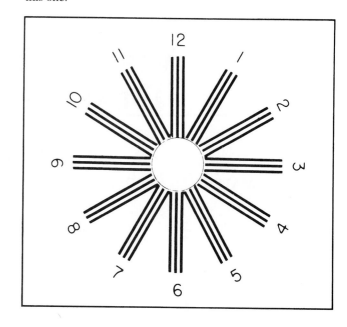

a. Hold the end of a meterstick against your partner's chin so that the stick extends outward at a right angle to the plane of the face (fig. 36.3).

b. Have your partner close the left eye.

c. Hold a 3″ × 5″ card with a word typed in the center at the distal end of the meterstick.

d. Slide the card along the stick toward your partner's open eye, and locate the *point closest to the eye* where your partner can still see the letters of the word sharply focused. This distance is called the *near point of accommodation*, and it tends to increase with age (table 36.1).

e. Repeat the procedure with the right eye closed.

f. Record the results in Part A of the laboratory report.

4. *Color vision test.* Some people exhibit defective color vision because they lack certain cones, usually those sensitive to the reds or greens. Because this trait is an X-linked (sex-linked) inheritance, the condition is more prevalent in males (7%) than in females (0.4%). People who lack or possess decreased sensitivity to the red-sensitive cones possess protanopia color blindness; those

b. Have your partner stand 20 feet in front of the chart, gently cover the left eye with a 3″ × 5″ card, focus on the spot in the center of the radiating lines, and report which lines, if any, appear more sharply focused and darker.

c. Repeat the procedure, using the left eye.

d. Record the results in Part A of the laboratory report.

3. *Accommodation test.* Accommodation is the changing of the shape of the lens that occurs when the normal eye is focused for close vision. It involves a reflex in which muscles of the ciliary body are stimulated to contract, releasing tension on the suspensory ligaments that are fastened to the lens capsule. This allows the capsule to rebound elastically, causing the surface of the lens to become more convex. The ability to accommodate is likely to decrease with age because the tissues involved tend to lose their elasticity.

To evaluate the ability to accommodate, follow these steps:

Table 36.1   Near Point of Accommodation

Age (years)	Average Near Point (cm)
10	7
20	10
30	13
40	20
50	45
60	90

Figure 36.3 To determine the near point of accommodation, slide the 3″ × 5″ card along the meterstick toward your partner's open eye until it reaches the closest location where your partner can still see the word sharply focused.

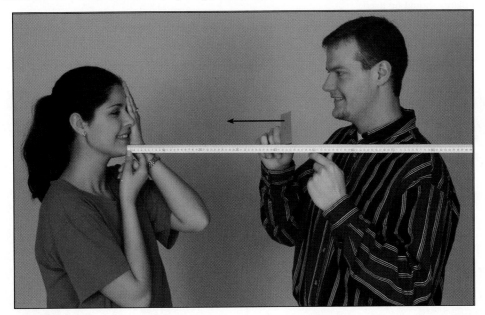

who lack or possess decreased sensitivity to green-sensitive cones possess deuteranopia color blindness. The color-blindness condition is often more of a deficiency or a weakness than one of blindness. Chapter 24 in the textbook describes the genetics for color blindness.

To conduct the color vision test, follow these steps:

a. Examine the color test plates in Ichikawa's or Ishihara's book to test for any red-green color vision deficiency. Also examine figure 36.4.

b. Hold the test plates approximately 30 inches from the subject in bright light. All responses should occur within 3 seconds.

c. Compare your responses with the correct answers in Ichikawa's or Ishihara's book. Determine the percentage of males and females in your class who exhibit any color-deficient vision. If an individual exhibits color-deficient vision, determine if the condition is protanopia or deuteranopia.

d. Record the class results in Part A of the laboratory report.

5. Complete Part A of the laboratory report.

PROCEDURE B—VISUAL DEMONSTRATIONS

Perform the following demonstrations with the help of your laboratory partner.

1. *Blind-spot demonstration.* There are no photoreceptors in the optic disc, which is located where the nerve fibers of the retina leave the eye and enter the optic nerve. Consequently, this region of the retina is commonly called the *blind spot.*

To demonstrate the blind spot, follow these steps:

a. Close your left eye, hold figure 36.5 about 35 cm away from your face, and stare at the + sign in the figure with your right eye.

b. Move the figure closer to your face as you continue to stare at the + until the dot on the figure suddenly disappears. This happens when the image of the dot is focused on the optic disc. Measure the distance using a metric ruler or a meterstick.

c. Repeat the procedures with your right eye closed. This time stare at the dot, and the + will disappear when the image falls on the optic disc. Measure the distance.

d. Record the results in Part B of the laboratory report.

Critical Thinking Application

Under normal visual circumstances, explain why small objects are not lost from our vision.

_____

_____

_____

_____

2. *Photopupillary reflex.* The smooth muscles of the iris control the size of the pupil. For example, when the intensity of light entering the eye increases, a photopupillary reflex is triggered, and the circular muscles of the iris are stimulated to contract. As a result, the size of the pupil decreases, and less light enters the eye.

Figure 36.4 Samples of Ishihara's color plates. These plates are reproduced from *Ishihara's Tests for Colour Blindness* published by Kanehara & Co., Ltd., Tokyo, Japan, but tests for color blindness cannot be conducted with this material. For accurate testing, the original plates should be used.

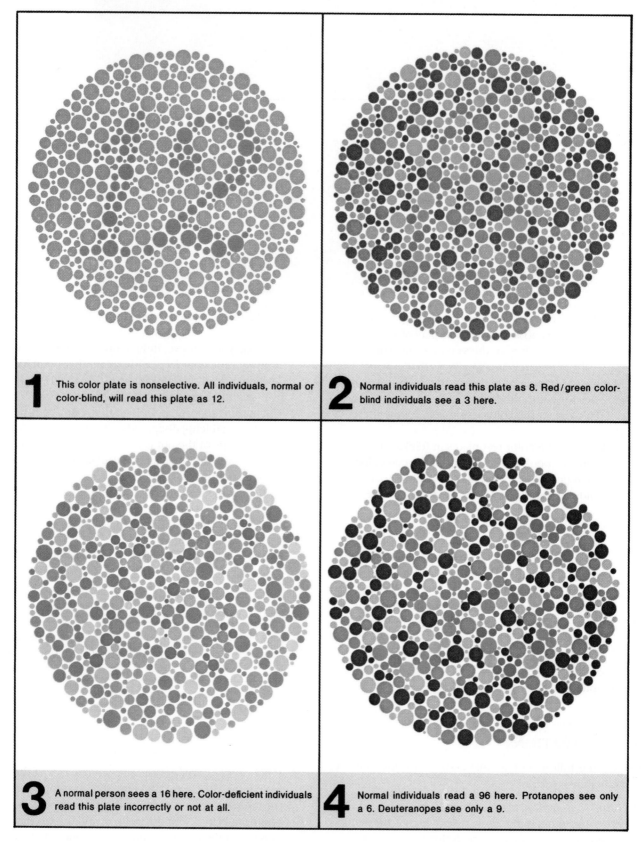

1 This color plate is nonselective. All individuals, normal or color-blind, will read this plate as 12.

2 Normal individuals read this plate as 8. Red/green color-blind individuals see a 3 here.

3 A normal person sees a 16 here. Color-deficient individuals read this plate incorrectly or not at all.

4 Normal individuals read a 96 here. Protanopes see only a 6. Deuteranopes see only a 9.

Figure 36.5 Blind-spot demonstration.

To demonstrate this reflex, follow these steps:

a. Ask your partner to sit with his or her hands thoroughly covering his or her eyes for 2 minutes.

b. Position a pen flashlight close to one eye with the light shining on the hand that covers the eye.

c. Ask your partner to remove the hand quickly.

d. Observe the pupil and note any change in its size.

e. Have your partner remove the other hand, but keep that uncovered eye shielded from extra light.

f. Observe both pupils and note any difference in their sizes.

3. *Accommodation pupillary reflex.* The pupil constricts as a normal accommodation reflex to focusing on close objects. To demonstrate the accommodation reflex, follow these steps:

a. Have your partner stare for several seconds at some dimly illuminated object in the room that is more than 20 feet away.

b. Observe the size of the pupil of one eye. Then hold a pencil about 25 cm in front of your partner's face, and have your partner stare at it.

c. Note any change in the size of the pupil.

4. *Convergence reflex.* The eyes converge as a normal convergence reflex to focusing on close objects. To demonstrate the convergence reflex, follow these steps:

a. Repeat the procedure outlined for the accommodation pupillary reflex.

b. Note any change in the position of the eyeballs as your partner changes focus from the distant object to the pencil.

5. Complete Part B of the laboratory report.

VISUAL TESTS AND DEMONSTRATIONS

PART A

1. Visual acuity test results:

Eye Tested	Acuity Values
Right eye	
Right eye with glasses (if applicable)	
Left eye	
Left eye with glasses (if applicable)	

2. Astigmatism test results:

Eye Tested	Darker Lines
Right eye	
Right eye with glasses (if applicable)	
Left eye	
Left eye with glasses (if applicable)	

3. Accommodation test results:

Eye Tested	Near Point (cm)
Right eye	
Right eye with glasses (if applicable)	
Left eye	
Left eye with glasses (if applicable)	

4. Color vision test results:

Condition	Males			Females		
	Class Number	Class Percentage	Expected Percentage	Class Number	Class Percentage	Expected Percentage
Normal color vision			93			99.6
Deficient red-green color vision			7			0.4
Protanopia (lack red-sensitive cones)			less-frequent type			less-frequent type
Deuteranopia (lack green-sensitive cones)			more-frequent type			more-frequent type

5. Complete the following:

    a. What is meant by 20/70 vision? _____

    _____

    b. What is meant by 20/10 vision? _____

    _____

    c. What visual problem is created by astigmatism? _____

    _____

    d. Why does the near point of accommodation often increase with age? _____

    _____

    e. Describe the eye defect that causes color-deficient vision. _____

    _____

PART B

1. Blind-spot results:
    a. Right eye distance _____
    b. Left eye distance _____

Complete the following:

2. Explain why an eye has a blind spot. _____

_____

3. Describe the photopupillary reflex. _____

_____

4. What difference did you note in the size of the pupils when one eye was exposed to bright light and the other eye was shielded from the light? _____

_____

5. Describe the accommodation pupillary reflex. _____

_____

6. Describe the convergence reflex. _____

_____

LABORATORY EXERCISE 37

ENDOCRINE SYSTEM

MATERIALS NEEDED

Textbook
Human torso
Compound light microscope
Prepared microscope slides of the following:
  Pituitary gland
  Thyroid gland
  Parathyroid gland
  Adrenal gland
  Pancreas

For Optional Activity:

Water bath equipped with temperature control
  mechanism set at 37°C (98.6°F)
Laboratory thermometer

The endocrine system consists of ductless glands that act together with parts of the nervous system to help control body activities. The endocrine glands secrete hormones that are transported in body fluids and affect cells possessing appropriate receptor molecules. In this way, hormones influence the rate of metabolic reactions, the transport of substances through cell membranes, and the regulation of water and electrolyte balances. By controlling cellular activities, endocrine glands play important roles in the maintenance of homeostasis.

PURPOSE OF THE EXERCISE

To review the structure and function of major endocrine glands and to examine microscopically the tissues of these glands.

LEARNING OBJECTIVES

After completing this exercise, you should be able to

1. Name and locate the major endocrine glands.
2. Name the hormones secreted by each of the major glands.
3. Describe the principal functions of each hormone.
4. Recognize tissue sections from the pituitary gland, thyroid gland, parathyroid glands, adrenal glands, and pancreas.

PROCEDURE

1. Review the sections entitled "Pituitary Gland," "Thyroid Gland," "Parathyroid Glands," "Adrenal Glands," "Pancreas," and "Other Endocrine Glands" in chapter 13 of the textbook.
2. As a review activity, label figures 37.1, 37.2, 37.3, 37.4, 37.5, and 37.6.
3. Complete Part A of Laboratory Report 37.
4. Examine the human torso and locate the following:

hypothalamus

pituitary stalk (infundibulum)

pituitary gland

  anterior lobe

  posterior lobe

thyroid gland

parathyroid glands

adrenal glands

  adrenal medulla

  adrenal cortex

pancreas

pineal gland

thymus gland

ovaries

testes

297

Figure 37.1   Label the major endocrine glands.

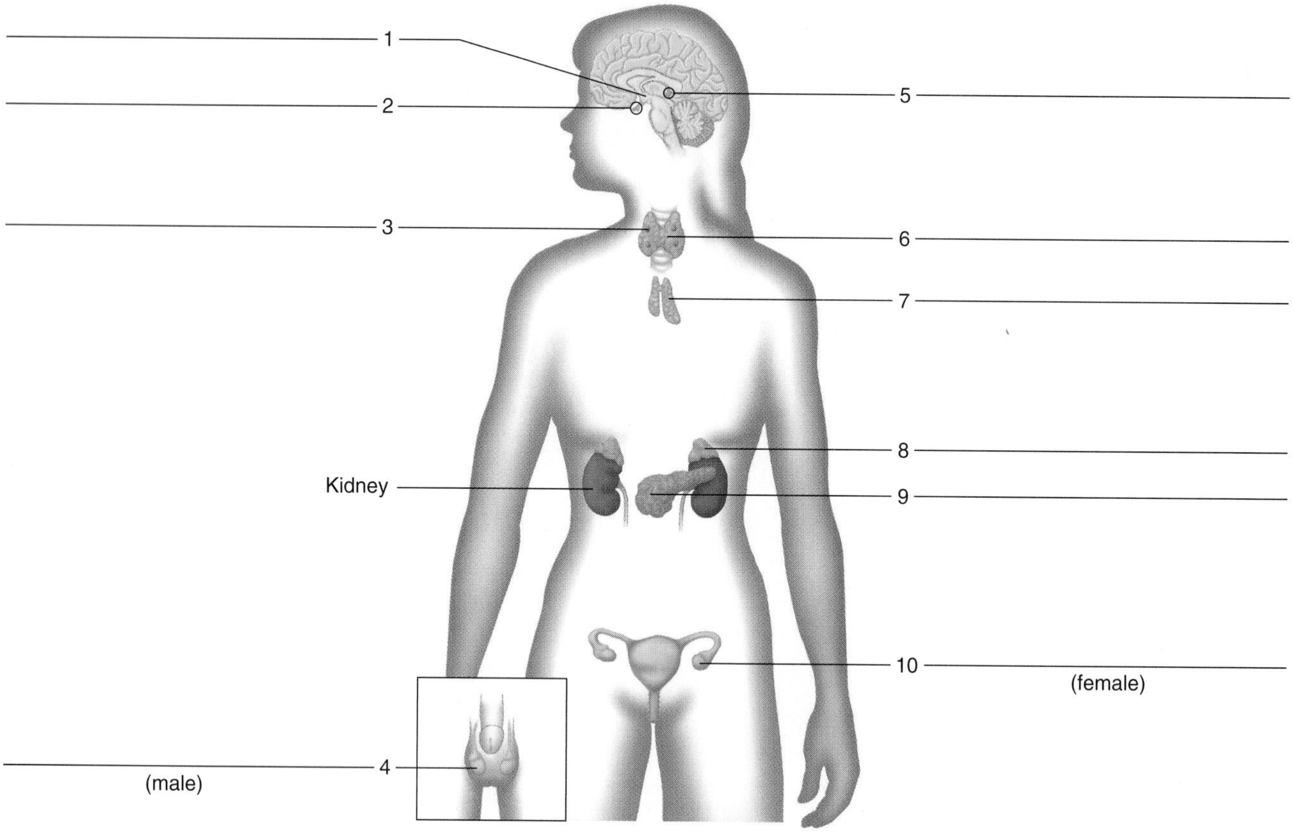

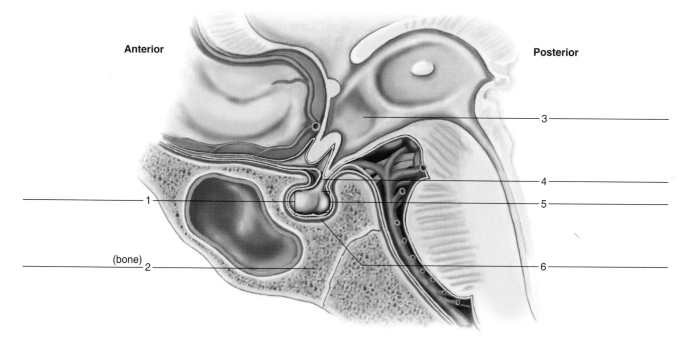

1	5
2	
3	6
	7
	8
Kidney	9
	10 (female)
(male) 4	

Figure 37.2   Label the features associated with the pituitary gland.

Anterior

Posterior

	3
	4
1	5
(bone) 2	6

Figure 37.3 Label the features associated with the thyroid gland (anterior view).

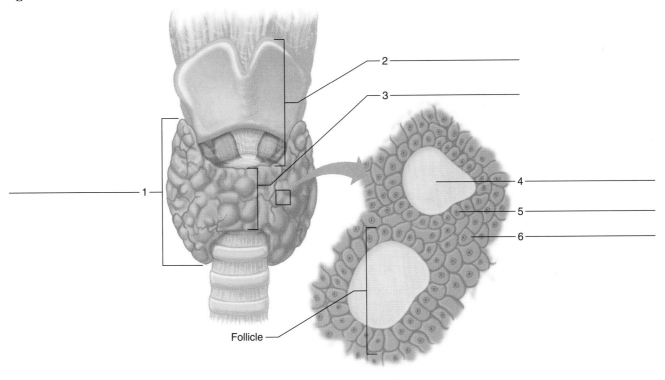

1 _____

2 _____

3 _____

4 _____

5 _____

6 _____

Follicle

Figure 37.4 Label the features associated with the parathyroid glands (posterior view).

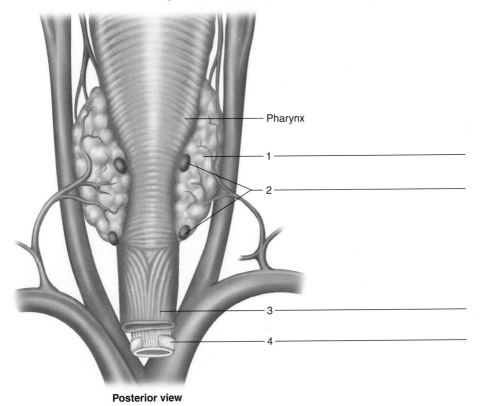

Pharynx

1 _____

2 _____

3 _____

4 _____

Posterior view

Figure 37.5   Label the features associated with the adrenal gland.

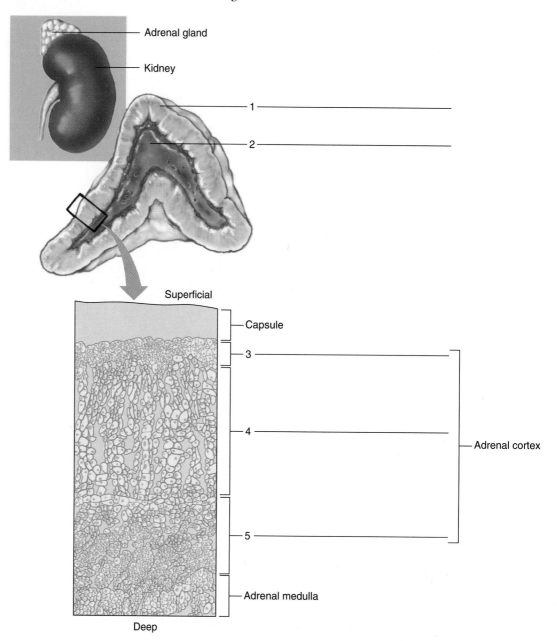

Adrenal gland

Kidney

1 _____

2 _____

Superficial

Capsule

3 _____

4 _____

Adrenal cortex

5 _____

Adrenal medulla

Deep

OPTIONAL ACTIVITY

The secretions of endocrine glands are usually controlled by negative feedback systems. As a result, the concentrations of hormones in body fluids remain relatively stable, although they will fluctuate slightly within a normal range. (See the section entitled "Control of Hormonal Secretions" in chapter 13 of the textbook.)

Similarly, the mechanism used to maintain the temperature of a laboratory water bath involves negative feedback. In this case, a temperature-sensitive thermostat in the water allows a water heater to operate whenever the water temperature drops below the thermostat's set point. Then, when the water temperature reaches the set point, the thermostat causes the water heater to turn off (a negative effect), and the water bath begins to cool again.

Use a laboratory thermometer to monitor the temperature of the water bath in the laboratory. Measure the temperature at regular intervals, until you have recorded ten readings. What was the lowest temperature you recorded? _____ The highest temperature? _____ What was the average temperature of the water bath? _____ How is the water bath temperature control mechanism similar to a hormonal control mechanism in the body?

_____

_____

_____

_____

_____

Figure 37.6 Label the features associated with the pancreas.

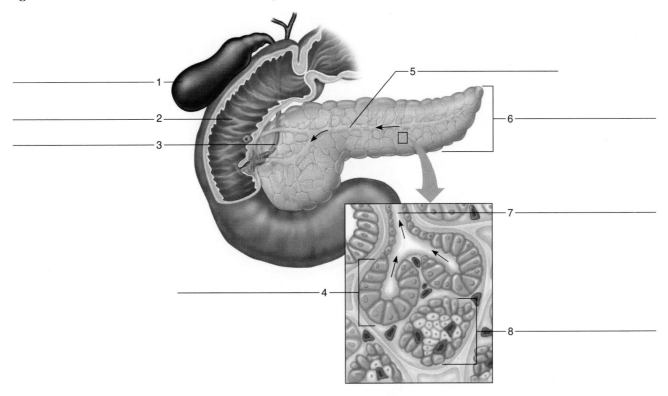

5. Examine the microscopic tissue sections of the following glands, and identify the features described:

   Pituitary gland. To examine the pituitary tissue, follow these steps:
   a. Observe the tissues using low-power magnification (fig. 37.7).
   b. Locate the *infundibulum (pituitary stalk)*, the *anterior lobe* (the largest part of the gland), and the *posterior lobe*.
   c. Observe an area of the anterior lobe with high-power magnification. Locate a cluster of relatively large cells and identify some *acidophil cells*, which contain pink-stained granules, and some *basophil cells*, which contain blue-stained granules. These acidophil and basophil cells are hormone-secreting cells.
   d. Observe an area of the posterior lobe with high-power magnification. Note the numerous unmyelinated nerve fibers present in this lobe. Also locate some *pituicytes*, a type of neuroglial cell, scattered among the nerve fibers.
   e. Prepare labeled sketches of representative portions of the anterior and posterior lobes of the pituitary gland in Part B of the laboratory report.

   Thyroid gland. To examine the thyroid tissue, follow these steps:
   a. Use low-power magnification to observe the tissue (fig. 37.8). Note the numerous *follicles*, each of which consists of a layer of cells surrounding a colloid-filled cavity.
   b. Observe the tissue using high-power magnification. Note that the cells forming the wall of a follicle are simple cuboidal epithelial cells.
   c. Prepare a labeled sketch of a representative portion of the thyroid gland in Part B of the laboratory report.

   Parathyroid gland. To examine the parathyroid tissue, follow these steps:
   a. Use low-power magnification to observe the tissue (fig. 37.9). Note that the gland consists of numerous tightly packed secretory cells.
   b. Switch to high-power magnification and locate two types of cells—a smaller form (chief cells) that are arranged in cordlike patterns and a larger form (oxyphil cells) that have distinct cell boundaries and are present in clusters. *Chief cells* secrete parathyroid hormone, whereas the function of *oxyphil cells* is not clearly understood.

Figure 37.7 Micrograph of the pituitary gland (6×).

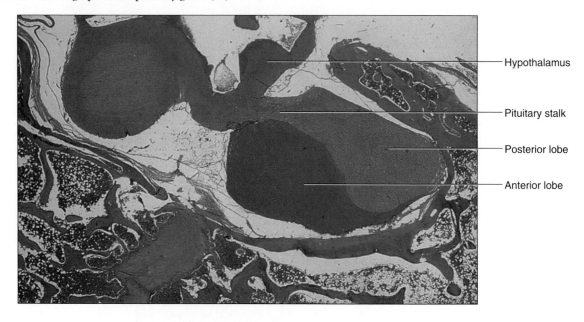

- Hypothalamus
- Pituitary stalk
- Posterior lobe
- Anterior lobe

Figure 37.8 Micrograph of the thyroid gland (100× micrograph enlarged to 300×).

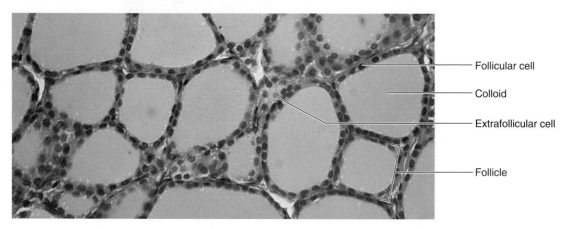

- Follicular cell
- Colloid
- Extrafollicular cell
- Follicle

c. Prepare a labeled sketch of a representative portion of the parathyroid gland in Part B of the laboratory report.

   Adrenal gland. To examine the adrenal tissue, follow these steps:

a. Use low-power magnification to observe the tissue (fig. 37.10). Note the thin capsule of connective tissue that covers the gland. Just beneath the capsule, there is a relatively thick *adrenal cortex.* The central portion of the gland is the *adrenal medulla.* The cells of the cortex are in three poorly defined layers. Those of the outer layer (zona glomerulosa) are arranged irregularly; those of the middle layer (zona fasciculata) are in long cords; and those of the inner layer (zona reticularis) are arranged in an interconnected network of cords. The cells of the medulla are relatively large and irregularly shaped, and they often occur in clusters.

Figure 37.9 Micrograph of the parathyroid gland (65×).

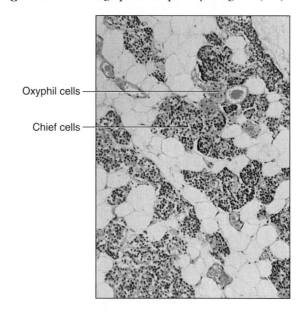

Oxyphil cells

Chief cells

Figure 37.10 Micrograph of the adrenal cortex and the adrenal medulla (75×).

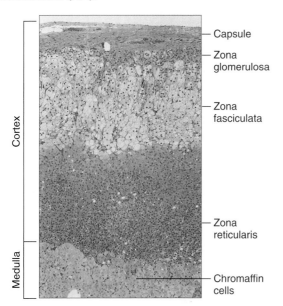

Cortex

— Capsule

— Zona glomerulosa

— Zona fasciculata

— Zona reticularis

Medulla

— Chromaffin cells

Figure 37.11 Micrograph of the pancreas (100× micrograph enlarged to 400×).

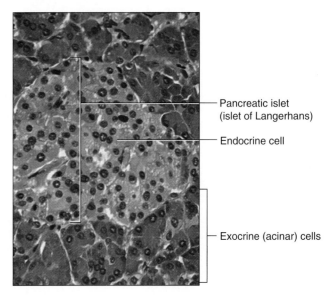

— Pancreatic islet (islet of Langerhans)

— Endocrine cell

— Exocrine (acinar) cells

b. Using high-power magnification, observe each of the layers of the cortex and the cells of the medulla.

c. Prepare labeled sketches of representative portions of the adrenal cortex and medulla in Part B of the laboratory report.

   Pancreas. To examine the pancreas tissue, follow these steps:

a. Use low-power magnification to observe the tissue (fig. 37.11). Note that the gland largely consists of deeply stained exocrine cells arranged in clusters around secretory ducts. These exocrine cells (acinar cells) secrete pancreatic juice rich in digestive enzymes. There are circular masses of lightly stained cells scattered throughout the gland. These clumps of cells constitute the *pancreatic islets (islets of Langerhans),* and they represent the endocrine portion of the pancreas.

b. Examine an islet, using high-power magnification.

c. Prepare a labeled sketch of a representative portion of the pancreas in Part B of the laboratory report.

Web Quest

What are common endocrine gland disorders? How can they be treated? Search these and review the endocrine system at
www.mhhe.com/shier10
Click on Student Edition. Click on Martin Lab Manual, Web Quest.

Laboratory Report 37

Name _____

Date _____

Section _____

ENDOCRINE SYSTEM

PART A

Complete the following:

1. Name six hormones secreted by the anterior lobe of the pituitary gland. _____
   _____
   _____
   _____

2. Name two hormones secreted by the posterior lobe of the pituitary gland. _____
   _____

3. Name the pituitary hormone responsible for the following actions:
   a. Stimulates ovarian follicle to secrete estrogen and egg development _____
   b. Causes kidneys to conserve water _____
   c. Stimulates cells to increase in size and divide more rapidly _____
   d. Essential for egg release from the ovary _____
   e. Stimulates secretion from thyroid gland _____
   f. Causes contraction of uterine wall muscles _____
   g. Stimulates secretion from adrenal cortex _____
   h. Stimulates milk production _____

4. Name two thyroid hormones that affect metabolic rate. _____
   _____

5. Name a hormone secreted by the thyroid gland that acts to lower blood calcium. _____

6. Name a hormone that acts to raise blood calcium. _____

7. Name three target organs of parathyroid hormone. _____
   _____

8. Name two hormones secreted by the adrenal medulla. _____
   _____

9. List five different effects produced by these medullary hormones. _____
   _____
   _____
   _____

10. Name the most important mineralocorticoid secreted by the adrenal cortex. _____

11. List three actions of this mineralocorticoid. _____
    _____
    _____

12. Name the most important glucocorticoid secreted by the adrenal cortex. _____

13. List three actions of this glucocorticoid. _____

_____

_____

14. Distinguish the hormones secreted by the alpha and beta cells of the pancreatic islets (islets of Langerhans).

_____

Critical Thinking Application

Briefly explain how the actions of pancreatic hormones complement one another.

_____

_____

_____

_____

Prepare labeled sketches to illustrate representative portions of the following endocrine glands:

Pituitary gland (_____×)
(anterior lobe)

Pituitary gland (_____×)
(posterior lobe)

Thyroid gland (_____×)

Parathyroid gland (_____×)

Adrenal gland (_____×)
(medulla)

Adrenal gland (_____×)
(cortex)

Pancreas (_____×)

BLOOD CELLS

SAFETY

- It is important that students learn and practice correct
  procedures for handling body fluids. Consider using
  either mammal blood other than human or
  contaminant-free blood that has been tested and is
  available from various laboratory supply houses. Some
  of the procedures might be accomplished as
  demonstrations only. If student blood is used, it is
  important that students handle only their own blood.
- Use an appropriate disinfectant to wash the
  laboratory tables before and after the procedures.
- Wear disposable gloves when handling blood
  samples.
- Clean end of a finger with 70% alcohol before the
  puncture is performed.
- The sterile blood lancet should be used only once.
- Dispose of used lancets and blood-contaminated items
  in an appropriate container (never use the
  wastebasket).
- Wash your hands before leaving the laboratory.

Blood is a type of connective tissue whose cells are
suspended in a liquid intercellular substance. These
cells are mainly formed in red bone marrow, and they
include *red blood cells, white blood cells,* and some cellular fragments called *platelets.*

Red blood cells transport gases between the body
cells and the lungs, white blood cells defend the body
against infections, and platelets play an important role
in stoppage of bleeding (hemostasis).

PURPOSE OF THE EXERCISE

To review the characteristics of blood cells, to examine
them microscopically, and to perform a differential
white blood cell count.

LEARNING OBJECTIVES

After completing this exercise, you should be able to

1. Describe the structure and function of red blood
   cells, white blood cells, and platelets.
2. Identify red blood cells, five types of white blood
   cells, and platelets on a stained blood slide.
3. Perform a differential white blood cell count.

WARNING

*Because of the possibility of blood infections
being transmitted from one student to another if
blood slides are prepared in the classroom, it is
suggested that commercially prepared blood slides
be used in this exercise. The instructor, however,
may wish to demonstrate the procedure for preparing such a slide. Observe all safety procedures for
this lab.*

DEMONSTRATION—BLOOD SLIDE
PREPARATION

To prepare a stained blood slide, follow these steps:

1. Obtain two precleaned microscope slides. Avoid
   touching their flat surfaces.
2. Thoroughly wash hands with soap and water and
   dry them with paper towels.
3. Cleanse the end of the middle finger with some
   sterile cotton moistened with 70% alcohol and let
   the finger dry in the air.
4. Remove a sterile disposable blood lancet from its
   package without touching the sharp end.

Figure 38.1 To prepare a blood smear: (*a*) place a drop of blood about 2 cm from the end of a clean slide; (*b*) hold a second slide at about a 45° angle to the first one, allowing the blood to spread along its edge; (*c*) push the second slide over the surface of the first so that it pulls the blood with it; (*d*) observe the completed blood smear. The ideal smear should be 1½ inches in length, be evenly distributed, and contain a smooth, feathered edge.

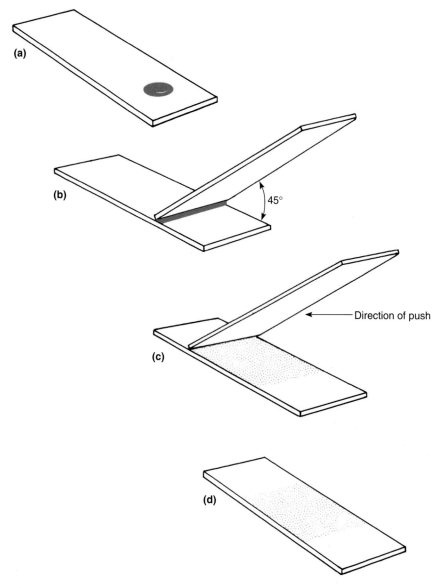

(a)

(b) 45°

Direction of push

(c)

(d)

5. Puncture the skin on the side near the tip of the middle finger with the lancet and properly discard the lancet.

6. Wipe away the first drop of blood with the cotton ball. Place a drop of blood about 2 cm from the end of a clean microscope slide.

7. Use a second slide to spread the blood across the first slide, as illustrated in figure 38.1. Discard the slide used for spreading the blood in the appropriate container.

8. Place the blood slide on a slide staining rack and let it dry in the air.

9. Put enough Wright's stain on the slide to cover the smear but not overflow the slide. Count the number of drops of stain that are used.

10. After 2–3 minutes, add an equal volume of distilled water to the stain and let the slide stand for 4 minutes.

From time to time, gently blow on the liquid to mix the water and stain.

11. Flood the slide with distilled water until the blood smear appears light blue.

12. Tilt the slide to pour off the water, and let the slide dry in the air.

13. Examine the blood smear with low-power magnification, and locate an area where the blood cells are well distributed. Observe these cells, using high-power magnification and then an oil immersion objective if one is available.

PROCEDURE A—TYPES OF BLOOD CELLS

1. Review the sections entitled "Characteristics of Red Blood Cells," "Types of White Blood Cells," and "Blood Platelets" in chapter 14 of the textbook.

Figure 38.2 Blood cells illustrating some of the numerous variations of each type.

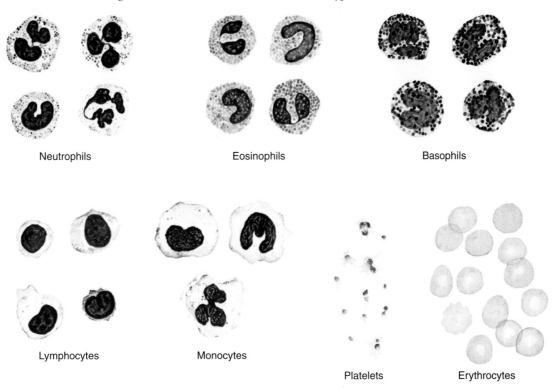

Neutrophils   Eosinophils   Basophils

Lymphocytes   Monocytes

Platelets   Erythrocytes

2. Complete Part A of Laboratory Report 38.
3. Refer to figures 14.9–14.13 in the textbook and figure 38.2 as an aid in identifying the various types of blood cells. Use the prepared slide of blood and locate each of the following:

red blood cell (erythrocyte)

white blood cell (leukocyte)

    granulocytes

        neutrophil

        eosinophil

        basophil

    agranulocytes

        lymphocyte

        monocyte

platelet (thrombocyte)

4. In Part B of the laboratory report, prepare sketches of single blood cells to illustrate each type. Pay particular attention to relative size, nuclear shape, and color of granules in the cytoplasm (if present).

PROCEDURE B—DIFFERENTIAL WHITE BLOOD CELL COUNT

A differential white blood cell count is performed to determine the percentage of each of the various types of white blood cells present in a blood sample. The test is useful because the relative proportions of white blood cells may change in particular diseases. Neutrophils, for example, usually increase during bacterial infections, whereas eosinophils may increase during certain parasitic infections and allergic reactions.

1. To make a differential white blood cell count, follow these steps:
   a. Using high-power magnification or an oil immersion objective, focus on the cells at one end of a prepared blood slide where the cells are well distributed.
   b. Slowly move the blood slide back and forth, following a path that avoids passing over the same cells twice (fig. 38.3).
   c. Each time you encounter a white blood cell, identify its type and record it in Part C of the laboratory report.
   d. Continue searching for and identifying white blood cells until you have recorded 100 cells in the data table. Because *percent* means "parts of 100," for each type of white blood cell, the total number observed is equal to its percentage in the blood sample.
2. Complete Part C of the laboratory report.

Figure 38.3 Move the blood slide back and forth to avoid passing the same cells twice.

Table 38.1 Differential White Blood Cell Count

Cell Type	Normal Value (percent)
Neutrophil	54–62
Eosinophil	1–3
Basophil	<1
Lymphocyte	25–33
Monocyte	3–9

OPTIONAL ACTIVITY

Obtain a prepared slide of pathological blood that has been stained with Wright's stain. Perform a differential white blood cell count using this slide, and compare the results with the values for normal blood listed in table 38.1.

What differences do you note? _____

_____

_____

_____

Web Quest

What are the functions of the various blood components? What do abnormal amounts indicate? Search these and review blood cell identification at
www.mhhe.com/shier10
Click on Student Edition. Click on Martin Lab Manual, Web Quest.

Laboratory Report 38

Name _____

Date _____

Section _____

BLOOD CELLS

PART A

Complete the following statements:

1. Mature red blood cells are also called _____ .

2. The shape of a red blood cell can be described as a _____ disc.

3. The function of red blood cells is _____ .

4. _____ is the oxygen-carrying substance in a red blood cell.

5. Red blood cells with high oxygen concentrations are bright red because of the presence of

   _____ .

6. Red blood cells cannot reproduce because they lack _____ when they are mature.

7. White blood cells are also called _____ .

8. White blood cells with granular cytoplasm are called _____ .

9. White blood cells lacking granular cytoplasm are called _____ .

10. Polymorphonuclear leukocyte is another name for a _____ with a segmented nucleus.

11. Normally, the most numerous white blood cells are _____ .

12. White blood cells whose cytoplasmic granules stain red in acid stain are called _____ .

13. _____ are normally the least abundant of the white blood cells.

14. _____ are the largest of the white blood cells.

15. _____ are small agranulocytes that have relatively large, round nuclei with thin rims of cytoplasm.

16. In red bone marrow, platelets develop from stem cells called _____ .

17. Upon an injury, platelets adhere to _____ found in connective tissue.

18. In the presence of damaged blood vessels, platelets release a substance called _____ ,which causes smooth muscle contraction.

PART B

Sketch a single blood cell of each type in the following spaces. Use colored pencils to represent the stained colors of the cells. Label any features that can be identified.

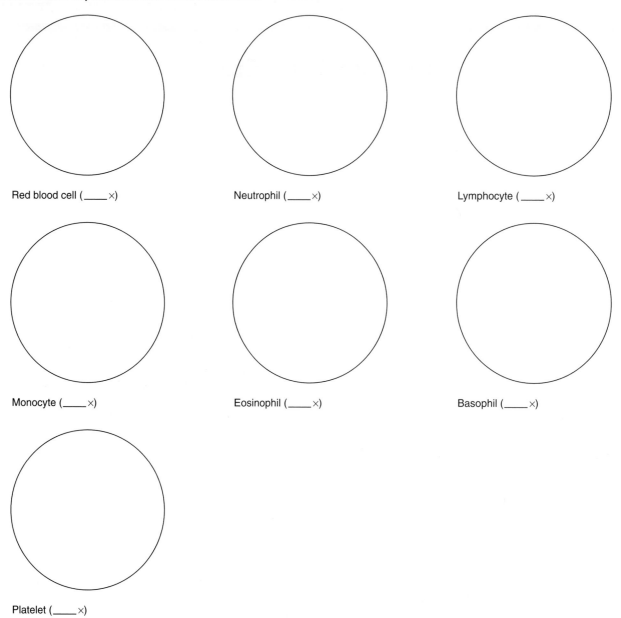

Red blood cell (_____ ×)

Neutrophil (_____ ×)

Lymphocyte (_____ ×)

Monocyte (_____ ×)

Eosinophil (_____ ×)

Basophil (_____ ×)

Platelet (_____ ×)

PART C

1. *Differential White Blood Cell Count Data Table.* As you identify white blood cells, record them on the table by using a tally system, such as ℍℍ Ⅱ. Place tally marks in the "Number Observed" column, and total each of the five WBCs when the differential count is completed. Obtain a total of all five WBCs counted to determine the percent of each WBC type.

Type of WBC	Number Observed	Total	Percent
Neutrophil			
Lymphocyte			
Monocyte			
Eosinophil			
Basophil			
		Total of column	

2. How do the results of your differential white blood cell count compare with the normal values listed in table 38.1? _____
_____
_____
_____
_____
_____

Critical Thinking Application

What is the difference between a differential white blood cell count and a total white blood cell count?

_____
_____
_____
_____
_____
_____

BLOOD TESTING— A DEMONSTRATION

SAFETY

- It is important that students learn and practice correct procedures for handling body fluids. Consider using either mammal blood other than human or contaminant-free blood that has been tested and is available from various laboratory supply houses. Some of the procedures might be accomplished as demonstrations only. If student blood is used, it is important that students handle only their own blood.
- Use an appropriate disinfectant to wash the laboratory tables before and after the procedures.
- Wear disposable gloves when handling blood samples.
- Clean the end of a finger with 70% alcohol before the puncture is performed.
- The sterile blood lancet should be used only once.
- Dispose of used lancets and blood-contaminated items into an appropriate container (never use the wastebasket).
- Wash your hands before leaving the laboratory.

As an aid in identifying various disease conditions, tests are often performed on blood to determine how its composition compares with normal values. These tests commonly include red blood cell percentage, hemoglobin content, red blood cell count, and total white blood cell count. Common tests performed on the blood can be found in Appendix B of the textbook. Use the clinical significance in Appendix B for the tests in this lab to determine what factors can influence the normal values.

PURPOSE OF THE EXERCISE

To observe the blood tests used to determine red blood cell percentage, hemoglobin content, red blood cell count, and total white blood cell count.

LEARNING OBJECTIVES

After completing this exercise, you should be able to

1. Determine the percentage of red blood cells in a blood sample.
2. Determine the hemoglobin content of a blood sample.
3. Describe how a red blood cell count is performed.
4. Describe how a total white blood cell count is performed.

WARNING

Because of the possibility of blood infections being transmitted from one student to another during blood-testing procedures, it is suggested that the following demonstrations be performed by the instructor. Observe all safety procedures listed for this lab.

DEMONSTRATION A—RED BLOOD CELL PERCENTAGE (HEMATOCRIT)

To determine the percentage of red blood cells in a whole blood sample, the cells must be separated from the liquid plasma. This separation can be rapidly accomplished by placing a tube of blood in a centrifuge. The force created by the spinning motion of the centrifuge causes the cells to be packed into the lower end of the tube. Then the quantities of cells and plasma can be measured, and the percentage of cells (hematocrit or packed cell volume) can be calculated.

1. To determine the percentage of red blood cells in a blood sample, follow these steps:
   a. Lance the end of a finger to obtain a drop of blood. *See the demonstration in Laboratory Exercise 38 for directions.*
   b. Touch the drop of blood with the colored end of a heparinized capillary tube. Hold the tube tilted slightly upward so that the blood will easily move into it by capillary action (fig. 39.1). To prevent an air bubble, keep the tip in the blood until filled.
   c. Allow the blood to fill about two-thirds of the length of the tube.
   d. Plug the blood end of the tube by pushing it with a rotating motion into sealing clay or by adding a plastic Critocap. Hold a finger over the tip of the dry end so that blood will not drain out while you seal the blood end.
   e. Place the sealed tube into one of the numbered grooves of a microhematocrit centrifuge. The tube's sealed end should point outward from the center and should touch the rubber lining on the rim of the centrifuge (fig. 39.1).
   f. The centrifuge should be balanced by placing specimen tubes on opposite sides of the moving head, the inside cover should be tightened with the lock wrench, and the outside cover should be securely fastened.
   g. Run the centrifuge for 3–5 minutes.
   h. After the centrifuge has stopped, remove the specimen tube and note that the red blood cells have been packed into the bottom of the tube. The clear liquid on top of the cells is plasma.
   i. Use a microhematocrit reader to determine the percentage of red blood cells in the tube. If a microhematocrit reader is not available, measure the total length of the blood column in millimeters (red cells plus plasma) and the length of the red blood cell column alone in millimeters. Divide the red blood cell length by the total blood column length and multiply the answer by 100 to calculate the percentage of red blood cells.
   j. Record the test result in Part A of Laboratory Report 39.
2. Complete Part B of the laboratory report.

318

Figure 39.1 Steps in the hematocrit (red blood cell percentage) procedure: (*a*) load a heparinized capillary tube with blood; (*b*) plug the blood end of the tube with sealing clay; (*c*) place the tube in a microhematocrit centrifuge.

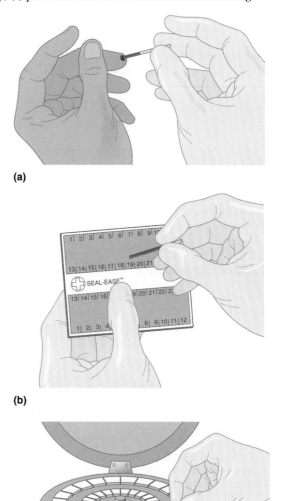

(a)

(b)

(c)

DEMONSTRATION B—HEMOGLOBIN CONTENT

Although the hemoglobin content of a blood sample can be measured in several ways, a common method uses a hemoglobinometer. This instrument is designed to compare the color of light passing through a hemolyzed blood sample with a standard color. The results of the test are expressed in grams of hemoglobin per 100 mL of blood or in percentage of normal values.

1. To measure the hemoglobin content of a blood sample, follow these steps:
   a. Obtain a hemoglobinometer and remove the blood chamber from the slot in its side.

b. Separate the pieces of glass from the metal clip and clean them with 70% alcohol and lens paper. Note that one of the pieces of glass has two broad, U-shaped areas surrounded by depressions. The other piece is flat on both sides.

c. Obtain a large drop of blood from a finger, as before.

d. Place the drop of blood on one of the U-shaped areas of the blood chamber glass.

e. Stir the blood with the tip of a hemolysis applicator until the blood appears clear rather than cloudy. This usually takes about 45 seconds.

f. Place the flat piece of glass on top of the blood plate and slide both into the metal clip of the blood chamber.

g. Push the blood chamber into the slot on the side of the hemoglobinometer, making sure that it is in all the way (fig. 39.2).

h. Hold the hemoglobinometer in the left hand with the thumb on the light switch on the underside.

i. Look into the eyepiece and note the green area that is split in half.

j. Slowly move the slide on the side of the instrument back and forth with the right hand until the two halves of the green area look the same.

k. Note the value in the upper scale (grams of hemoglobin per 100 mL of blood), indicated by the mark in the center of the movable slide.

l. Record the test result in Part A of the laboratory report.

2. Complete Part C of the laboratory report.

DEMONSTRATION C—RED BLOOD CELL COUNT

Although modern clinical laboratories commonly use electronic instruments to obtain blood cell counts, a hemocytometer provides a convenient and inexpensive way to count both red and white blood cells. This instrument is a special microscope slide on which there are two counting areas (fig. 39.3). Each counting area contains a grid of tiny lines forming nine large squares that are further subdivided into smaller squares (fig. 39.4).

The squares on the counting area have known dimensions. When a coverslip is placed over the area, it is

Figure 39.2 Steps in the hemoglobin content procedure: (*a*) load the blood chamber with blood; (*b*) stir the blood with a hemolysis applicator; (*c*) place the blood chamber in the slot of the hemoglobinometer; (*d*) match the colors in the green area by moving the slide on the side of the instrument

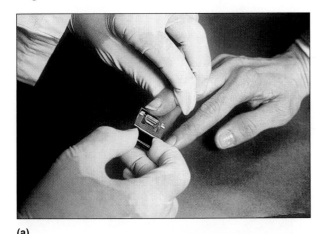

(a)

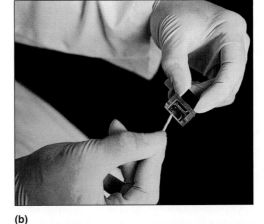

(b)

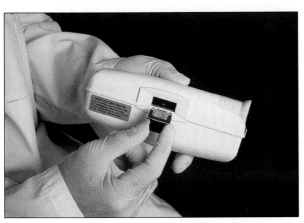

(c)

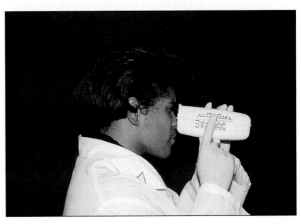

(d)

319

Figure 39.3 Location of the counting areas of a hemocytometer.

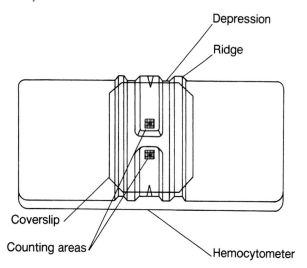

Figure 39.4 The pattern of lines in a counting area of a hemocytometer. The squares marked with *R* are used to count red blood cells, whereas the squares marked with *W* are used to count white blood cells.

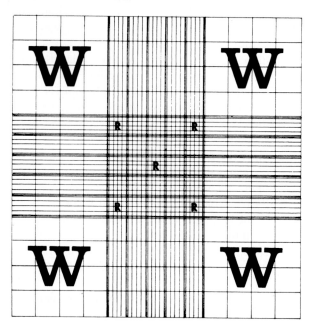

held a known distance above the grid by glass ridges of the hemocytometer. Thus, when a liquid is placed under the coverslip, the volume of liquid covering each part of the grid can be calculated. Also, if the number of blood cells in a tiny volume of blood can be counted, it is possible to calculate the number that must be present in any larger volume.

When red blood cells are counted using a hemocytometer, a small sample of blood is drawn into a special pipette, and the blood is diluted to reduce the number of cells that must be counted. Some of the diluted blood is spread over the counting area, and with the aid of a microscope, all of the cells in the grid areas marked *R* are counted (fig. 39.4). The total count is multiplied by 10,000 to correct for the dilution and for the fact that only a small volume of blood was observed. The final result provides the number of red blood cells per cubic millimeter in the original blood sample.

1. To perform a red blood cell count, follow these steps:
   a. Examine a hemocytometer, and use a microscope to locate the grid of a counting area. Focus on the large central square of the grid with low-power and then with high-power magnification. Adjust the light intensity so that the lines are clear and sharp.
   b. Wash the hemocytometer with soap and water, and dry it. Place the coverslip over the counting areas and set the instrument aside.
   c. Examine the Unopette system for counting red blood cells, which consists of a reservoir containing blood cell diluting fluid and a plastic capillary tube assembly within a plastic shield (fig. 39.5).
   d. Place the Unopette reservoir on the table and gently force the pointed tip of the capillary tube shield through the thin diaphragm at the top of the reservoir.

Figure 39.5 The Unopette system consists of a reservoir containing blood cell diluting fluid (*left*) and a plastic capillary tube within a plastic shield (*right*).

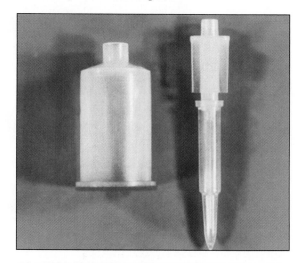

   e. Lance the tip of a finger, as before, to obtain a drop of blood.
   f. Remove the Unopette capillary tube from its shield and, holding the tube horizontally, touch the tip of the tube to the drop of blood. Allow the tube to fill completely with blood by capillary action (fig. 39.6).
   g. Gently squeeze the reservoir, taking care not to expel any of its fluid content, and while maintaining pressure on the sides of the reservoir, insert the capillary tube through the punctured diaphragm of the reservoir. Release the pressure on the sides of the reservoir and allow the blood to be drawn into its chamber.

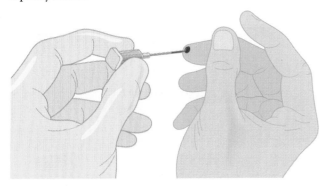

Figure 39.6 The Unopette capillary tube fills with blood by capillary action.

Figure 39.7 The Unopette dropper system being used to fill a hemocytometer with blood. When you squeeze the reservoir, a drop of diluted blood enters the counting areas under the coverslip by capillary action.

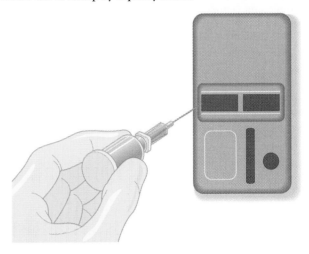

h. Hold the capillary tube assembly in the reservoir and gently squeeze the sides of the reservoir several times to mix the blood with the diluting fluid inside. Also, invert the reservoir a few times to aid this mixing process.

i. Remove the capillary tube assembly from the reservoir and insert the opposite end of the assembly into the top of the reservoir, thus converting the parts into a dropper system (fig. 39.7).

j. Gently squeeze the sides of the reservoir to expel some of the diluted blood from the capillary tube. Discard the first four drops of this mixture, and place the next drop of diluted blood at the edge of the coverslip near a counting area of the hemocytometer (fig. 39.7). If the hemocytometer is properly charged with diluted blood, the space between the counting area and the coverslip will be filled and will lack air bubbles, but no fluid will spill over into the depression on either side.

k. Place the hemocytometer on the microscope stage and focus on the large, central square of the counting area with the low-power objective and then the high-power objective. Adjust the light so that the grid lines and blood cells are clearly visible.

l. Count all the cells in the five areas corresponding to those marked with *R* in figure 39.4. To obtain an accurate count, include cells that are touching the lines at the tops and left sides of the squares, but do not count those touching the bottoms and right sides of the squares. The use of a hand counter facilitates the counting task.

m. Multiply the total count by 10,000, and record the result (cells per cubic millimeter of blood) in Part A of the laboratory report.

n. Discard or clean the materials as directed by the laboratory instructor.

2. Complete Part D of the laboratory report.

DEMONSTRATION D—TOTAL WHITE BLOOD CELL COUNT

A hemocytometer is used to make a total white blood cell count in much the same way that it was used to count red blood cells. However, in the case of white cell counting, a diluting fluid is used that destroys red blood cells. Also, in making the white cell count, all of the cells in the four large squares marked *W* in figure 39.4 are included. The total count is multiplied by 50 to calculate the total number of white blood cells in a cubic millimeter of the blood sample.

1. To perform a total white blood cell count, follow these steps:
   a. Clean the hemocytometer as before.
   b. Lance the tip of a finger, as before, to obtain a drop of blood.
   c. Repeat steps 1c–1g of the procedure for counting red blood cells, but use a Unopette system for counting white blood cells.
   d. After mixing the blood with the diluting fluid in the reservoir, discard the first four drops of the mixture and charge the hemocytometer with diluted blood as before.
   e. Use low-power magnification to locate the areas of the grid corresponding to those marked with *W* in figure 39.4.
   f. Count all of the cells in the four large squares (remember that the red blood cells were destroyed by the white blood cell diluting fluid), following the same counting rules as before.
   g. Multiply the total by 50, and record the result in Part A of the laboratory report.
   h. Discard or clean the materials as directed by the laboratory instructor.

2. Complete Part E of the laboratory report.

ALTERNATIVE PROCEDURE

Laboratories in modern hospitals and clinics use updated hematology analyzers (fig. 39.8) for evaluations of the blood factors that are described in laboratory exercise 39. The more traditional procedures performed in these laboratory activities will help you to better understand each blood characteristic.

Figure 39.8
Modern hematology analyzer being used in the laboratory of a clinic.

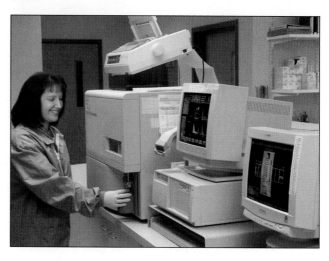

Laboratory Report **39**

Name _____

Date _____

Section _____

BLOOD TESTING— A DEMONSTRATION

PART A

Blood test data:

Blood Test	Test Results	Normal Values
Hematocrit (mL per 100 mL blood)		Men: 40–54 Women: 37–47
Hemoglobin content (g per 100 mL blood)		Men: 14–18 Women: 12–16
Red blood cell count (cells per mm³ blood)		Men: 4,600,000–6,200,000 Women: 4,200,000–5,400,000
White blood cell count (cells per mm³ blood)		5,000–10,000

PART B

Complete the following:

1. How does the hematocrit from the demonstration blood test compare with the normal value?

   _____

   _____

2. What conditions might produce a decreased red blood cell percentage?_____

   _____

   _____

3. What conditions might produce an increased red blood cell percentage? _____

   _____

   _____

PART C

Complete the following:

1. How does the hemoglobin content from the demonstration blood test compare with the normal value?_____

   _____

   _____

2. What conditions might produce a decreased hemoglobin content? _____

   _____

   _____

   _____

3. What conditions might produce an increased hemoglobin content? _____

_____

_____

_____

PART D

Complete the following:

1. How does the red blood cell count from the demonstration blood test compare with the normal value? _____

_____

_____

2. What conditions might produce a decreased red blood cell count? _____

_____

_____

_____

3. What conditions might produce an increased red blood cell count? _____

_____

_____

_____

PART E

Complete the following:

1. How does the white blood cell count from the demonstration blood test compare with the normal value? _____

_____

_____

2. What conditions might produce a decreased white blood cell count? _____

_____

_____

_____

3. What conditions might produce an increased white blood cell count? _____

_____

_____

_____

Critical Thinking Application

Which blood tests performed in this lab could be used to determine possible anemia?

_____

_____

_____

BLOOD TYPING

SAFETY

- It is important that students learn and practice correct procedures for handling body fluids. Consider using contaminant-free blood that has been tested and is available from various laboratory supply houses. Some of the procedures might be accomplished as demonstrations only. If student blood is used, it is important that students handle only their own blood.
- Use an appropriate disinfectant to wash the laboratory tables before and after the procedures.
- Wear disposable gloves when handling blood samples.
- Clean the end of a finger with 70% alcohol before the puncture is performed.
- The sterile blood lancet should be used only once.
- Dispose of used lancets and blood-contaminated items into an appropriate container (never use the wastebasket).
- Wash your hands before leaving the laboratory.

Blood typing involves identifying protein substances called *antigens* that are present in red blood cell membranes. Although there are many different antigens associated with human red blood cells, only a few of them are of clinical importance. These include the antigens of the ABO group and those of the Rh group.

To determine which antigens are present, a blood sample is mixed with blood-typing sera that contain known types of antibodies. If a particular antibody contacts a corresponding antigen, a reaction occurs, and the red blood cells clump together (agglutination).

Thus, if blood cells are mixed with serum containing antibodies that react with antigen A and the cells clump together, antigen A must be present in those cells.

PURPOSE OF THE EXERCISE

To determine the ABO blood type of a blood sample and to observe an Rh blood-typing test.

LEARNING OBJECTIVES

After completing this exercise, you should be able to

1. Explain the basis of ABO blood typing.
2. Determine the ABO type of a blood sample.
3. Explain the basis of Rh blood typing.
4. Describe how the Rh type of a blood sample is determined.

WARNING

Because of the possibility of blood infections being transmitted from one student to another if blood testing is performed in the classroom, it is suggested that commercially prepared blood-typing kits containing virus-free human blood be used for ABO blood typing. The instructor may wish to demonstrate Rh blood typing. Observe all of the safety procedures listed for this lab.

ABO BLOOD TYPING

1. Review the section entitled "ABO Blood Group" in chapter 14 of the textbook.
2. Compete Part A of Laboratory Report 40.
3. Perform the ABO blood type test using the blood-typing kit. To do this, follow these steps:
   a. Obtain a clean microscope slide and mark across its center with a wax pencil to divide it into right and left halves. Also write "Anti-A" near the edge of the left half and "Anti-B" near the edge of the right half (fig. 40.1).
   b. Place a small drop of blood on each half of the microscope slide. Work quickly so that the blood will not have time to clot.

Figure 40.1 Slide prepared for ABO blood typing.

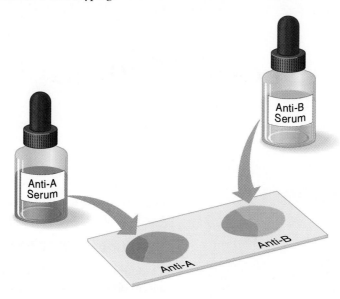

c. Add a drop of anti-A serum to the blood on the left half and a drop of anti-B serum to the blood on the right half. Note the color coding of the anti-A and anti-B typing sera. To avoid contaminating the serum, avoid touching the blood with the serum while it is in the dropper; instead allow the serum to fall from the dropper onto the blood.

d. Use separate toothpicks to stir each sample of serum and blood together, and spread each over an area about as large as a quarter. Dispose of toothpicks in an appropriate container.

e. Examine the samples for clumping of blood cells (agglutination) after 2 minutes.

f. See table 40.1 and figure 40.2 for aid in interpreting the test results.

g. Discard contaminated materials as directed by the laboratory instructor.

4. Complete Part B of the laboratory report.

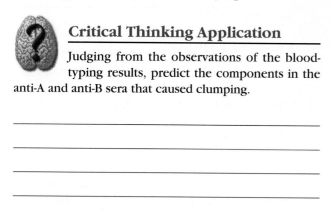

Critical Thinking Application

Judging from the observations of the blood-typing results, predict the components in the anti-A and anti-B sera that caused clumping.

_____

_____

_____

_____

DEMONSTRATION—RH BLOOD TYPING

1. Review the section entitled "Rh Blood Group" in chapter 14 of the textbook.

Table 40.1 Possible Reactions of ABO Blood-Typing Sera

Reactions		Blood Type
Anti-A Serum	**Anti-B Serum**	
Clumping	No clumping	Type A
No clumping	Clumping	Type B
Clumping	Clumping	Type AB
No clumping	No clumping	Type O

Note: The inheritance of the ABO blood groups is described in chapter 24 of the textbook.

2. Complete Part C of the laboratory report.

3. To determine the Rh blood type of a blood sample, follow these steps:

a. Lance the tip of a finger. (See the demonstration procedures in Laboratory Exercise 38 for directions.) Place a small drop of blood in the center of a clean microscope slide.

b. Add a drop of anti-D serum to the blood and mix them together with a clean toothpick.

c. Place the slide on the plate of a warming box (Rh blood-typing box or Rh view box) that has been prewarmed to 45°C (113°F) (fig. 40.3).

d. Slowly rock the box back and forth to keep the mixture moving, and watch for clumping of the blood cells. When clumping occurs in anti-D serum, the clumps usually are smaller than those that appear in anti-A or anti-B sera, so they may be less obvious. However, if clumping occurs, the blood is called Rh positive; if no clumping occurs *within 2 minutes,* the blood is called Rh negative.

e. Discard all contaminated materials in appropriate containers.

4. Complete Part D of the laboratory report.

326

Figure 40.2 Four possible results of the ABO test.

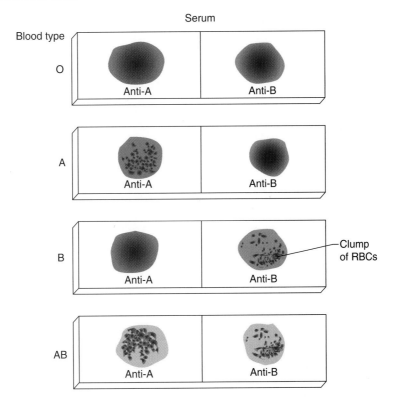

Figure 40.3 Slide warming box used for Rh blood typing.

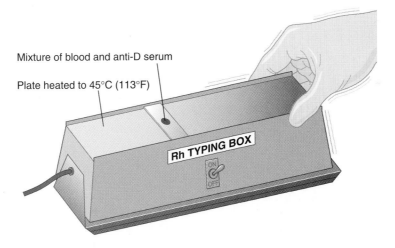

Name _____

Date _____

Section _____

BLOOD TYPING

PART A

Complete the following statements:

1. The antigens of the ABO blood group are located in the _____ .

2. The blood of every person contains one of (how many possible?) _____ combinations of antigens.

3. Type A blood contains antigen _____ .

4. Type B blood contains antigen _____ .

5. Type A blood contains antibody _____ in the plasma.

6. Type B blood contains antibody _____ in the plasma.

7. Persons with ABO blood type _____ are often called universal recipients.

8. Persons with ABO blood type _____ are often called universal donors.

9. Although antigens are present at the time of birth, antibodies usually do not appear in the blood until the age of _____ .

PART B

Complete the following:

1. What was the ABO type of the blood tested? _____

2. What ABO antigens are present in the red blood cells of this type of blood? _____

3. What ABO antibodies are present in the plasma of this type of blood? _____

4. If a person with this blood type needed a blood transfusion, what ABO type(s) of blood could be received safely? _____

5. If a person with this blood type was serving as a blood donor, what ABO blood type(s) could receive the blood safely? _____

PART C

Complete the following statements:

1. The Rh blood group was named after the _____ .
2. Of the antigens in the Rh group, the most important is _____ .
3. If red blood cells lack Rh antigens, the blood is called _____ .
4. Rh antibodies form only in persons with _____ type blood in response to special stimulation.
5. If an Rh-negative person who is sensitive to Rh-positive blood receives a transfusion of Rh-positive blood, the donor's cells are likely to _____ .
6. An Rh-negative woman, who might be carrying an _____ fetus, is given an injection of RhoGAM to prevent erythroblastosis fetalis.

PART D

Complete the following:

1. What was the Rh type of the blood tested in the demonstration? _____
2. What Rh antigen is present in the red blood cells of this type of blood? _____
3. What Rh antibody is normally present in the plasma of this type of blood? _____
4. If a person with this blood type needed a blood transfusion, what type of blood could be received safely?

   _____
5. If a person with this blood type was serving as a blood donor, a person with what type of blood could receive the blood safely? _____

STRUCTURE OF THE HEART

MATERIALS NEEDED

Textbook
Dissectible human heart model
Preserved sheep or other mammalian heart
Dissecting tray
Dissecting instruments

For Optional Activity:

Colored pencils

SAFETY

- Wear disposable gloves when working on the heart dissection.
- Save or dispose of the dissected heart as instructed.
- Wash the dissecting tray and instruments as instructed.
- Wash your laboratory table.
- Wash your hands before leaving the laboratory.

The heart is a muscular pump located within the mediastinum and resting upon the diaphragm. It is enclosed by the lungs, thoracic vertebrae, and sternum, and attached at its top (the base) are several large blood vessels. Its distal end extends downward to the left and terminates as a bluntly pointed apex.

The heart and the proximal ends of the attached blood vessels are enclosed by a double-layered pericardium. The inner layer of this membrane (visceral pericardium) consists of a thin covering that is closely applied to the surface of the heart, whereas the outer layer (parietal pericardium with fibrous pericardium) forms a tough, protective sac surrounding the heart.

PURPOSE OF THE EXERCISE

To review the structural characteristics of the human heart and to examine the major features of a mammalian heart.

LEARNING OBJECTIVES

After completing this exercise, you should be able to

1. Identify the major structural features of the human heart.

2. Compare the features of the human heart with those of another mammal.

PROCEDURE A—THE HUMAN HEART

1. Review the section entitled "Structure of the Heart" in chapter 15 of the textbook.
2. As a review activity, label figures 41.1, 41.2, and 41.3.
3. Complete Part A of Laboratory Report 41.
4. Examine the human heart model, and locate the following features:

heart

  base

  apex

pericardium (pericardial sac)

  fibrous pericardium (outer layer)

  parietal pericardium (inner lining of fibrous pericardium)

  visceral pericardium (epicardium)

pericardial cavity (between parietal and visceral pericardial membranes)

myocardium

endocardium

atria

  right atrium

  left atrium

  auricles

ventricles

  right ventricle

  left ventricle

atrioventricular orifices

atrioventricular valves (A-V valves)

  tricuspid valve (right atrioventricular valve)

  mitral valve (bicuspid valve; left atrioventricular valve)

chordae tendineae

331

Figure 41.1 Label this anterior view of the human heart.

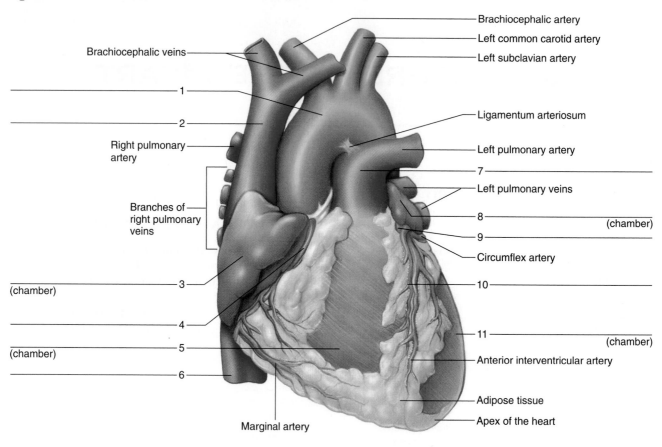

Brachiocephalic veins

Right pulmonary artery

Branches of right pulmonary veins

(chamber)

(chamber)

Marginal artery

Brachiocephalic artery
Left common carotid artery
Left subclavian artery

Ligamentum arteriosum

Left pulmonary artery

Left pulmonary veins

(chamber)

Circumflex artery

(chamber)

Anterior interventricular artery

Adipose tissue
Apex of the heart

1
2
3
4
5
6
7
8
9
10
11

Figure 41.2 Label this posterior view of the human heart.

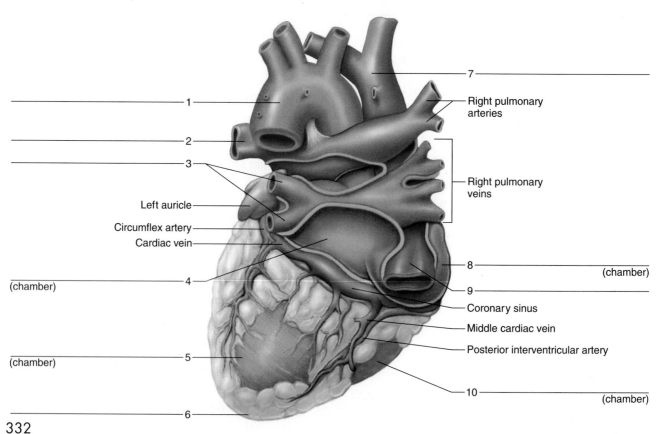

Left auricle
Circumflex artery
Cardiac vein

(chamber)

(chamber)

Right pulmonary arteries

Right pulmonary veins

(chamber)

Coronary sinus
Middle cardiac vein
Posterior interventricular artery

(chamber)

1
2
3
4
5
6
7
8
9
10

Figure 41.3 Label this anterior view of a coronal section of the human heart.

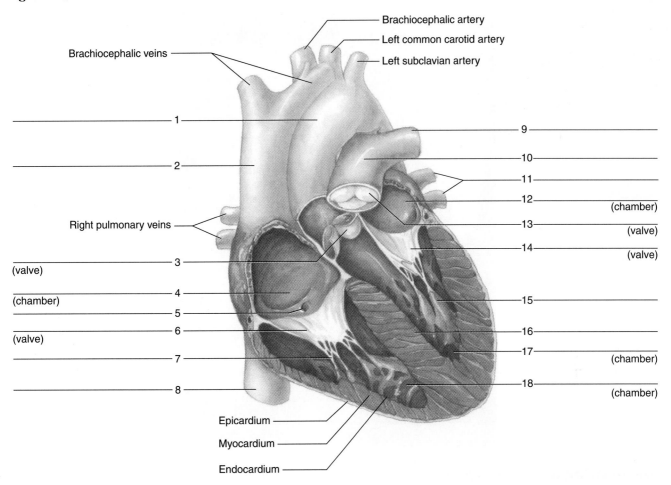

Brachiocephalic artery
Left common carotid artery
Left subclavian artery

Brachiocephalic veins

Right pulmonary veins

1
2

9
10
11
12
(chamber)
13
(valve)
14
(valve)

3
(valve)

4
(chamber)

5
6
(valve)

7

8

15

16
17
(chamber)

18
(chamber)

Epicardium
Myocardium
Endocardium

papillary muscles

atrioventricular sulcus

interventricular sulci

   anterior sulcus

   posterior sulcus

superior vena cava

inferior vena cava

pulmonary trunk

pulmonary arteries

pulmonary veins

aorta

semilunar valves

   pulmonary valve

   aortic valve

left coronary artery

   circumflex artery

   anterior interventricular artery

right coronary artery

   posterior interventricular artery

   marginal artery

cardiac veins

   great cardiac vein

   middle cardiac vein

   small cardiac vein

coronary sinus

5. Complete Part B of the laboratory report.

OPTIONAL ACTIVITY

Use red and blue colored pencils to color the blood vessels in figure 41.3. Use red to illustrate a blood vessel high in oxygen, and use blue to illustrate a blood vessel low in oxygen. You can check your work by referring to the corresponding figures in the textbook, which are presented in full color.

Figure 41.4 Ventral surface of sheep heart.

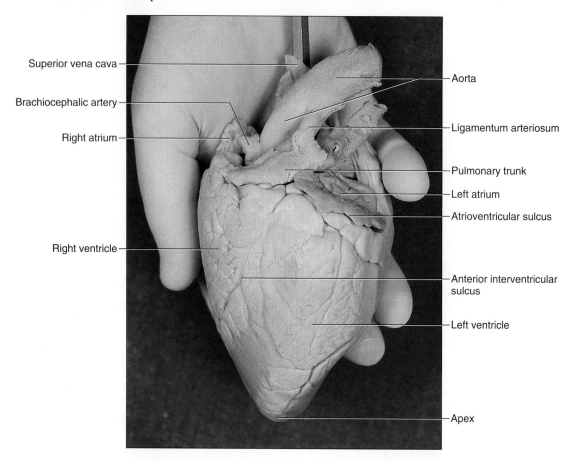

Superior vena cava

Brachiocephalic artery

Right atrium

Right ventricle

Aorta

Ligamentum arteriosum

Pulmonary trunk

Left atrium

Atrioventricular sulcus

Anterior interventricular sulcus

Left ventricle

Apex

PROCEDURE B—DISSECTION OF A SHEEP HEART

1. Obtain a preserved sheep heart. Rinse it in water thoroughly to remove as much of the preservative as possible. Also run water into the large blood vessels to force any blood clots out of the heart chambers.
2. Place the heart in a dissecting tray with its ventral surface up (fig. 41.4), and proceed as follows:
   a. Although the relatively thick *pericardial sac* probably is missing, look for traces of this membrane around the origins of the large blood vessels.
   b. Locate the *visceral pericardium,* which appears as a thin, transparent layer on the surface of the heart. Use a scalpel to remove a portion of this layer and expose the *myocardium* beneath. Also note the abundance of fat along the paths of various blood vessels. This adipose tissue occurs in the loose connective tissue that underlies the visceral pericardium.
   c. Identify the following:

      right atrium

      right ventricle

      left atrium

      left ventricle

      atrioventricular sulcus

      anterior interventricular sulcus

   d. Carefully remove the fat from the anterior interventricular sulcus, and expose the blood vessels that pass along this groove. They include a branch of the *left coronary artery* (anterior interventricular artery) and a *cardiac vein.*
3. Examine the dorsal surface of the heart (fig. 41.5), and proceed as follows:
   a. Identify the *atrioventricular sulcus* and the *posterior interventricular sulcus.*
   b. Locate the stumps of two relatively thin-walled veins that enter the right atrium. Demonstrate this connection by passing a slender probe through them. The upper vessel is the *superior vena cava,* and the lower one is the *inferior vena cava.*
4. Open the right atrium. To do this, follow these steps:
   a. Insert a blade of the scissors into the superior vena cava and cut downward through the atrial wall (fig. 41.5).
   b. Open the chamber, locate the *tricuspid valve (right atrioventricular valve),* and examine its cusps.

Figure 41.5 Dorsal surface of sheep heart. To open the right atrium, insert a blade of the scissors into the superior (anterior in sheep) vena cava and cut downward.

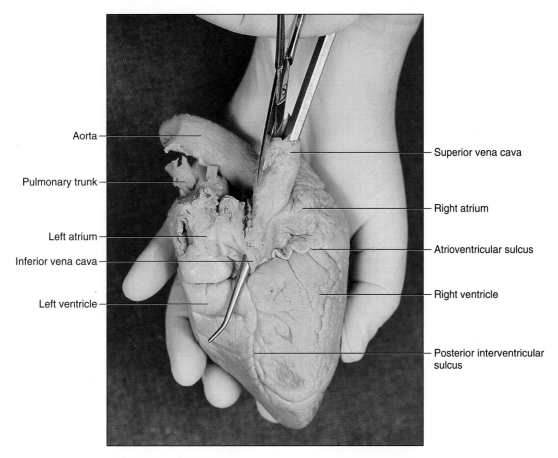

Aorta —
Pulmonary trunk —
Left atrium —
Inferior vena cava —
Left ventricle —

— Superior vena cava
— Right atrium
— Atrioventricular sulcus
— Right ventricle
— Posterior interventricular sulcus

c. Also locate the opening to the *coronary sinus* between the valve and the inferior vena cava.

d. Run some water through the tricuspid valve to fill the chamber of the right ventricle.

e. Gently squeeze the ventricles, and watch the cusps of the valve as the water moves up against them.

5. Open the right ventricle as follows:

a. Continue cutting downward through the tricuspid valve and the right ventricular wall until you reach the apex of the heart.

b. Locate the *chordae tendineae* and the *papillary muscles.*

c. Find the opening to the *pulmonary trunk,* and use the scissors to cut upward through the wall of the right ventricle. Follow the pulmonary trunk until you have exposed the *pulmonary valve.*

d. Examine the valve and its cusps.

6. Open the left side of the heart. To do this, follow these steps:

a. Insert the blade of the scissors through the wall of the left atrium and cut downward to the apex of the heart.

b. Open the left atrium, and locate the four openings of the *pulmonary veins.* Pass a slender probe through each opening, and locate the stump of its vessel.

c. Examine the *bicuspid valve (left atrioventricular valve)* and its cusps.

d. Also examine the left ventricle, and compare the thickness of its wall with that of the right ventricle.

7. Locate the aorta, which leads away from the left ventricle, and proceed as follows:

a. Compare the thickness of the aortic wall with that of a pulmonary artery.

b. Use scissors to cut along the length of the aorta to expose the *aortic valve* at its base.

c. Examine the cusps of the valve, and locate the openings of the *coronary arteries* just distal to them.

8. As a review, locate and identify the stumps of each of the major blood vessels associated with the heart.

9. Discard or save the specimen as directed by the laboratory instructor.

10. Complete Part C of the laboratory report.

Web Quest

Trace blood flow through an animated heart at various rates. Identify heart structures and take an animated tour of the heart. Search these at www.mhhe.com/shier10

Click on Student Edition. Click on Martin Lab Manual, Web Quest.

STRUCTURE OF THE HEART

PART A

Match the terms in column A with the descriptions in column B. Place the letter of your choice in the space provided.

Column A	Column B
a. Aorta	_____ 1. Upper chamber of the heart
b. Atrioventricular sulcus	_____ 2. Structure from which chordae tendineae originate
c. Atrium	_____ 3. Prevents blood movement from right ventricle to right atrium
d. Cardiac vein	
e. Coronary artery	_____ 4. Membranes around heart
f. Coronary sinus	_____ 5. Groove separating left and right ventricles
g. Endocardium	_____ 6. Prevents blood movement from left ventricle to left atrium
h. Interventricular sulcus	_____ 7. Gives rise to left and right pulmonary arteries
i. Mitral (bicuspid) valve	
j. Myocardium	_____ 8. Drains blood from myocardium into right atrium
k. Papillary muscle	_____ 9. Inner lining of heart chamber
l. Pericardial cavity	_____ 10. Layer largely composed of cardiac muscle tissue
m. Pericardial sac	
n. Pulmonary trunk	_____ 11. Space containing serous fluid to reduce friction during heartbeats
o. Tricuspid valve	_____ 12. Drains blood from myocardial capillaries
	_____ 13. Supplies blood to heart muscle
	_____ 14. Distributes blood to body organs (systemic circuit) except lungs
	_____ 15. Groove separating atrial and ventricular portions of heart

Figure 41.6 Identify the features indicated on this anterior view of a coronal section of a human heart model, using the terms provided. (*Note:* The pulmonary valve is not shown on the portion of the model photographed.)

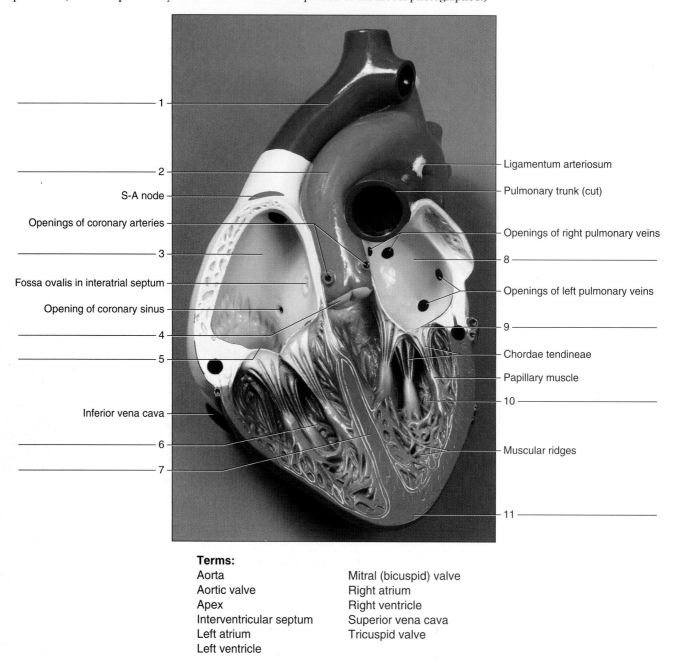

Terms:

Aorta	Mitral (bicuspid) valve
Aortic valve	Right atrium
Apex	Right ventricle
Interventricular septum	Superior vena cava
Left atrium	Tricuspid valve
Left ventricle	

PART B

Locate the labeled features and identify the numbered features in figure 41.6 of a dissectible human heart model (coronal section).

PART C

Complete the following:

1. Compare the structure of the tricuspid valve with that of the pulmonary valve._____

_____

_____

_____

_____

2. Describe the action of the tricuspid valve when you squeezed the water-filled right ventricle. _____

_____

_____

_____

_____

3. Describe the function of the chordae tendineae and the papillary muscles. _____

_____

_____

_____

_____

4. What is the significance of the difference in thickness between the wall of the aorta and the wall of the

   pulmonary trunk? _____

_____

_____

_____

_____

5. List in order the major blood vessels, chambers, and valves through which blood must pass in traveling from a

   vena cava to the aorta. _____

_____

_____

_____

_____

_____

Critical Thinking Application

What is the significance of the difference in thickness of the ventricular walls?

_____

_____

_____

_____

THE CARDIAC CYCLE

A set of atrial contractions while the ventricular walls relax, followed by ventricular contractions while the atrial walls relax, constitutes a cardiac cycle. Such a cycle is accompanied by blood pressure changes within the heart chambers, opening and closing of heart valves, and movement of blood in and out of the chambers. These valves closing produce vibrations in the tissues and thus create the sounds associated with the heartbeat.

A number of electrical changes also occur in the myocardium as it contracts and relaxes. These changes can be detected by using metal electrodes and an instrument called an *electrocardiograph.* The recording produced by the instrument is an *electrocardiogram,* or *ECG (EKG).*

PURPOSE OF THE EXERCISE

To review the events of a cardiac cycle, to become acquainted with normal heart sounds, and to record an electrocardiogram.

LEARNING OBJECTIVES

After completing this exercise, you should be able to

1. Describe the major events of a cardiac cycle.
2. Identify the sounds produced during a cardiac cycle.
3. Record an electrocardiogram.
4. Identify the components of a normal ECG pattern.
5. Describe the phases of a cardiac cycle represented by each part of a normal ECG pattern.

PROCEDURE A—HEART SOUNDS

1. Review the sections entitled "Cardiac Cycle" and "Heart Sounds" in chapter 15 of the textbook.
2. Complete Part A of Laboratory Report 42.
3. Listen to your own heart sounds. To do this, follow these steps:
    a. Obtain a stethoscope, and clean its earpieces and the diaphragm by using cotton moistened with alcohol.
    b. Fit the earpieces into your ear canals so that the angles are positioned in the forward direction.
    c. Firmly place the diaphragm (or bell) of the stethoscope on the chest over the apex of the heart (fig. 42.1) and listen to the sounds. This is a good location to hear the first sound (*lubb*) of a cardiac cycle when the A-V valves close.
    d. Move the diaphragm to the second intercostal space, just to the left of the sternum, and listen to the sounds from this region. You should be able to hear the second sound (*dupp*) of the cardiac cycle clearly when the semilunar valves close.
4. Inhale slowly and deeply, and exhale slowly while you listen to the heart sounds from each of the locations as before. Note any changes that have occurred in the sounds.
5. Exercise vigorously outside the laboratory for a few minutes so that other students listening to heart sounds will not be disturbed. After the exercise period, listen to the heart sounds and note any changes that have occurred in them. *This exercise should be avoided by anyone with health risks.*
6. Complete Part B of the laboratory report.

PROCEDURE B—THE ELECTROCARDIOGRAM

1. Review sections entitled "Cardiac Conduction System," "Electrocardiogram," and "Arrhythmias—Clinical Application" in chapter 15 of the textbook.
2. Complete Part C of the laboratory report.

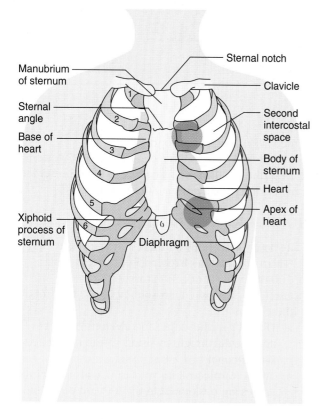

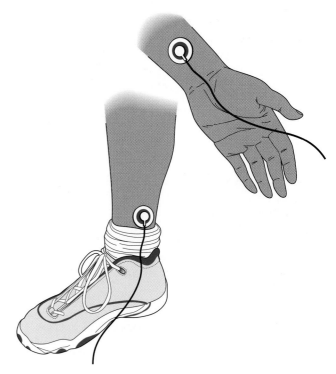

3. The laboratory instructor will demonstrate the proper adjustment and use of the instrument available to record an electrocardiogram.

4. Record your laboratory partner's ECG. To do this, follow these steps:

   a. Have your partner lie on a cot or table close to the electrocardiograph, remaining as relaxed and still as possible.

   b. Scrub the electrode placement locations with cotton moistened with alcohol (fig. 42.2). Apply a small quantity of electrode cream to the skin on the insides of the wrists and ankles. (Any jewelry on the wrists or ankles should be removed.)

   c. Spread some electrode cream over the inner surfaces of four plate electrodes and attach one to each of the prepared skin areas (fig. 42.2). Make sure there is good contact between the skin and the metal of the electrodes. The electrode plate on the right ankle is the grounding system.

   d. Attach the plate electrodes to the corresponding cables of a lead selector switch. When an ECG recording is made, only two electrodes are used at a time, and the selector switch allows various combinations of electrodes (leads) to be activated. Three standard limb leads placed on the two wrists and the left ankle are used for an ECG. This arrangement has become known as *Einthoven's triangle,*[1] which enables the recording of the potential difference between any two of the electrodes.

   The standard Leads I, II, and III are called bipolar leads because they are the potential difference between two electrodes (a positive and a negative). Lead I measures the potential difference between the right wrist (negative) and the left wrist (positive). Lead II measures the potential difference between the right wrist and the left ankle, and Lead III measures the potential difference between the left wrist and the left ankle. The right ankle is always the ground.

   e. Turn on the recording instrument and adjust it as previously demonstrated by the laboratory instructor. The paper speed should be set at 2.5 cm/second. This is the standard speed for ECG recordings.

   f. Set the lead selector switch to Lead I (right wrist, left wrist electrodes), and record the ECG for 1 minute.

   g. Set the lead selector switch to Lead II (right wrist, left ankle electrodes), and record the ECG for 1 minute.

   h. Set the lead selector switch to Lead III (left wrist, left ankle electrodes), and record the ECG for 1 minute.

---

[1]Willem Einthoven (1860–1927), a Dutch physiologist, received the Nobel prize for physiology or medicine for his work with electrocardiograms.

i. Remove the electrodes and clean the cream from the metal and skin.

j. Use figure 42.3 to label the ECG components of the results from Leads I, II, and III. The P-Q interval is often called the P-R interval because the Q wave is often small or absent. The normal P-Q interval is 0.12–0.20 seconds. The normal QRS complex duration is less than 0.10 seconds.

5. Complete Part D of the laboratory report.

Web Quest

Identify normal heart sounds and a murmur. What is the purpose of stress electrocardiography? Compare normal and abnormal ECGs.

Search these at

www.mhhe.com/shier10

Click on Student Edition. Click on Martin Lab Manual, Web Quest.

Figure 42.3 Components of a normal ECG pattern with a time scale.

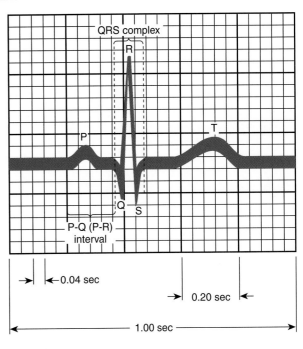

Laboratory Report 42

Name _____

Date _____

Section _____

THE CARDIAC CYCLE

PART A

Complete the following statements:

1. About _____ % of the blood from the atria passes into the ventricles before the atrial walls contract.
2. The period during which a heart chamber is contracting is called _____ .
3. The period during which a heart chamber is relaxing is called _____ .
4. During ventricular contraction, the A-V valves (tricuspid and mitral valves) remain _____ .
5. During ventricular relaxation, the A-V valves remain _____ .
6. The pulmonary and aortic valves open when the pressure in the _____ exceeds the pressure in the pulmonary trunk and aorta.
7. Heart sounds are due to _____ in heart tissues created by changes in blood flow.
8. The first sound of a cardiac cycle occurs when the _____ are closing.
9. The second sound of a cardiac cycle occurs when the _____ are closing.
10. The sound created when blood leaks back through an incompletely closed valve is called a _____ .

PART B

Complete the following:

1. What changes did you note in the heart sounds when you inhaled deeply?

    _____

    _____

2. What changes did you note in the heart sounds following the exercise period?

    _____

    _____

PART C

Complete the following statements:

1. The fibers of the cardiac conduction system are specialized _____ tissue.
2. Normally, the _____ node serves as the pacemaker of the heart.
3. The _____ node is located in the inferior portion of the interatrial septum.
4. The large fibers on the distal side of the A-V node make up the _____ .
5. The fibers that carry cardiac impulses from the interventricular septum into the myocardium are called _____ .
6. An _____ is a recording of electrical changes occurring in the myocardium during the cardiac cycle.

7. Between cardiac cycles, cardiac muscle fibers remain _____ with no detectable electrical changes.

8. The P wave corresponds to depolarization of the muscle fibers of the _____ .

9. The QRS complex corresponds to depolarization of the muscle fibers of the _____ .

10. The T wave corresponds to repolarization of the muscle fibers of the _____ .

11. Why is atrial repolarization not observed in the ECG? _____
_____

12. A rapid heartbeat is called _____ ; a slow heartbeat is called _____ .

PART D

1. Attach a short segment of the ECG recording from each of the three leads you used, and label the waves of each.

   Lead I

   Lead II

   Lead III

2. What differences do you find in the ECG patterns of these leads?_____
_____
_____
_____
_____
_____

3. How much time passed from the beginning of the P wave to the beginning of the QRS complex (P-Q interval, or P-R interval) in the ECG from Lead I? _____

4. What is the significance of this P-Q (P-R) interval? _____

_____

_____

_____

_____

5. How can you determine the heart rate from an electrocardiogram? _____

_____

_____

_____

6. What was your heart rate as determined from the ECG? _____

Critical Thinking Application

If a person's heart rate is 72 beats per minute, determine the number of QRS complexes that would have appeared on an ECG during the first 30 seconds.

_____

_____

_____

FACTORS AFFECTING THE CARDIAC CYCLE

MATERIALS NEEDED

Textbook
Physiological recording apparatus such as a kymograph
    or Physiograph
Live frog
Dissecting tray
Dissecting instruments
Dissecting pins
Frog Ringer's solution in plastic squeeze bottle
Thread
Small hook
Medicine dropper
Thermometer
Ice
Hot plate
Calcium chloride, 2% solution
Potassium chloride, 5% solution

For Optional Activity:

Epinephrine, 1:10,000 solution
Acetylcholine, 1:10,000 solution
Caffeine, 0.2% solution

SAFETY

- Wear disposable gloves when handling the frogs.
- Dispose of the frogs according to your laboratory
  instructor.
- Wash your hands before leaving the laboratory.

Although the cardiac cycle is controlled by the S-A node serving as the pacemaker, the rate of heart action can be altered by various other factors. These factors include parasympathetic and sympathetic nerve impulses that originate in the cardiac center of the medulla oblongata, changes in body temperature, and concentrations of certain ions.

PURPOSE OF THE EXERCISE

To review the mechanism by which the heartbeat is regulated, to observe the action of a frog heart, and to investigate the effects of various factors on the frog heartbeat.

LEARNING OBJECTIVES

After completing this exercise, you should be able to

1. Describe the mechanism by which the human cardiac cycle is controlled.
2. List several factors that affect the rate of the heartbeat.
3. Identify the atrial and ventricular contractions and determine the heart rate from a recording of a frog heartbeat.
4. Test the effects of various factors on the action of a frog heart.

PROCEDURE

1. Review the section entitled "Regulation of the Cardiac Cycle" in chapter 15 of the textbook.
2. Complete Part A of Laboratory Report 43.

GENERAL SUGGESTION

Try to become familiar with the content and organization of this lab before you pith a frog. If you work quickly, one pithed frog should last for all of the experimental steps.

3. Observe the normal action of a frog heart. To do this, follow these steps:
   a. Obtain a live frog, and pith it according to the directions in Procedure C of Laboratory Exercise 19.
   b. Place the frog in a dissecting tray with its ventral surface up, and pin its jaw and legs to the tray with dissecting pins.
   c. Use scissors to make a midline incision through the skin from the pelvis to the jaw.
   d. Cut the skin laterally on each side in the pelvic and pectoral regions, and pin the resulting flaps of skin to the tray (fig. 43.1).
   e. Remove the exposed pectoral muscles and the sternum, being careful not to injure the underlying organs.
   f. Note the beating heart surrounded by the thin-walled pericardium. Use forceps to lift the pericardium upward, and carefully slit it open with scissors, thus exposing the heart.

Figure 43.1 Pin the frog to the dissecting tray and make incisions through the skin as indicated.

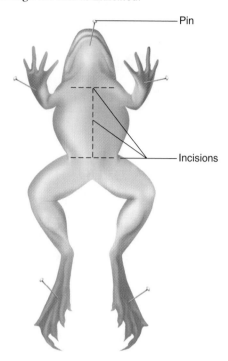

Pin

Incisions

Figure 43.2 Attach a hook and thread to the tip of the ventricle.

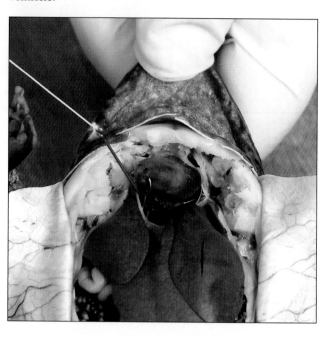

g. Flood the heart with frog Ringer's solution, and keep it moist throughout this exercise.

h. Note that the frog heart has only three chambers—two atria and a ventricle. Watch the heart carefully as it beats, and note the sequence of chamber movements during a cardiac cycle.

4. Tie a piece of thread about 45 cm long to a small metal hook, and insert the hook into the tip (apex) of the ventricle without penetrating the chamber (fig. 43.2). The laboratory instructor will demonstrate how to connect the thread to a physiological recording apparatus so that you can record the frog heart movements. The thread should be adjusted so that there is no slack in it, but at the same time, it should not be so taut that it pulls the heart out of its normal position (fig. 43.3).

5. Record the movements of the frog heart for 2–3 minutes. Identify on the recording the smaller atrial contraction waves and the larger ventricular contraction waves. Also, determine the heart rate (beats per minute) for each minute of recording, and calculate the average rate. Enter the results in Part B of the laboratory report.

6. Test the effect of temperature change on the frog's heart rate. To do this, follow these steps:

a. Remove as much as possible of the Ringer's solution from around the heart, using a medicine dropper.

b. Flood the heart with fresh Ringer's solution that has been cooled in an ice water bath to about 10°C (50°F).

c. Record the heart movements, and determine the heart rate as before.

d. Remove the cool liquid from around the heart, and replace it with room temperature Ringer's solution.

e. After the heart is beating at its normal rate again, flood it with Ringer's solution that has been heated on a hot plate to about 35°C (95°F).

f. Record the heart movements, and determine the heart rate as before.

g. Enter the results in Part B of the laboratory report.

7. Complete Part B of the laboratory report.

8. Test the effect of an increased concentration of calcium ions on the frog heart. If the frog heart from the previous experiment is still beating, replace the fluid around it with room temperature Ringer's solution, and wait until its rate is normal. Otherwise, prepare a fresh specimen, and determine its normal rate as before. To perform the test, follow these steps:

a. Flood the frog heart with 2% calcium chloride. (This solution of calcium chloride will allow ionization to occur, providing Ca^{++}.)

b. Record the heartbeat for about 5 minutes, and note any change in rate.

c. Flood the heart with fresh Ringer's solution until heart rate returns to normal.

9. Test the effect of an increased concentration of potassium ions on the frog heart. To do this, follow these steps:

a. Flood the heart with 5% potassium chloride. (This solution of potassium chloride will allow ionization to occur, providing K^+.)

b. Record the heartbeat for about 5 minutes, and note any change in rate.

10. Complete Part C of the laboratory report.

Figure 43.3   Attach the thread from the heart to the recording apparatus so that there is no slack in the thread.

Name _____

Date _____

Section _____

FACTORS AFFECTING THE CARDIAC CYCLE

PART A

Complete the following statements:

1. The primary function of the heart is to _____ .
2. The _____ normally controls the heart rate.
3. Parasympathetic nerve fibers that supply the heart make up part of the _____ nerve.
4. Endings of parasympathetic nerve fibers secrete _____ , which causes the heart rate to decrease.
5. Sympathetic nerve fibers reach the heart by means of _____ nerves.
6. Endings of sympathetic nerve fibers secrete _____ , which causes the heart rate to increase.
7. The cardiac control center is located in the _____ of the brainstem.
8. Baroreceptors (pressoreceptors) located in the walls of the aorta and carotid arteries are sensitive to changes in _____ .
9. If baroreceptors (pressoreceptors) in the walls of the venae cavae are stimulated by stretching, the cardioaccelerator center sends _____ impulses to the heart.
10. Rising body temperature usually causes the heart rate to _____ .
11. Of the ions that affect heart action, the most important ions are calcium and _____ .

PART B

1. Describe the actions of the frog heart chambers during a cardiac cycle.

   _____

   _____

   _____

   _____

   _____

2. Attach a short segment of the normal frog heart recording in the following space. Label the atrial and ventricular waves of one cardiac cycle.

3. Temperature effect results:

Temperature	Heart Rate
10°C (50°F)	
Room temperature	
35°C (95°F)	

4. Summarize the effect of temperature on the frog's heart action that was demonstrated by this experiment. _____

_____
_____
_____
_____
_____

PART C

Complete the following:

1. Describe the effect of an increased calcium ion (Ca^{++}) concentration on the frog's heart rate._____

_____
_____
_____
_____

2. Describe the effect of an increased potassium ion (K^+) concentration on the frog's heart rate. _____

_____
_____
_____
_____

Critical Thinking Application

In testing the effects of different ions on heart action, why were chlorides used in each case?

_____
_____
_____
_____

BLOOD VESSELS

MATERIALS NEEDED

Textbook
Compound light microscope
Prepared microscope slides:
  artery cross section
  vein cross section
Live frog
Frog Ringer's solution
Paper towel
Rubber bands
Frog board or heavy cardboard (with a 1-inch hole cut
  in one corner)
Dissecting pins
Thread
Masking tape

For Optional Activity:

Ice
Hot plate
Thermometer

SAFETY

- Wear disposable gloves when handling the live frogs.
- Return the frogs to the location indicated after the experiment.
- Wash your hands before leaving the laboratory.

The blood vessels form a closed system of tubes that carry blood to and from the heart, lungs, and body cells. These tubes include arteries and arterioles that conduct blood away from the heart; capillaries in which exchanges of substances occur between the blood and surrounding tissues; and venules and veins that return blood to the heart.

PURPOSE OF THE EXERCISE

To review the structure and functions of blood vessels and to observe examples of blood vessels microscopically.

LEARNING OBJECTIVES

After completing this exercise, you should be able to

1. Describe the structure and functions of arteries, capillaries, and veins.

2. Distinguish cross sections of arteries and veins microscopically.
3. Identify the three major layers in the wall of an artery or vein.
4. Identify the types of blood vessels in the web of a frog's foot.

PROCEDURE

1. Review the section entitled "Blood Vessels" in chapter 15 of the textbook.
2. As a review activity, label figures 44.1 and 44.2.
3. Complete Part A of Laboratory Report 44.
4. Obtain a microscope slide of an artery cross section and examine it using low-power and high-power magnification. Identify the three distinct layers (tunics) of the arterial wall. The inner layer (*tunica interna*) is composed of an endothelium (simple squamous epithelium) and appears as a wavy line due to an abundance of elastic fibers that have recoiled just beneath it. The middle layer (*tunica media*) consists of numerous concentrically arranged smooth muscle cells with elastic fibers scattered among them. The outer layer (*tunica externa*) contains connective tissue that is rich in collagenous fibers (fig. 44.3).
5. Prepare a labeled sketch of the arterial wall in Part B of the laboratory report.
6. Obtain a slide of a vein cross section and examine it as you did the artery cross section. Note the thinner wall and larger lumen relative to an artery of comparable size. Identify the three layers of the wall, and prepare a labeled sketch in Part B of the laboratory report.
7. Complete Part B of the laboratory report.
8. Observe the blood vessels in the webbing of a frog's foot. To do this, follow these steps:
   a. Obtain a live frog. Wrap its body in a moist paper towel, leaving one foot extending outward. Secure the towel with rubber bands, but be careful not to wrap the animal so tightly that it could be injured. Try to keep the nostrils exposed.
   b. Place the frog on a frog board or on a piece of heavy cardboard with the foot near the hole in one corner.

Figure 44.1 Label the tunics of the wall of this artery and vein.

Artery

Vein

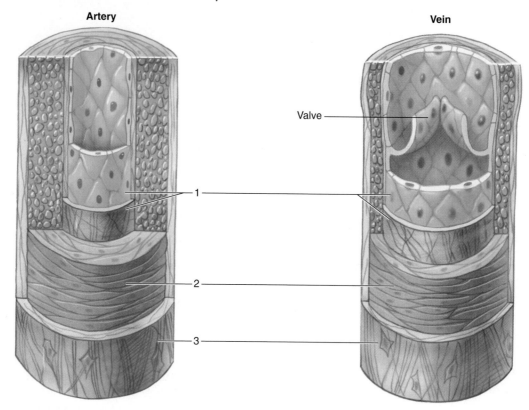

Valve

1

2

3

Figure 44.2 Label this arteriole by placing the correct numbers in the spaces provided.

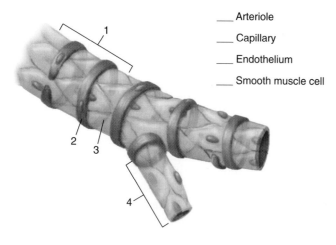

1

2

3

4

___ Arteriole

___ Capillary

___ Endothelium

___ Smooth muscle cell

c. Fasten the wrapped body to the board with masking tape.

d. Carefully spread the web of the foot over the hole and secure it to the board with dissecting pins and thread (fig. 44.4). Keep the web moist with frog Ringer's solution.

e. Secure the board on the stage of a microscope with heavy rubber bands and position it so that the web is beneath the objective lens.

f. Focus on the web, using low-power magnification, and locate some blood vessels. Note the movement of the blood cells and the direction of the blood flow. You might notice that red blood cells of frogs are nucleated. Identify an arteriole, a capillary, and a venule.

g. Examine each of these vessels with high-power magnification.

OPTIONAL ACTIVITY

Investigate the effect of temperature change on the blood vessels of the frog's foot by flooding the web with a small quantity of ice water. Observe the blood vessels with low-power magnification and note any changes in their diameters or the rate of blood flow. Remove the ice water and replace it with water heated to about 35°C (95°F). Repeat your observations. What do you conclude from this experiment?

_____

_____

_____

_____

h. When finished, return the frog to the location indicated by your instructor. The microscope lenses and stage will likely need cleaning after the experiment.

9. Complete Part C of the laboratory report.

Figure 44.3 Cross section of an artery and a vein (5×).

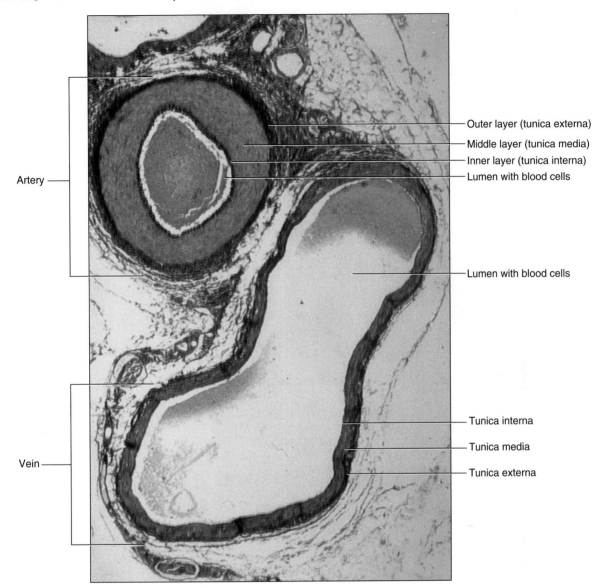

Outer layer (tunica externa)

Middle layer (tunica media)

Inner layer (tunica interna)

Lumen with blood cells

Artery

Lumen with blood cells

Tunica interna

Tunica media

Vein

Tunica externa

Figure 44.4   Spread the web of the foot over the hole and secure it to the board with pins and thread.

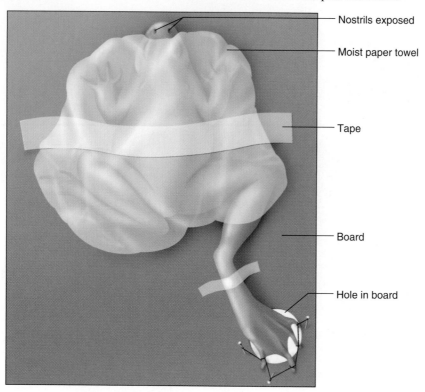

Nostrils exposed

Moist paper towel

Tape

Board

Hole in board

Name _____

Date _____

Section _____

BLOOD VESSELS

PART A

Complete the following statements:

1. Simple squamous epithelial tissue called _____ forms the inner linings of the tunica interna of blood vessels.

2. The _____ of an artery wall contains many smooth muscle cells.

3. The _____ of an artery wall is largely composed of connective tissue.

4. When contraction of the smooth muscle in a blood vessel wall occurs, the vessel is referred to as being in a condition of _____ .

5. Relaxation of the smooth muscle in a blood vessel wall results in the vessel being in a condition of _____ .

6. A vessel (metarteriole) that connects an arteriole directly to a venule, thus bypassing the capillaries, is called a (an) _____ .

7. The smallest blood vessels are called _____ .

8. The openings or intercellular channels of capillary walls within muscle tissues are relatively _____ .

9. The protective tight arrangement between the capillaries and tissues of the brain is called the _____ .

10. Tissues with high rates of metabolism tend to have _____ densities of capillaries.

11. _____ are composed of smooth muscles that encircle the entrances to capillaries and thus can control the distribution of blood within tissues.

12. The process called _____ provides the most important means of transfer of biochemicals through capillary walls.

13. Oxygen and carbon dioxide move easily through most areas of cell membranes because they are soluble in _____ .

14. Water moves through capillary walls by passing through _____ in the cell membranes and through slitlike openings between the endothelial cells.

15. Filtration results when substances are forced through capillary walls by _____ pressure.

16. The presence of plasma proteins in blood increases its _____ pressure as compared to tissue fluids.

17. Excess tissue fluid is returned to the venous circulation by means of _____ vessels.

18. The substance called _____ causes an increase in capillary membrane permeability.

19. _____ in certain veins close if blood begins to back up in the vein.

20. _____ are vessels with large lumens that serve as blood reservoirs.

PART B

1. Sketch and label a section of an arterial wall.

2. Sketch and label a section of a venous wall.

3. Describe the differences you noted in the structures of the arterial and venous walls. Mention each of the three layers of the wall. _____

_____

_____

_____

_____

Critical Thinking Application

Explain the functional significance of the differences you noted in the structures of the arterial and venous walls.

_____

_____

_____

_____

PART C

Complete the following:

1. How did you distinguish between arterioles and venules when you observed the vessels in the web of the frog's foot? _____

   _____

   _____

   _____

   _____

   _____

2. How did you recognize capillaries in the web? _____

   _____

   _____

   _____

   _____

3. What differences did you note in the rate of blood flow through the arterioles, capillaries, and venules?_____

   _____

   _____

   _____

   _____

4. Did you observe any evidence of precapillary sphincter activity? Explain your answer. _____

   _____

   _____

   _____

PULSE RATE AND BLOOD PRESSURE

The surge of blood that enters the arteries each time the ventricles of the heart contract causes the elastic walls of these vessels to swell. Then, as the ventricles relax, the walls recoil. This alternate expanding and recoiling of an arterial wall can be felt as a pulse in vessels that run close to the surface of the body.

The force exerted by the blood pressing against the inner walls of arteries also creates blood pressure. This pressure reaches a maximum during ventricular contraction and then drops to its lowest level while the ventricles are relaxed.

PURPOSE OF THE EXERCISE

To examine the pulse, determine the pulse rate, measure blood pressure, and investigate the effects of body position and exercise on pulse rate and blood pressure.

LEARNING OBJECTIVES

After completing this exercise, you should be able to

1. Determine pulse rate.
2. Test the effects of various factors on pulse rate.
3. Measure blood pressure using a sphygmomanometer.
4. Test the effects of various factors on blood pressure.
5. Calculate pulse pressure and mean arterial pressure from blood pressure readings.

PROCEDURE A—PULSE RATE

1. Review the sections entitled "Arterial Blood Pressure" and "Measurement of Arterial Blood Pressure— Clinical Application" in chapter 15 of the textbook.
2. Complete Part A of Laboratory Report 45.
3. Examine your laboratory partner's radial pulse. To do this, follow these steps:
   a. Have your partner sit quietly, remaining as relaxed as possible.
   b. Locate the pulse by placing your index and middle fingers over the radial artery on the anterior surface of the wrist. Do not use your thumb for sensing the pulse because you may feel a pulse coming from an artery in the thumb itself.
   c. Note the characteristics of the pulse. That is, could it be described as regular or irregular, strong or weak, hard or soft?
   d. To determine the pulse rate, count the number of pulses that occur in 1 minute. This can be accomplished by counting pulses in 30 seconds and multiplying that number by 2.
4. Repeat the procedure and determine the pulse rate in each of the following conditions:
   a. immediately after lying down;
   b. 5 minutes after lying down;
   c. immediately after standing;
   d. 5 minutes after standing quietly;
   e. immediately after 3 minutes of moderate exercise *(omit if the person has health problems);*
   f. 5 minutes after exercise has ended.
5. Complete Part B of the laboratory report.

DEMONSTRATION

If the equipment is available, the laboratory instructor will demonstrate how a photoelectric pulse pickup transducer or plethysmogram can be used together with a physiological recording apparatus to record the pulse. Such a recording allows an investigator to analyze certain characteristics of the pulse more precisely than is possible using a finger to examine the pulse. For example, the pulse rate can be determined very accurately from a recording, and the heights of the pulse waves provide information about the blood pressure.

PROCEDURE B—BLOOD PRESSURE

1. Measure your laboratory partner's arterial blood pressure. To do this, follow these steps:
    a. Obtain a sphygmomanometer and a stethoscope.
    b. Clean the earpieces and the diaphragm of the stethoscope with cotton moistened with 70% alcohol.
    c. Have your partner sit quietly with bare upper limb resting on a table at heart level. Have the person remain as relaxed as possible.
    d. Locate the brachial artery at the antecubital space. Wrap the cuff of the sphygmomanometer around the arm so that its lower border is about 2.5 cm above the bend of the elbow. Center the bladder of the cuff in line with the *brachial pulse* (fig. 45.1).
    e. Palpate the *radial pulse.* Close the valve on the neck of the rubber bulb connected to the cuff, and pump air from the bulb into the cuff. Inflate the cuff while watching the sphygmomanometer, and note the pressure when the pulse disappears. (This is a rough estimate of the systolic pressure.) Immediately deflate the cuff.
    f. Position the stethoscope over the brachial artery. Reinflate the cuff to a level 30 mm Hg higher than the point where the pulse disappeared during palpation.
    g. Slowly open the valve of the bulb until the pressure in the cuff drops at a rate of about 2 or 3 mm Hg per second.
    h. Listen for sounds (Korotkoff sounds) from the brachial artery. When the first loud tapping sound is heard, record the reading as the systolic pressure. This indicates the pressure exerted against the arterial wall during systole.
    i. Continue to listen to the sounds as the pressure drops, and note the level when the last sound is heard. Record this reading as the diastolic pressure, which measures the constant arterial resistance.
    j. Release all of the pressure from the cuff.
    k. Repeat the procedure until you have two blood pressure measurements from each arm, allowing 2–3 minutes of rest between readings.
    l. Average your readings and enter them in the table in Part C of the laboratory report.
2. Measure your partner's blood pressure in each of the following conditions:
    a. immediately after lying down;
    b. 5 minutes after lying down;
    c. immediately after standing;
    d. 5 minutes after standing quietly;
    e. immediately after 3 minutes of moderate exercise (*omit if the person has health problems*);
    f. 5 minutes after exercise has ended.
3. Complete Part C of the laboratory report.

Figure 45.1 Blood pressure is commonly measured by using a sphygmomanometer (blood pressure cuff).

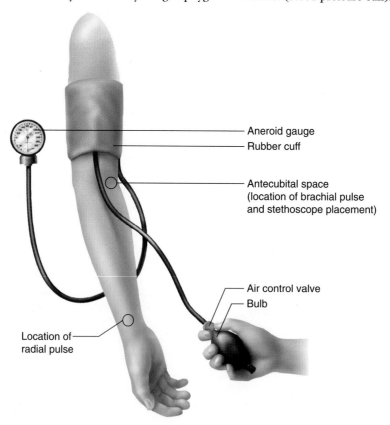

Aneroid gauge

Rubber cuff

Antecubital space (location of brachial pulse and stethoscope placement)

Air control valve

Bulb

Location of radial pulse

Name _____

Date _____

Section _____

PULSE RATE AND BLOOD PRESSURE

PART A

Complete the following statements:

1. The term *blood pressure* most commonly is used to refer to systemic _____ pressure.

2. The maximum pressure achieved during ventricular contraction is called _____ pressure.

3. The lowest pressure that remains in the arterial system during ventricular relaxation is called _____ pressure.

4. The pulse rate is equal to the _____ rate.

5. A pulse that feels full and is not easily compressed is produced by an elevated _____ .

6. The instrument commonly used to measure systemic arterial blood pressure is called a (an) _____ .

7. Blood pressure is expressed in units of _____ .

8. The upper number of the fraction used to record blood pressure indicates the _____ pressure.

9. The difference between the systolic and diastolic pressure is called the _____ .

10. The mean arterial pressure is approximated by adding the _____ pressure to one-third of the pulse pressure.

11. The _____ artery in the arm is the standard systemic artery in which blood pressure is measured.

PART B

1. Enter your observations of pulse characteristics and pulse rates in the table.

Test Subject	Pulse Characteristics	Pulse Rate
Sitting		
Lying down		
5 minutes later		
Standing		
5 minutes later		
After exercise		
5 minutes later		

2. Summarize the effects of body position and exercise on the characteristics and rates of the pulse. _____

_____

_____

_____

_____

PART C

1. Enter the initial measurements of blood pressure in the table.

Reading	Blood Pressure in Right Arm	Blood Pressure in Left Arm
First		
Second		
Average		

2. Enter your test results in the table.

Test Subject	Blood Pressure
Lying down	
5 minutes later	
Standing	
5 minutes later	
After exercise	
5 minutes later	

3. Summarize the effects of body position and exercise on blood pressure. _____

_____

_____

_____

_____

_____

4. Summarize any correlations between pulse rate and blood pressure from any of the experimental conditions.

_____

_____

_____

_____

Critical Thinking Application

When a pulse is palpated and counted, which blood pressure (systolic or diastolic) would be characteristic at that moment? Explain your answer.

_____

_____

_____

LABORATORY EXERCISE 46

MAJOR ARTERIES AND VEINS

MATERIALS NEEDED

Textbook
Human torso
Anatomical charts of the cardiovascular system

The blood vessels of the cardiovascular system can be divided into two major pathways—the pulmonary circuit and the systemic circuit. Within each circuit, arteries transport blood away from the heart. After exchanges of gases, nutrients, and wastes have occurred between the blood and the surrounding tissues, veins return the blood to the heart.

PURPOSE OF THE EXERCISE

To review the major circulatory pathways and to locate the major arteries and veins on anatomical charts and in the torso.

LEARNING OBJECTIVES

After completing this exercise, you should be able to

1. Distinguish and trace the pulmonary and systemic circuits.
2. Locate the major arteries in these circuits on a chart or model.
3. Locate the major veins in these circuits on a chart or model.

PROCEDURE A—PATHS OF CIRCULATION

1. Review the sections entitled "Pulmonary Circuit" and "Systemic Circuit" in chapter 15 of the textbook.
2. As a review activity, label figure 46.1.
3. Locate the following blood vessels on the available anatomic charts and the human torso:

pulmonary trunk	1
pulmonary arteries	2
pulmonary veins	4
aorta	1
superior vena cava	1
inferior vena cava	1

 Critical Thinking Application

Why is the left ventricle wall thicker than the right ventricle wall?

_____

_____

_____

_____

Figure 46.1 Label the major blood vessels associated with the pulmonary and systemic circuits, using the terms provided.

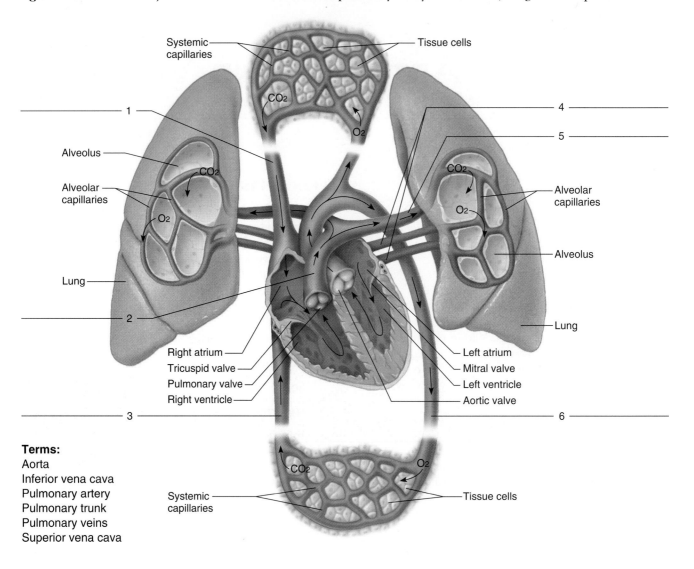

Systemic capillaries

Tissue cells

1

4

5

Alveolus

CO_2

CO_2

Alveolar capillaries

O_2

O_2

Alveolar capillaries

Alveolus

Lung

Lung

2

Right atrium

Tricuspid valve

Pulmonary valve

Right ventricle

Left atrium

Mitral valve

Left ventricle

Aortic valve

3

6

CO_2

O_2

Systemic capillaries

Tissue cells

Terms:
Aorta
Inferior vena cava
Pulmonary artery
Pulmonary trunk
Pulmonary veins
Superior vena cava

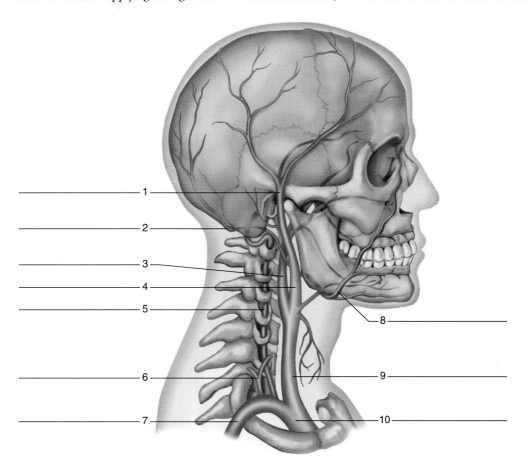

PROCEDURE B—THE ARTERIAL SYSTEM

1. Review the section entitled "Arterial System" in chapter 15 of the textbook.
2. As a review activity, label figures 46.2, 46.3, 46.4, and 46.5.
3. Locate the following arteries of the systemic circuit on the charts and torso:

aorta

    ascending aorta

    aortic sinus

    aortic arch (arch of the aorta)

    thoracic aorta

    abdominal aorta

branches of the aorta

    coronary arteries

    brachiocephalic artery

left common carotid artery

left subclavian artery

celiac artery

    gastric artery

    splenic artery

    hepatic artery

superior mesenteric artery

suprarenal arteries

renal arteries

gonadal arteries

    ovarian arteries

    testicular arteries

inferior mesenteric artery

middle sacral artery

common iliac arteries

Figure 46.3 Label the major arteries of the shoulder and upper limb.

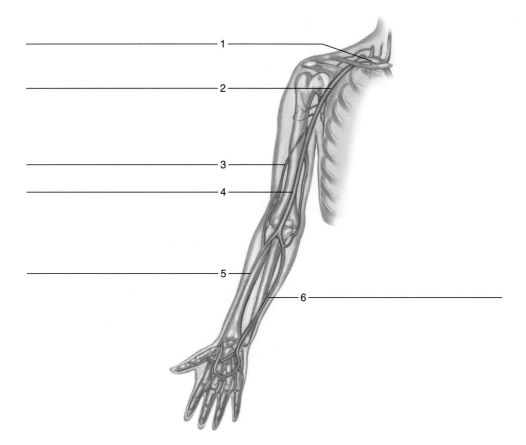

arteries to neck, head, and brain

   vertebral arteries

   thyrocervical arteries

   common carotid arteries

   external carotid arteries

      superior thyroid artery

      facial artery

      superficial temporal artery

   internal carotid arteries

arteries to shoulder and upper limb

   axillary artery

   brachial artery

   deep brachial artery

   ulnar artery

   radial artery

arteries to the thoracic and abdominal walls

   internal thoracic artery

   anterior intercostal artery

   posterior intercostal artery

arteries to pelvis and lower limb

   common iliac artery

   internal iliac artery

   external iliac artery

   femoral artery

   deep femoral artery

   popliteal artery

   anterior tibial artery

   dorsalis pedis artery (dorsal pedis artery)

   posterior tibial artery

4. Complete Parts A and B of Laboratory Report 46.

Figure 46.4 Label the arteries associated with the abdominal aorta.

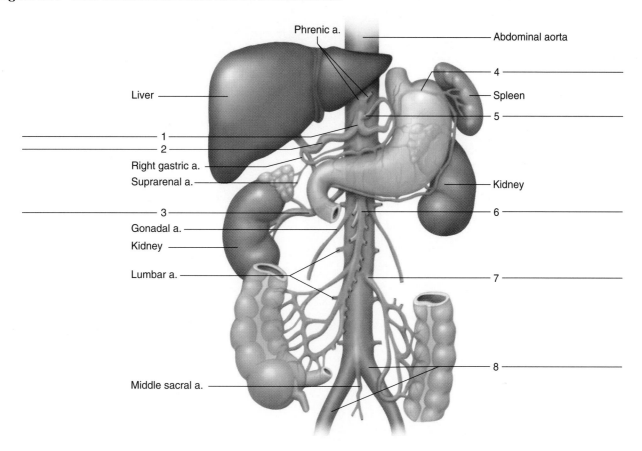

Phrenic a.

Abdominal aorta

Liver

4

Spleen

5

1

2

Right gastric a.

Suprarenal a.

Kidney

3

6

Gonadal a.

Kidney

Lumbar a.

7

8

Middle sacral a.

PROCEDURE C—THE VENOUS SYSTEM

1. Review the section entitled "Venous System" in chapter 15 of the textbook.
2. As a review activity, label figures 46.6, 46.7, 46.8, 46.9, and 46.10.
3. Locate the following veins of the systemic circuit on the charts and the torso:

veins from brain, head, and neck

external jugular veins

internal jugular veins

subclavian veins

brachiocephalic veins

superior vena cava

veins from upper limb and shoulder

radial vein

ulnar vein

brachial vein

basilic vein

axillary vein

cephalic vein

median cubital vein

subclavian vein

veins of the abdominal and thoracic walls

internal thoracic veins

intercostal veins

azygos vein

posterior intercostal veins

superior hemiazygos vein

inferior hemiazygos vein

Figure 46.5 Label the major arteries supplying the pelvis and lower limb, using the terms provided.

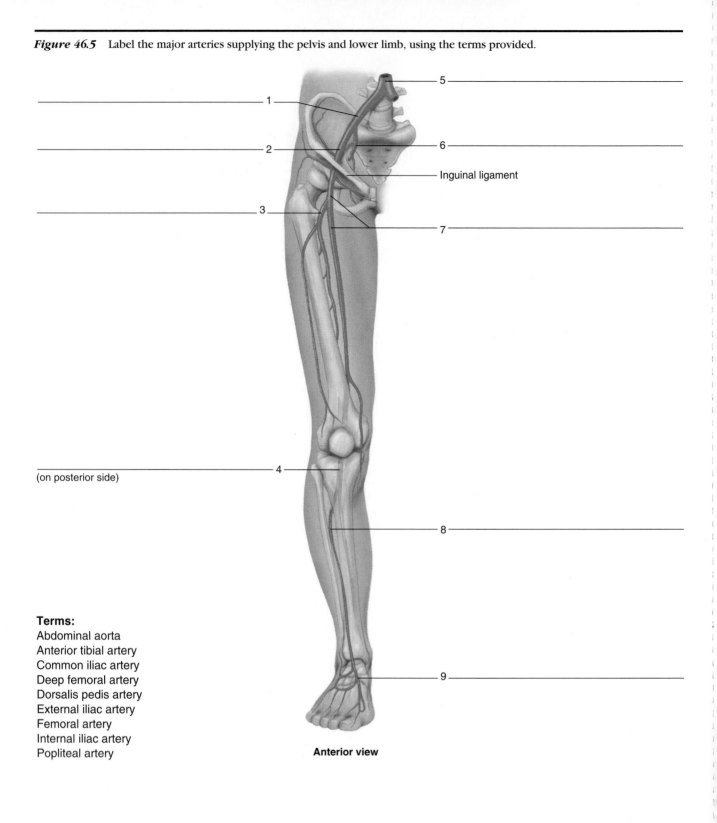

1 _____

2 _____

3 _____

_____
(on posterior side)

4 _____

5 _____

6 _____

_____ Inguinal ligament

7 _____

8 _____

9 _____

Anterior view

Terms:
Abdominal aorta
Anterior tibial artery
Common iliac artery
Deep femoral artery
Dorsalis pedis artery
External iliac artery
Femoral artery
Internal iliac artery
Popliteal artery

Figure 46.6 Label the major veins associated with the head and neck. (Note that the clavicle has been removed.)

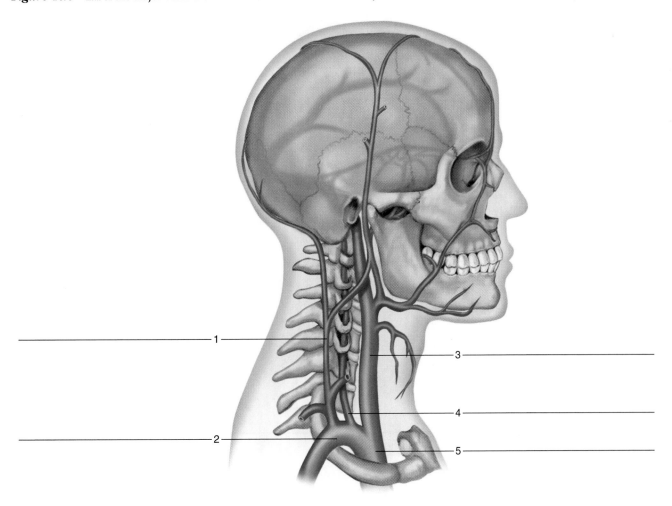

veins of the abdominal viscera

    hepatic portal vein

    gastric vein

    superior mesenteric vein

    splenic vein

    inferior mesenteric vein

    hepatic vein

    gonadal vein

    renal vein

    suprarenal vein

veins from lower limb and pelvis

    anterior tibial vein

    posterior tibial vein

    popliteal vein

    femoral vein

    great saphenous vein

    small saphenous vein

    external iliac vein

    internal iliac vein

    common iliac vein

    inferior vena cava

4. Complete Parts C, D, and E of the laboratory report.

Figure 46.7 Label the veins associated with the thoracic wall, using the terms provided.

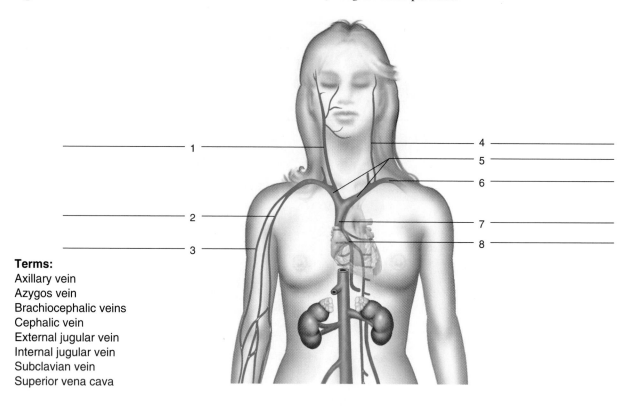

1 _____

2 _____

3 _____

4 _____

5 _____

6 _____

7 _____

8 _____

Terms:
Axillary vein
Azygos vein
Brachiocephalic veins
Cephalic vein
External jugular vein
Internal jugular vein
Subclavian vein
Superior vena cava

Figure 46.8 Label the veins of the upper limb and shoulder.

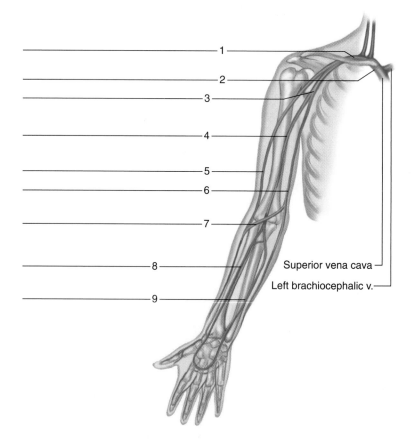

1 _____

2 _____

3 _____

4 _____

5 _____

6 _____

7 _____

8 _____

9 _____

Superior vena cava

Left brachiocephalic v.

Figure 46.9 Label the veins that drain the abdominal viscera (hepatic portal system).

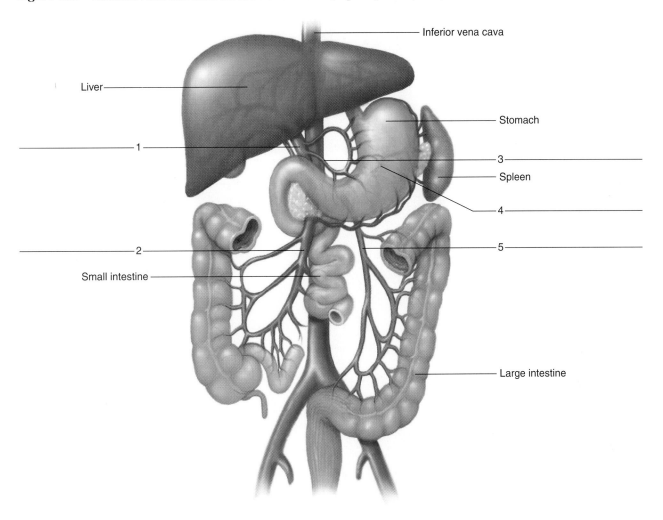

Inferior vena cava

Liver

Stomach

1

3

Spleen

4

2

5

Small intestine

Large intestine

Web Quest

Identify the major arteries and veins, and describe their functions. Observe an animation of the exchanges in a capillary. Search these at www.mhhe.com/shier10
Click on Student Edition. Click on Martin Lab Manual, Web Quest.

Figure 46.10 Label the veins of the pelvis and lower limb, using the terms provided.

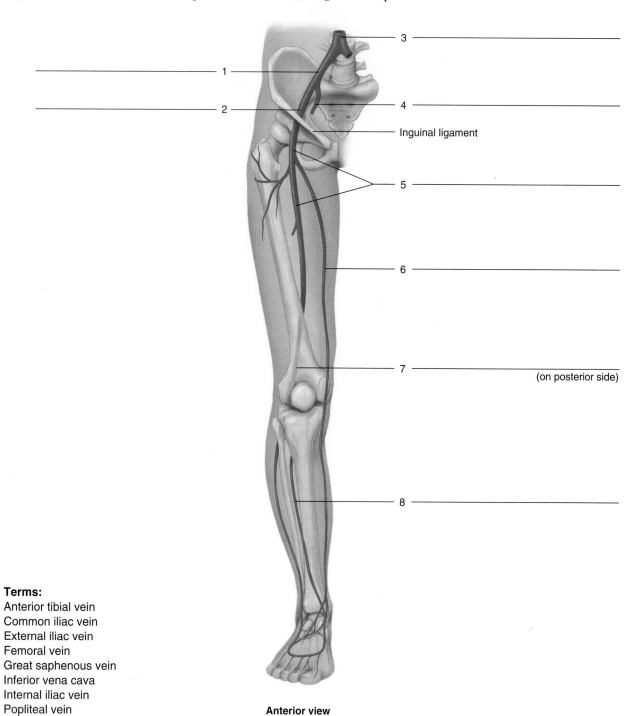

1 _____

2 _____

3 _____

4 _____

Inguinal ligament

5 _____

6 _____

7 _____ (on posterior side)

8 _____

Terms:
Anterior tibial vein
Common iliac vein
External iliac vein
Femoral vein
Great saphenous vein
Inferior vena cava
Internal iliac vein
Popliteal vein

Anterior view

MAJOR ARTERIES AND VEINS

PART A

Match the arteries in column A with the regions supplied in column B. Place the letter of your choice in the space provided.

	Column A		Column B
a.	Anterior tibial	_____ 1.	Jaw, teeth, and face
b.	Celiac	_____ 2.	Larynx, trachea, thyroid gland
c.	Costocervical	_____ 3.	Kidney
d.	Deep brachial	_____ 4.	Upper digestive tract, spleen, and liver
e.	External carotid	_____ 5.	Foot and toes
f.	Inferior mesenteric	_____ 6.	Gluteal muscles
g.	Internal carotid	_____ 7.	Triceps muscle
h.	Internal iliac	_____ 8.	Thoracic wall
i.	Lumbar	_____ 9.	Posterior abdominal wall
j.	Phrenic	_____ 10.	Adrenal gland
k.	Popliteal	_____ 11.	Diaphragm
l.	Renal	_____ 12.	Lower colon
m.	Suprarenal	_____ 13.	Brain
n.	Thyrocervical	_____ 14.	Forearm muscles
o.	Ulnar	_____ 15.	Knee joint

PART B

Provide the name of the missing artery in each of the following sequences:

1. Brachiocephalic artery, _____ , right axillary artery

2. Ascending aorta, _____ , thoracic aorta

3. Abdominal aorta, _____ , ascending colon

4. Abdominal aorta, _____ , descending colon

5. Brachiocephalic artery, _____ , right external carotid artery

6. Subclavian artery, _____ , basilar artery

7. External carotid artery, _____ , lips

8. Axillary artery, _____ , radial artery

9. Common iliac artery, _____ , femoral artery

10. Pulmonary trunk, _____ , lungs

PART C

Match each vein in column A with the vein it drains into from column B. Place the letter of your choice in the space provided.

Column A

a. Anterior tibial
b. Basilic
c. Brachiocephalic
d. Common iliac
e. External jugular
f. Femoral
g. Popliteal
h. Radial

Column B

_____ 1. Popliteal

_____ 2. Axillary

_____ 3. Inferior vena cava

_____ 4. Subclavian

_____ 5. Brachial

_____ 6. Superior vena cava

_____ 7. Femoral

_____ 8. External iliac

PART D

Provide the name of the missing vein or veins in each of the following sequences:

1. Right subclavian vein, _____ , superior vena cava

2. Posterior tibial vein, _____ , femoral vein

3. Internal iliac vein, _____ , inferior vena cava

4. Medial cubital vein, _____ , axillary vein

5. Dorsalis pedis vein, _____ , popliteal vein

6. Great saphenous vein, _____ , external iliac vein

7. Liver, _____ , inferior vena cava

8. Lungs, _____ , left atrium

9. Kidney, _____ , inferior vena cava

PART E

Label the major arteries and veins indicated in figure 46.11.

Figure 46.11 Label the major arteries and veins of the systemic and pulmonary circuits, using the terms provided.

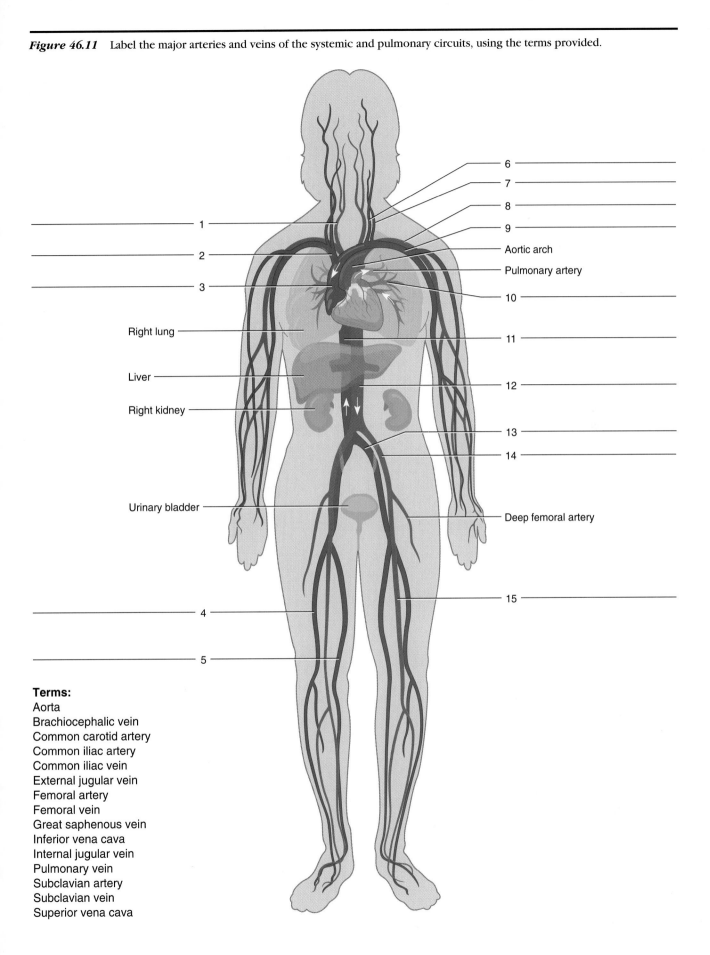

Right lung

Liver

Right kidney

Urinary bladder

1

2

3

4

5

6

7

8

9

Aortic arch

Pulmonary artery

10

11

12

13

14

Deep femoral artery

15

Terms:
Aorta
Brachiocephalic vein
Common carotid artery
Common iliac artery
Common iliac vein
External jugular vein
Femoral artery
Femoral vein
Great saphenous vein
Inferior vena cava
Internal jugular vein
Pulmonary vein
Subclavian artery
Subclavian vein
Superior vena cava

CAT DISSECTION: CARDIOVASCULAR SYSTEM

In this laboratory exercise, you will dissect the major organs of the cardiovascular system of the cat. As before, while you are examining the organs of the cat, compare them with the corresponding organs of the human torso.

If the cardiovascular system of the cat has been injected, the arteries will be filled with red latex, and the veins will be filled with blue latex. This will make it easier for you to trace the vessels as you dissect them.

PURPOSE OF THE EXERCISE

To examine the major organs of the cardiovascular system of the cat and to compare them with the corresponding organs of the human torso.

LEARNING OBJECTIVES

After completing this exercise, you should be able to

1. Locate and identify the major organs of the cardiovascular system of the cat.
2. Identify the corresponding organs in the human torso.
3. Compare the features of the cardiovascular system of the cat with those of the human.

PROCEDURE A—THE ARTERIAL SYSTEM

1. Place the embalmed cat in the dissecting tray with its ventral surface up.
2. Open the thoracic cavity, and expose its contents. To do this, follow these steps:
   a. Make a longitudinal incision passing anteriorly from the diaphragm along one side of the sternum. Continue the incision through the neck muscles to the mandible. Try to avoid damaging the internal organs as you cut.
   b. Make a lateral cut on each side along the anterior surface of the diaphragm, and cut the diaphragm loose from the thoracic wall.
   c. Spread the sides of the thoracic wall outward, and use a scalpel to make a longitudinal cut along each side of the inner wall of the rib cage to weaken the ribs. Continue to spread the thoracic wall laterally to break the ribs so that the flaps of the wall will remain open (figs. 47.1 and 47.2).
3. Note the location of the heart and the large blood vessels associated with it. Slit the outer *parietal pericardium* that surrounds the heart by cutting with scissors along the midventral line. Note how this membrane is connected to the *visceral pericardium* that is attached to the surface of the heart. Locate the *pericardial cavity,* the space between the two layers of the pericardium.
4. Examine the heart (see figs. 41.1, 41.2, and 41.3, as well as figs. 47.2 and 47.3). It should be noted that the arrangement of blood vessels coming off the aortic arch is different in the cat. Locate the following:

right atrium

left atrium

right ventricle

left ventricle

pulmonary trunk

aorta

coronary arteries

Figure 47.1 Arteries of the cat's thorax, neck, and forelimb.

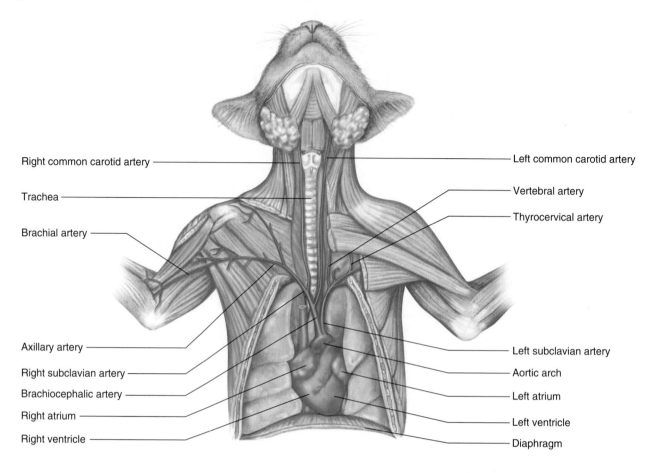

Right common carotid artery

Trachea

Brachial artery

Axillary artery

Right subclavian artery

Brachiocephalic artery

Right atrium

Right ventricle

Left common carotid artery

Vertebral artery

Thyrocervical artery

Left subclavian artery

Aortic arch

Left atrium

Left ventricle

Diaphragm

5. Use a scalpel to open the heart chambers by making a cut along the frontal plane from its apex to its base. Remove any remaining latex from the chambers. Examine the valves between the chambers, and note the relative thicknesses of the chamber walls (fig. 47.4). Do not remove the heart as it is needed for future examination of the relationship between major blood vessels.

6. Using figures 47.1, 47.3, and 47.4 as guides, locate and dissect the following arteries of the thorax and neck:

aortic arch

brachiocephalic artery (on right only)

right subclavian artery

left subclavian artery

right common carotid artery

left common carotid artery

7. Trace the right subclavian artery into the forelimb, and locate the following arteries:

vertebral artery

thyrocervical artery

subscapular artery

axillary artery

brachial artery

radial artery

ulnar artery

Figure 47.2 Features of the cat's neck and thoracic cavity.

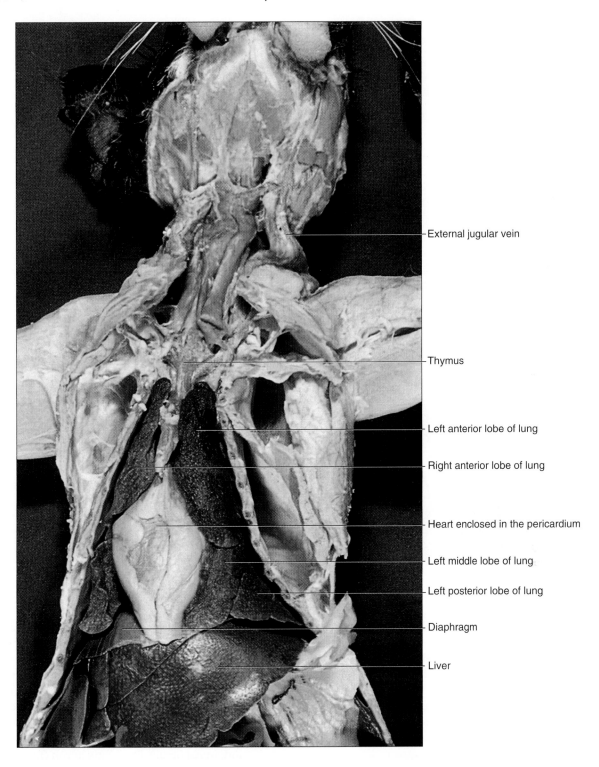

External jugular vein

Thymus

Left anterior lobe of lung

Right anterior lobe of lung

Heart enclosed in the pericardium

Left middle lobe of lung

Left posterior lobe of lung

Diaphragm

Liver

Figure 47.3 Blood vessels of the cat's upper forelimb and neck.

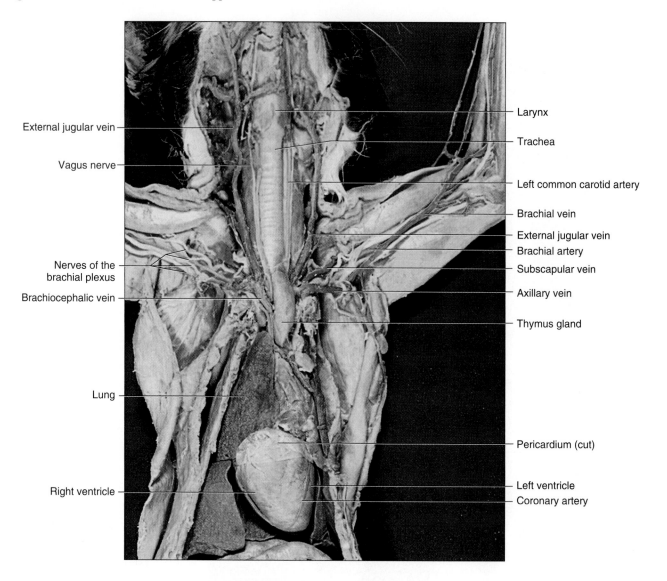

External jugular vein

Vagus nerve

Nerves of the
brachial plexus

Brachiocephalic vein

Lung

Right ventricle

Larynx

Trachea

Left common carotid artery

Brachial vein

External jugular vein

Brachial artery

Subscapular vein

Axillary vein

Thymus gland

Pericardium (cut)

Left ventricle

Coronary artery

8. Open the abdominal cavity. To do this, follow these steps:
   a. Use scissors to make an incision through the body wall along the midline from the symphysis pubis to the diaphragm.
   b. Make a lateral incision through the body wall along either side of the inferior border of the diaphragm and along the bases of the thighs.

c. Reflect the flaps created in the body wall as you would open a book, and expose the contents of the abdominal cavity.
d. Note the *parietal peritoneum* that forms the inner lining of the abdominal wall. Also note the *greater omentum,* a structure composed of a double layer of peritoneum that hangs from the border of the stomach and covers the lower abdominal organs like a fatty apron (fig. 47.5).

Figure 47.4   Blood vessels of the cat's neck and thorax.

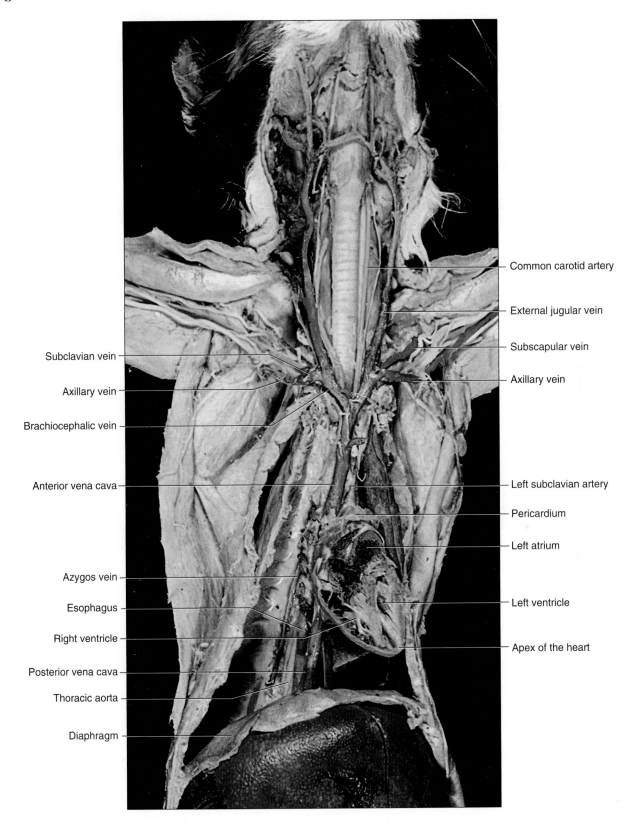

Common carotid artery

External jugular vein

Subscapular vein

Axillary vein

Subclavian vein

Axillary vein

Brachiocephalic vein

Anterior vena cava

Left subclavian artery

Pericardium

Left atrium

Azygos vein

Esophagus

Left ventricle

Right ventricle

Posterior vena cava

Apex of the heart

Thoracic aorta

Diaphragm

Figure 47.5 Organs of the cat's thoracic and abdominal cavities.

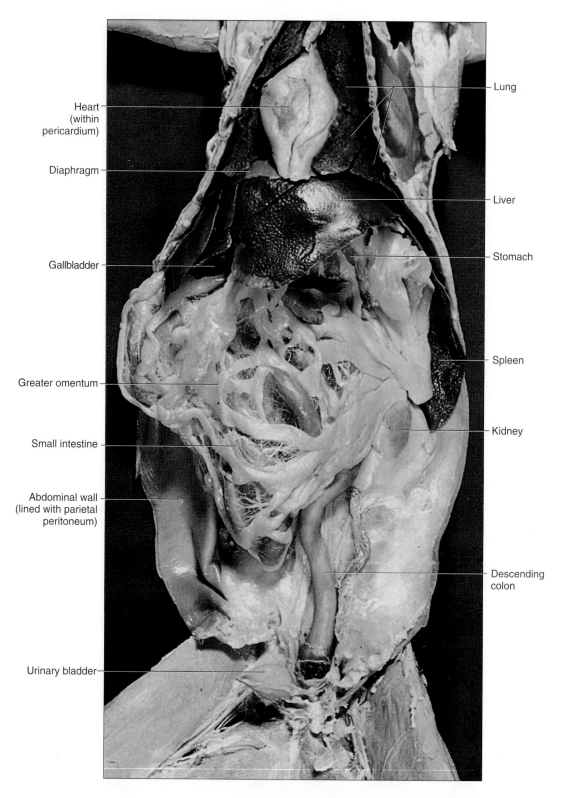

Heart
(within
pericardium)

Diaphragm

Gallbladder

Greater omentum

Small intestine

Abdominal wall
(lined with parietal
peritoneum)

Urinary bladder

Lung

Liver

Stomach

Spleen

Kidney

Descending
colon

Figure 47.6 Arteries of the cat's abdomen and hindlimb.

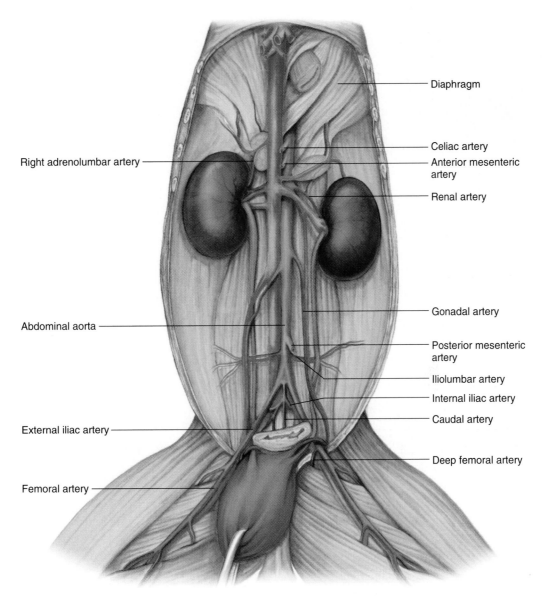

9. As you expose and dissect blood vessels, try not to destroy other visceral organs needed for future studies.

Using figures 47.6, 47.7, and 47.8 as guides, locate and dissect the following arteries of the abdomen:

abdominal aorta (unpaired)

celiac artery (unpaired)

hepatic artery (unpaired)

gastric artery (unpaired)

splenic artery (unpaired)

anterior mesenteric artery (corresponds to superior mesenteric artery)

renal arteries (paired)

adrenolumbar arteries (paired)

gonadal arteries (paired)

posterior mesenteric artery (corresponds to inferior mesenteric artery)

iliolumbar arteries (paired)

external iliac arteries (paired)

internal iliac arteries (paired)

caudal artery (unpaired)

Figure 47.7   Arteries of the cat's abdomen, left lateral view.

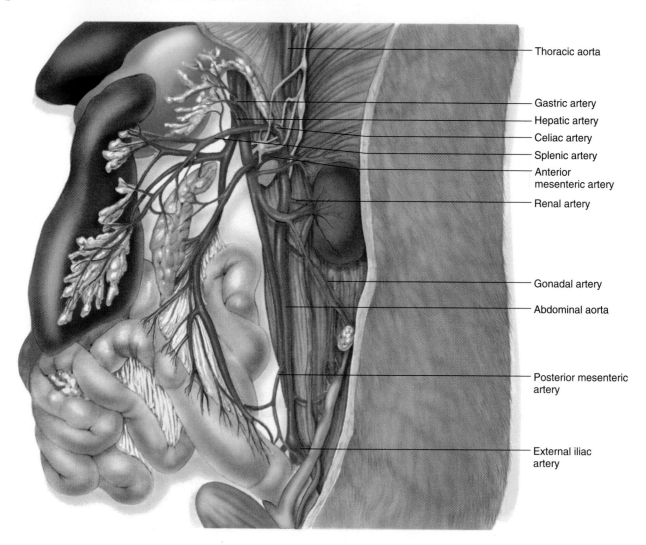

Thoracic aorta

Gastric artery

Hepatic artery

Celiac artery

Splenic artery

Anterior
mesenteric artery

Renal artery

Gonadal artery

Abdominal aorta

Posterior mesenteric
artery

External iliac
artery

10. Trace the external iliac artery into the right
    hindlimb (fig. 47.9), and locate the following:

    femoral artery

    deep femoral artery

11. Complete Part A of Laboratory Report 47.

PROCEDURE B—THE VENOUS SYSTEM

1. Examine the heart again, and locate the following
   veins:

   anterior vena cava (corresponds to superior vena
   cava)

   posterior vena cava (corresponds to inferior vena
   cava)

   pulmonary veins

Figure 47.8 Blood vessels of the cat's abdominal cavity, with the digestive organs removed.

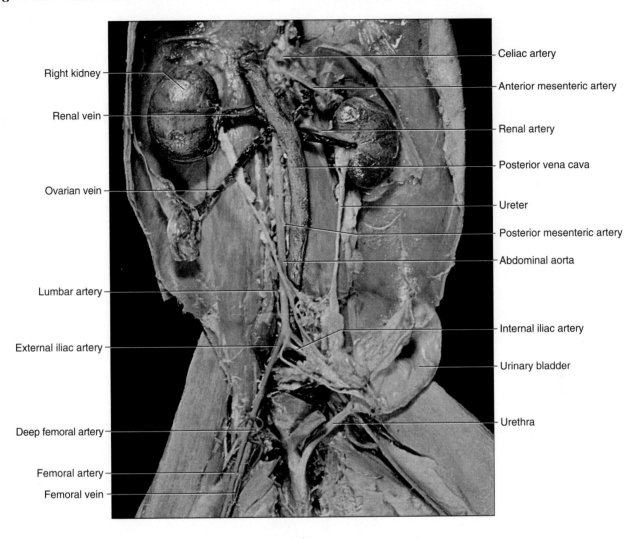

Right kidney

Renal vein

Ovarian vein

Lumbar artery

External iliac artery

Deep femoral artery

Femoral artery
Femoral vein

Celiac artery

Anterior mesenteric artery

Renal artery

Posterior vena cava

Ureter

Posterior mesenteric artery

Abdominal aorta

Internal iliac artery

Urinary bladder

Urethra

2. Using figures 47.3, 47.4, and 47.10 as guides, locate and dissect the following veins in the thorax and neck:

right brachiocephalic vein

left brachiocephalic vein

right subclavian vein

left subclavian vein

internal jugular vein

external jugular vein

3. Trace the right subclavian vein into the forelimb, and locate the following veins:

axillary vein

subscapular vein

brachial vein

long thoracic vein

Figure 47.9   Blood vessels of the cat's upper hindlimb.

External iliac vein

Femoral vein

External iliac artery

Deep femoral artery

Femoral artery

Figure 47.10   Veins of the cat's thorax, neck, and forelimb.

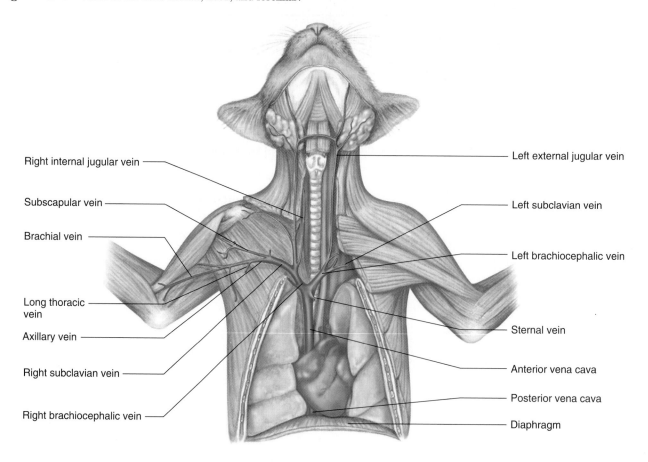

Right internal jugular vein

Subscapular vein

Brachial vein

Long thoracic vein

Axillary vein

Right subclavian vein

Right brachiocephalic vein

Left external jugular vein

Left subclavian vein

Left brachiocephalic vein

Sternal vein

Anterior vena cava

Posterior vena cava

Diaphragm

Figure 47.11 Veins of the cat's abdomen and hindlimb.

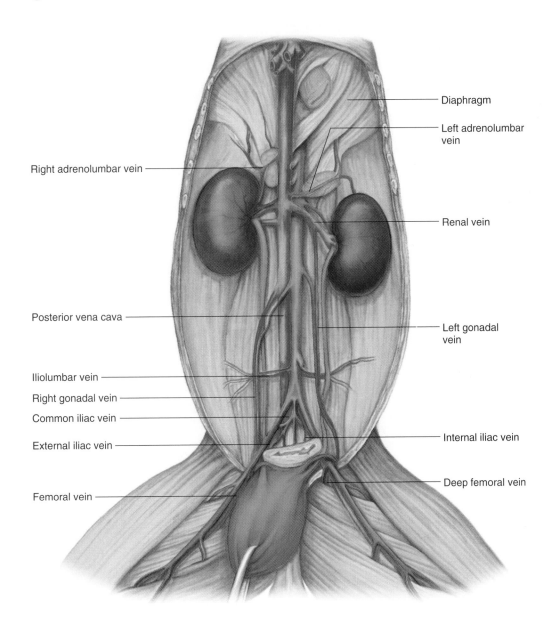

Diaphragm

Left adrenolumbar vein

Right adrenolumbar vein

Renal vein

Posterior vena cava

Left gonadal vein

Iliolumbar vein

Right gonadal vein

Common iliac vein

Internal iliac vein

External iliac vein

Deep femoral vein

Femoral vein

4. Using figures 47.8, 47.11, and 47.12 as guides, locate and dissect the following veins in the abdomen:

posterior vena cava

adrenolumbar vein

anterior mesenteric vein (corresponds to superior mesenteric vein)

posterior mesenteric vein (corresponds to inferior mesenteric vein)

hepatic portal vein

splenic vein

renal vein

gonadal vein

iliolumbar vein

common iliac vein

internal iliac vein

external iliac vein

Figure 47.12 Veins of the cat's abdomen, left lateral view.

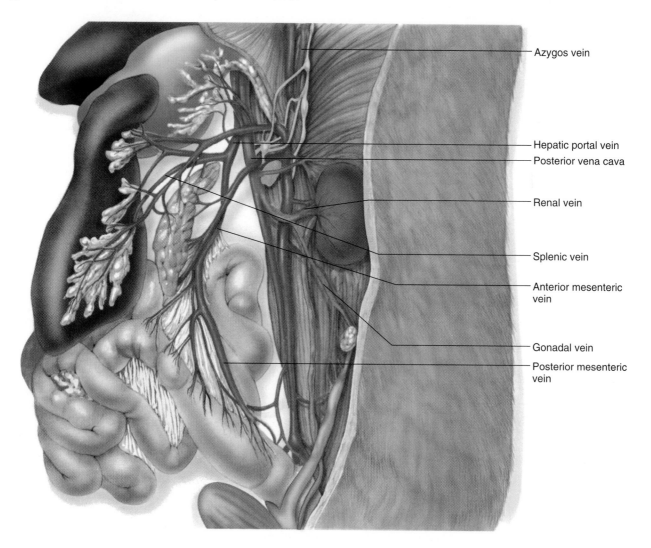

Azygos vein

Hepatic portal vein

Posterior vena cava

Renal vein

Splenic vein

Anterior mesenteric vein

Gonadal vein

Posterior mesenteric vein

5. Trace the external iliac vein into the hindlimb (fig. 47.9), and locate the following veins:

femoral vein

deep femoral vein

6. Complete Part B of the laboratory report.

CAT DISSECTION: CARDIOVASCULAR SYSTEM

PART A

Complete the following:

1. Describe the position and attachments of the parietal pericardium of the heart of the cat. _____
   _____
   _____
   _____

2. Describe the relative thicknesses of the walls of the heart chambers of the cat. _____
   _____
   _____
   _____

3. Explain how the wall thicknesses are related to the functions of the chambers. _____
   _____
   _____
   _____
   _____
   _____

4. Compare the origins of the common carotid arteries of the cat with those of the human. _____
   _____
   _____
   _____
   _____

5. Compare the origins of the iliac arteries of the cat with those of the human. _____
   _____
   _____
   _____
   _____

PART B

Complete the following:

1. Compare the origins of the brachiocephalic veins of the cat with those of the human. _____
   _____
   _____
   _____

2. Compare the relative sizes of the external and internal jugular veins of the cat with those of the human. _____
   _____
   _____
   _____
   _____
   _____

3. List twelve veins that cats and humans have in common.
   _____
   _____
   _____
   _____
   _____
   _____

LABORATORY EXERCISE 48

LYMPHATIC SYSTEM

MATERIALS NEEDED

Textbook
Human torso
Anatomical chart of the lymphatic system
Compound light microscope
Prepared microscope slides:
    Lymph node section
    Human thymus section
    Human spleen section

For Demonstration:

Embalmed cat with collecting ducts dissected

SAFETY

• Wear disposable gloves when working on the cat demonstration.
• Wash your hands before leaving the laboratory.

The lymphatic system is closely associated with the cardiovascular system and includes a network of capillaries and vessels that assist in the circulation of body fluids. These lymphatic capillaries and vessels provide pathways through which excess fluid can be transported away from intercellular spaces within tissues and returned to the bloodstream.

The organs of the lymphatic system also help to protect the tissues against infections by filtering particles from lymph and by supporting the activities of lymphocytes that furnish immunity against specific disease-causing agents.

PURPOSE OF THE EXERCISE

To review the structure of the lymphatic system and to observe the microscopic structure of a lymph node, thymus, and spleen.

LEARNING OBJECTIVES

After completing this exercise, you should be able to

1. Locate and identify the major lymphatic pathways in an anatomical chart or model.
2. Locate and identify the major chains of lymph nodes in an anatomical chart or model.
3. Describe the structure of a lymph node.
4. Identify the major microscopic structures of a lymph node, thymus, and spleen.

PROCEDURE A— LYMPHATIC PATHWAYS

1. Review the section entitled "Lymphatic Pathways" in chapter 16 of the textbook.
2. As a review activity, label figure 48.1.
3. Complete Part A of Laboratory Report 48.
4. Observe the human torso and the anatomical chart of the lymphatic system, and locate the following features:

lymphatic vessels

lymph nodes

lymphatic trunks

    lumbar trunk

    intestinal trunk

    intercostal trunk

    bronchomediastinal trunk

    subclavian trunk

    jugular trunk

collecting ducts

    thoracic (left lymphatic) duct

    right lymphatic duct

internal jugular veins

subclavian veins

Figure 48.1 Label the diagram by placing the correct numbers in the spaces provided.

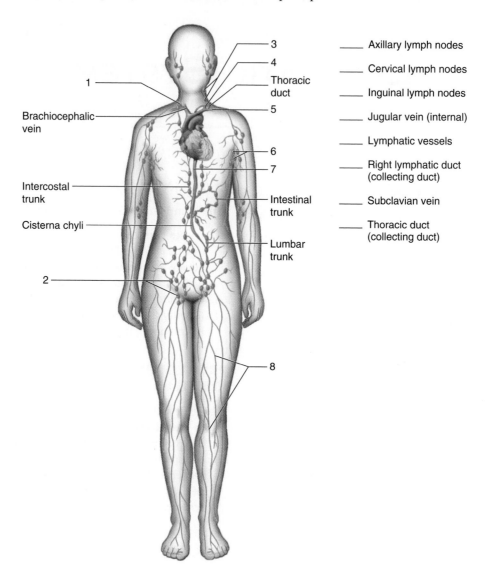

Brachiocephalic vein

Intercostal trunk

Cisterna chyli

Thoracic duct

Intestinal trunk

Lumbar trunk

_____ Axillary lymph nodes

_____ Cervical lymph nodes

_____ Inguinal lymph nodes

_____ Jugular vein (internal)

_____ Lymphatic vessels

_____ Right lymphatic duct (collecting duct)

_____ Subclavian vein

_____ Thoracic duct (collecting duct)

DEMONSTRATION

E xamine the collecting ducts in the thorax of the embalmed cat. The *thoracic duct* joins the venous system near the junction of the left subclavian vein and the left external jugular vein (see fig. 47.10). Trace this duct posteriorly, and note its beaded appearance, which is due to the presence of valves in the vessel. These valves prevent the backflow of lymph.

Look for the short *right lymphatic duct,* which joins the venous system near the junction of the right subclavian vein and the right external jugular vein. How do the collecting ducts compare in size?

_____

_____

What is the significance of this observation?

_____

_____

PROCEDURE B—LYMPH NODES

1. Review the section entitled "Lymph Nodes" in chapter 16 of the textbook.
2. As a review activity, label figure 48.2.
3. Complete Part B of the laboratory report.
4. Observe the anatomical chart of the lymphatic system and the human torso, and locate the clusters of lymph nodes in the following regions:

cervical region

axillary region

inguinal region

pelvic cavity

abdominal cavity

thoracic cavity

Figure 48.2 Label this diagram of a lymph node by placing the correct numbers in the spaces provided.

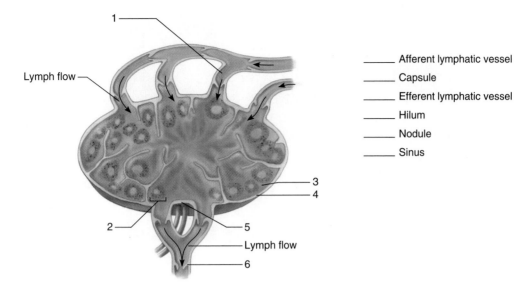

Lymph flow

1

_____ Afferent lymphatic vessel
_____ Capsule
_____ Efferent lymphatic vessel
_____ Hilum
_____ Nodule
_____ Sinus

3
4
2
5
Lymph flow
6

5. Palpate the lymph nodes in your cervical region. They are located along the lower border of the mandible and between the ramus of the mandible and the sternocleidomastoid muscle. They feel like small, firm lumps.

6. Obtain a prepared microscope slide of a lymph node and observe it, using low-power magnification. Identify the *capsule* that surrounds the node and is mainly composed of collagenous fibers, the *lymph nodules* that appear as dense masses near the surface of the node, and the *lymph sinus* that appears as narrow space between the nodules and the capsule.

7. Using high-power magnification, examine a nodule within the lymph node. Note that the nodule contains densely packed *lymphocytes.*

8. Prepare a labeled sketch of a representative section of a lymph node in Part D of the laboratory report.

PROCEDURE C—THYMUS AND SPLEEN

1. Review the section entitled "Thymus and Spleen" in chapter 16 of the textbook.

2. Locate the thymus and spleen in the anatomical chart of the lymphatic system and on the human torso.

3. Complete Part C of the laboratory report.

4. Obtain a prepared microscope slide of human thymus and observe it, using low-power magnification (fig. 48.3). Note how the thymus is subdivided into *lobules* by *septa* of connective tissue that contain blood vessels. Identify the *capsule* of loose connective tissue that surrounds the thymus, the outer *cortex* of a lobule that is composed of densely packed cells and is deeply stained, and the inner *medulla* of a lobule that is composed of loosely packed lymphocytes and epithelial cells and is lightly stained.

5. Examine the cortex tissue of a lobule using high-power magnification. The cells of the cortex are composed of densely packed *lymphocytes* among some epithelial cells and macrophages. Some of these cortical cells may be undergoing mitosis, so their chromosomes may be visible.

6. Prepare a labeled sketch of a representative section of the thymus in Part D of the laboratory report.

7. Obtain a prepared slide of the human spleen and observe it, using low-power magnification (fig. 48.4). Identify the *capsule* of dense connective tissue that surrounds the spleen. Note that the tissues of the spleen include circular *nodules of white* (in unstained tissue) *pulp* that are enclosed in a matrix of *red pulp.*

8. Using high-power magnification, examine a nodule of white pulp and red pulp. The cells of the white pulp are mainly *lymphocytes.* Also, there may be an arteriole centrally located in the nodule. The cells of the red pulp are mostly red blood cells with many lymphocytes and macrophages.

9. Prepare a labeled sketch of a representative section of the spleen in Part D of the laboratory report.

Figure 48.3 A section of the thymus gland (10× micrograph enlarged to 20×).

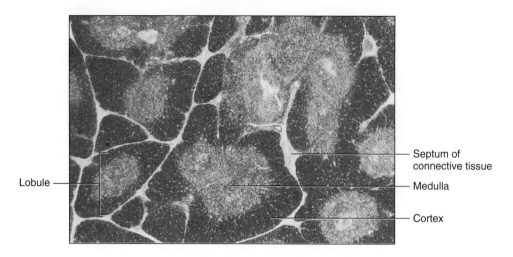

Lobule

Septum of
connective tissue

Medulla

Cortex

Figure 48.4 Micrograph of a section of the spleen (15×).

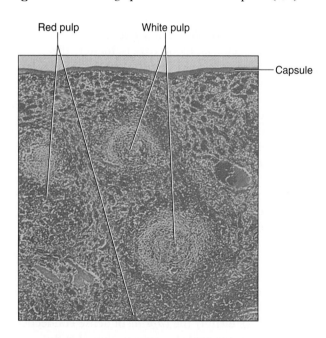

Red pulp

White pulp

Capsule

Web Quest

Locate the six major areas of lymph nodes, and identify the components of the lymphatic system. Search these at www.mhhe.com/shier10 Click on Student Edition. Click on Martin Lab Manual, Web Quest.

Name _____

Date _____

Section _____

LYMPHATIC SYSTEM

PART A

Complete the following statements:

1. Lymphatic pathways begin as _____ that merge to form lymphatic vessels.

2. The wall of a lymphatic capillary consists of a single layer of _____ cells.

3. Once tissue (interstitial) fluid is inside a lymph capillary, the fluid is called _____ .

4. Lymphatic vessels have walls that are similar to those of _____ but thinner.

5. Lymphatic vessels contain _____ that help prevent the backflow of lymph.

6. Lymphatic vessels usually lead to _____ that filter the fluid being transported.

7. The lymphatic trunk that drains the abdominal viscera is called the _____ trunk.

8. The lymphatic trunk that drains the head and neck is called the _____ trunk.

9. The _____ is the larger and longer of the two lymphatic collecting ducts.

PART B

Complete the following statements:

1. Lymph nodes contain large numbers of white blood cells called _____ and macrophages that fight invading microorganisms.

2. The indented region of a bean-shaped lymph node is called the _____ .

3. _____ that contain germinal centers are the structural units of a lymph node.

4. The spaces within a lymph node are called _____ through which lymph circulates.

5. Lymph enters a node through a(an) _____ lymphatic vessel.

6. The partially encapsulated lymph nodes in the pharynx are called _____ .

7. The aggregations of lymph nodules found within the mucosal lining of the small intestine are called

   _____ .

8. The lymph nodes in the cervical region are associated with the lymphatic vessels that drain the

   _____ .

9. The lymph nodes associated with the lymphatic vessels that drain the lower limbs are located in the

   _____ region.

PART C

Complete the following statements:

1. The thymus is located in the _____ , anterior to the aortic arch.

2. The thymus reaches its greatest size during infancy and early childhood and tends to decrease in size following _____ .

3. As a person ages, the thymus tissue is often replaced by _____ .

4. The lymphocytes of the thymus develop from precursor cells that originated in the _____ .

5. The hormones secreted by the thymus are called _____ .

6. The _____ is the largest of the lymphatic organs.

7. Blood vessels enter the spleen through the region called the _____ .

8. The sinuses within the spleen contain _____ .

9. The tiny islands (nodules) of tissue within the spleen that contain many lymphocytes comprise the _____ .

10. The _____ of the spleen contains large numbers of red blood cells, lymphocytes, and macrophages.

11. _____ within the spleen engulf and destroy foreign particles and cellular debris.

PART D

Lymph node sketch (_____×)	Thymus sketch (_____×)
Spleen sketch (_____×)	

ORGANS OF THE DIGESTIVE SYSTEM

MATERIALS NEEDED

Textbook
Human torso
Sagittal head section model
Skull with teeth
Teeth, sectioned
Tooth model, sectioned
Paper cup
Compound light microscope
Prepared microscope slides of the following:
    Parotid gland (salivary gland)
    Esophagus
    Stomach (fundus)
    Pancreas (exocrine portion)
    Small intestine (jejunum)
    Large intestine

The digestive system includes the organs associated with the alimentary canal and several accessory structures. The alimentary canal, which is a muscular tube, passes through the body from the opening of the mouth to the anus. It includes the mouth, pharynx, esophagus, stomach, small intestine, and large intestine. The canal is adapted to move substances throughout its length. It is specialized in various regions to store, digest, and absorb food materials and to eliminate the residues. The accessory organs, which include the salivary glands, liver, gallbladder, and pancreas, secrete products into the alimentary canal that aid digestive functions.

PURPOSE OF THE EXERCISE

To review the structure and function of the digestive organs and to examine the tissues of these organs microscopically.

LEARNING OBJECTIVES

After completing this exercise, you should be able to

1. Locate the major digestive organs.
2. Describe the functions of these organs.
3. Recognize tissue sections of these organs.
4. Identify the major features of each tissue section.

PROCEDURE A—MOUTH AND SALIVARY GLANDS

1. Review the sections entitled "Mouth" and "Salivary Glands" in chapter 17 of the textbook.
2. As a review activity, label figures 49.1, 49.2, and 49.3.
3. Examine the mouth of the torso, the sagittal head section model, and a skull. Locate the following structures:

oral cavity

vestibule

tongue

    frenulum (lingual)

    papillae

lingual tonsils

palate

    hard palate

    soft palate

    uvula

palatine tonsils

pharyngeal tonsils (adenoids)

gums (gingivae)

teeth

    incisors

    cuspids (canines)

    bicuspids (premolars)

    molars

Figure 49.1 Label the major features of the mouth.

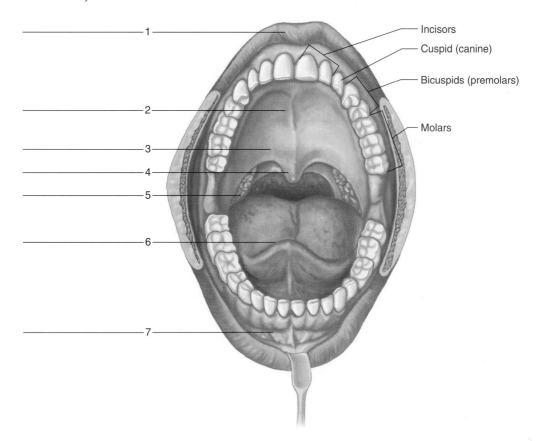

Incisors

Cuspid (canine)

Bicuspids (premolars)

Molars

1

2

3

4

5

6

7

4. Examine a sectioned tooth and a tooth model. Locate the following features:

crown

   enamel

   dentin

neck

root

   pulp cavity

   cementum

   root canal

5. Observe the head of the torso, and locate the following:

parotid salivary gland

parotid duct (Stensen's duct)

submandibular salivary gland

submandibular duct (Wharton's duct)

sublingual salivary gland

6. Examine a microscopic section of a parotid gland, using low- and high-power magnification. Note the numerous glandular cells arranged in clusters around small ducts. Also note a larger secretory duct surrounded by lightly stained cuboidal epithelial cells (fig. 49.4).

7. Complete Part A of Laboratory Report 49.

Figure 49.2 Label the features associated with the major salivary glands.

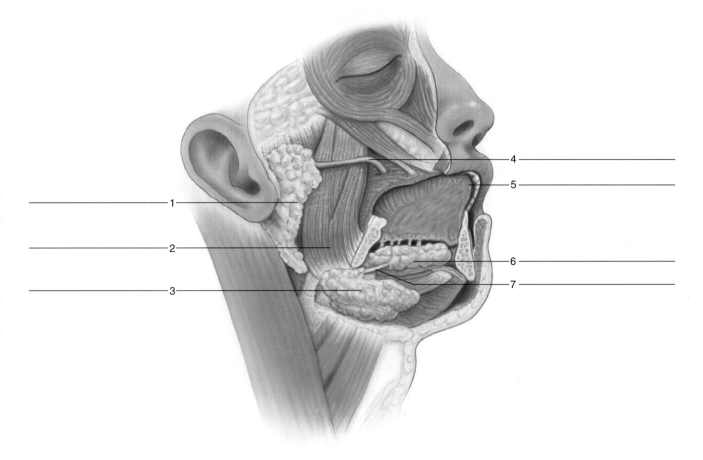

PROCEDURE B—PHARYNX AND ESOPHAGUS

1. Review the section entitled "Pharynx and Esophagus" in chapter 17 of the textbook.
2. As a review activity, label figure 49.5.
3. Observe the torso, and locate the following features:

   pharynx

       nasopharynx

       opening to auditory tube (eustachian tube)

       oropharynx

       laryngopharynx

       constrictor muscles

   epiglottis

   esophagus

       esophageal hiatus

   lower esophageal sphincter (cardiac sphincter)

4. Have your partner take a swallow from a cup of water. Carefully watch the movements in the anterior region of the neck. What steps in the swallowing process did you observe?

   _____

   _____

   _____

   _____

5. Examine a microscopic section of esophagus wall, using low-power magnification. Note that the inner lining is composed of stratified squamous epithelium and that there are layers of muscle tissue in the wall. Locate some mucous glands in the submucosa. They appear as clusters of lightly stained cells.
6. Complete Part B of the laboratory report.

Figure 49.3   Label the features of this cuspid tooth.

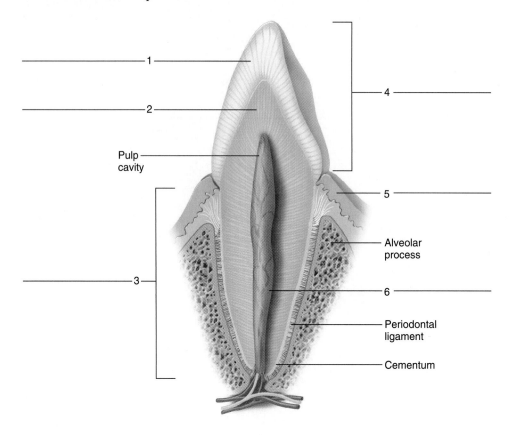

Pulp
cavity

Alveolar
process

Periodontal
ligament

Cementum

Figure 49.4   Micrograph of the parotid salivary gland (300×).

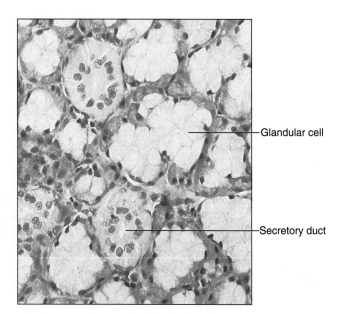

Glandular cell

Secretory duct

Figure 49.5 Label the features associated with the pharynx.

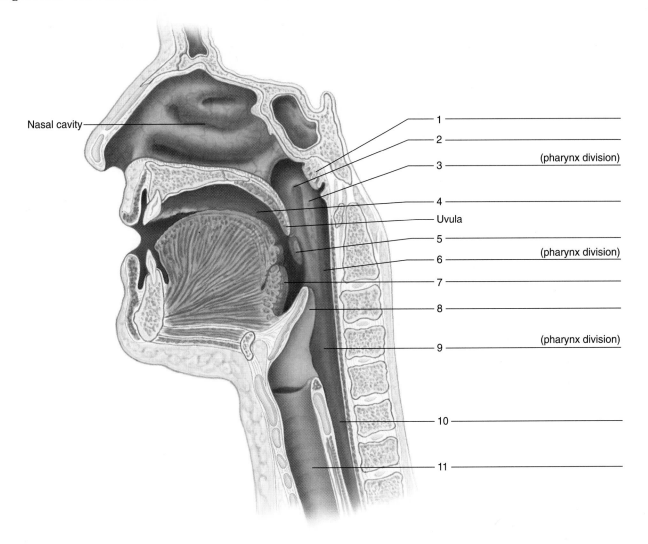

Nasal cavity

1 _____

2 _____

(pharynx division)

3 _____

4 _____

Uvula

5 _____

(pharynx division)

6 _____

7 _____

8 _____

(pharynx division)

9 _____

10 _____

11 _____

PROCEDURE C—THE STOMACH

1. Review the section entitled "Stomach" in chapter 17 of the textbook.
2. As a review activity, label figures 49.6 and 49.7.
3. Observe the torso, and locate the following features of the stomach:

rugae	**pyloric canal**
cardiac region	**pyloric sphincter (valve)**
fundic region	**lesser curvature**
body region	**greater curvature**
pyloric region	

4. Examine a microscopic section of stomach wall, using low-power magnification. Note how the inner lining of simple columnar epithelium dips inward to form gastric pits. The gastric glands are tubular structures that open into the gastric pits. Near the deep ends of these glands, you should be able to locate some intensely stained (bluish) chief cells and some lightly stained (pinkish) parietal cells (fig. 49.8). What are the functions of these cells?

_____

_____

_____

_____

_____

5. Complete Part C of the laboratory report.

Figure 49.6   Label the major regions of the stomach and associated structures.

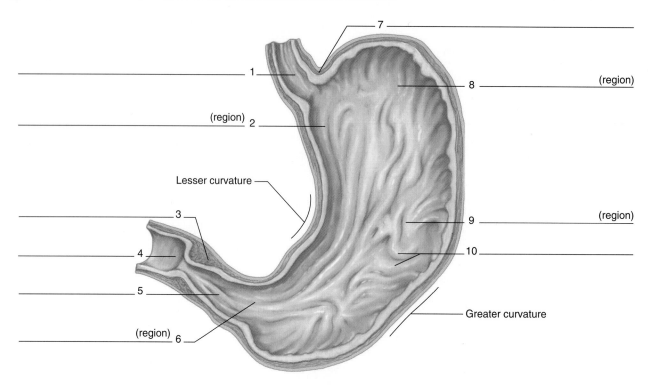

Lesser curvature

Greater curvature

(region)
(region)
(region)
(region)

Figure 49.7   Label the lining of the stomach by placing the correct numbers in the spaces provided.

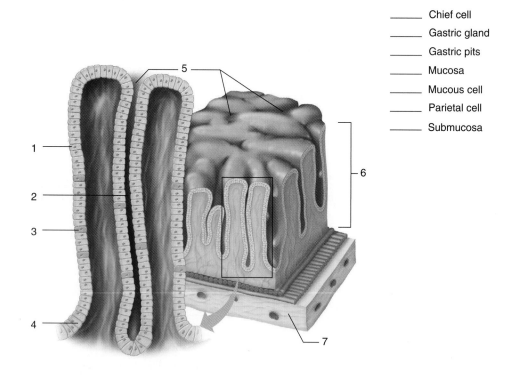

_____ Chief cell

_____ Gastric gland

_____ Gastric pits

_____ Mucosa

_____ Mucous cell

_____ Parietal cell

_____ Submucosa

Figure 49.8 Micrograph of the mucosa of the stomach wall (50× micrograph enlarged to 100×).

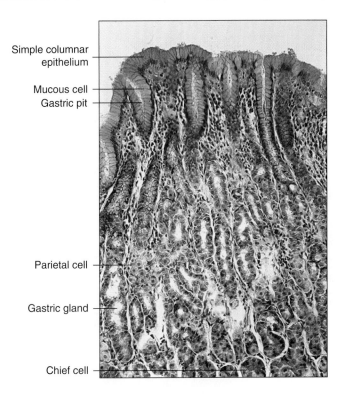

Simple columnar
epithelium

Mucous cell
Gastric pit

Parietal cell

Gastric gland

Chief cell

PROCEDURE D—PANCREAS AND LIVER

1. Review the sections entitled "Pancreas," "Liver," and "Small Intestine" in chapter 17 of the textbook.
2. As a review activity, label figure 49.9.
3. Observe the torso, and locate the following structures:

   pancreas

   pancreatic duct

   liver

      right lobe

      quadrate lobe

      caudate lobe

      left lobe

   round ligament (ligamentum teres)

   falciform ligament

   coronary ligament

   gallbladder

   hepatic ducts

   common hepatic duct

   cystic duct

   common bile duct

   hepatopancreatic sphincter (sphincter of Oddi)

4. Examine the pancreas slide, using low-power magnification. Observe the exocrine (acinar) cells that secrete pancreatic juice. See figure 37.11 in Laboratory Exercise 37 for a micrograph of the pancreas.
5. Complete Part D of the laboratory report.

Figure 49.9 Label the features associated with the liver and pancreas.

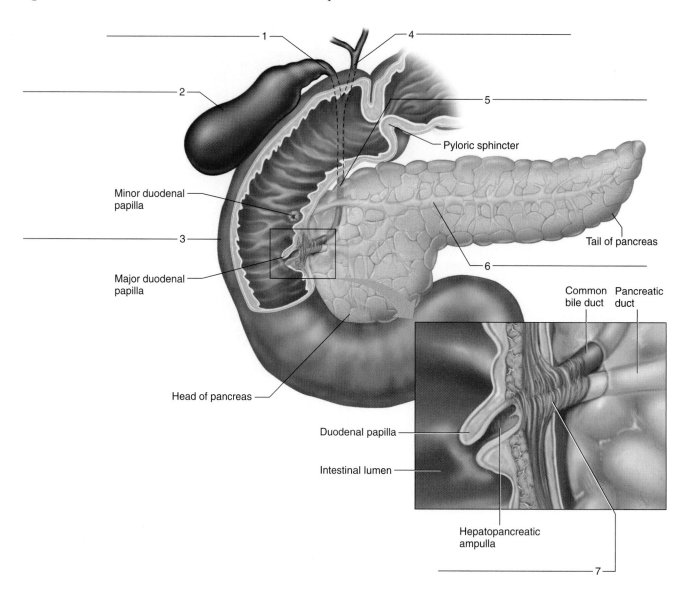

Minor duodenal papilla

Pyloric sphincter

Tail of pancreas

Major duodenal papilla

Head of pancreas

Common bile duct   Pancreatic duct

Duodenal papilla

Intestinal lumen

Hepatopancreatic ampulla

PROCEDURE E—SMALL AND LARGE INTESTINES

1. Review the sections entitled "Small Intestine" and "Large Intestine" in chapter 17 of the textbook.
2. As a review activity, label figure 49.10.
3. Observe the torso, and locate each of the following features:

small intestine

    duodenum

    jejunum

    ileum

mesentery

ileocecal sphincter (valve)

large intestine

    large intestinal wall

        haustra

        teniae coli

        epiploic appendages

    cecum

    vermiform appendix

    ascending colon

    right colic (hepatic) flexure

    transverse colon

Figure 49.10 Label this diagram of the small and large intestines by placing the correct numbers in the spaces provided.

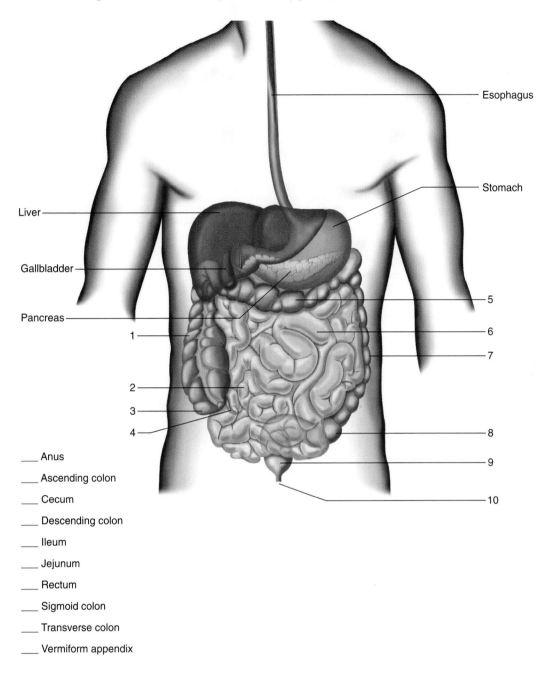

Esophagus

Stomach

Liver

Gallbladder

5

Pancreas

6

1

7

2

3

4

8

9

10

___ Anus

___ Ascending colon

___ Cecum

___ Descending colon

___ Ileum

___ Jejunum

___ Rectum

___ Sigmoid colon

___ Transverse colon

___ Vermiform appendix

left colic (splenic) flexure

descending colon

sigmoid colon

rectum

anal canal

anal columns

anal sphincter muscles

internal anal sphincter

external anal sphincter

anus

4. Using low-power magnification, examine a microscopic section of small intestine wall. Identify the mucosa, submucosa, muscular layer, and serosa. Note the villi that extend into the lumen of the tube. Study a single villus, using high-power magnification. Note the core of connective tissue and the covering of simple columnar epithelium that contains some lightly stained goblet cells (fig. 49.11). What is the function of these villi?

_____

_____

_____

5. Examine a microscopic section of large intestine wall. Note the lack of villi. Also note the tubular mucous glands that open on the surface of the inner lining and the numerous lightly stained goblet cells. Locate the four layers of the wall (fig. 49.12). What is the function of the mucus secreted by these glands?

_____

_____

_____

6. Complete Part E of the laboratory report.

 Critical Thinking Application

How is the structure of the small intestine better adapted for absorption than the large intestine?

_____

_____

_____

_____

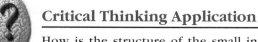 **Web Quest**

Summarize the functions of the organs of the digestive system. Search these at www.mhhe.com/shier10 Click on Student Edition. Click on Martin Lab Manual, Web Quest.

Figure 49.11 Micrograph of the small intestine wall (40×).

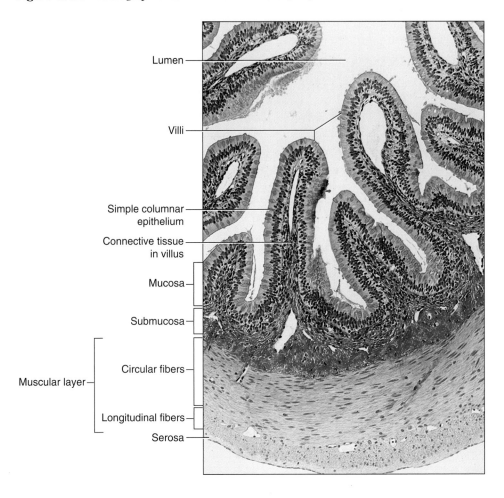

Lumen

Villi

Simple columnar
epithelium

Connective tissue
in villus

Mucosa

Submucosa

Muscular layer

Circular fibers

Longitudinal fibers

Serosa

Figure 49.12 Micrograph of the large intestine wall (64×).

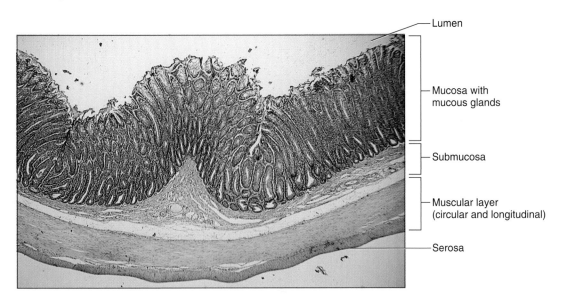

Lumen

Mucosa with
mucous glands

Submucosa

Muscular layer
(circular and longitudinal)

Serosa

Laboratory Report **49**

Name _____

Date _____

Section _____

ORGANS OF THE DIGESTIVE SYSTEM

PART A

Match the terms in column A with the descriptions in column B. Place the letter of your choice in the space provided.

Column A

a. Adenoids (pharyngeal tonsils)
b. Amylase
c. Crown
d. Dentin
e. Frenulum (lingual)
f. Incisor
g. Molar
h. Oral cavity
i. Palate
j. Papillae
k. Periodontal ligament
l. Serous cell
m. Sublingual gland
n. Uvula
o. Vestibule

Column B

_____ 1. Bonelike substance beneath tooth enamel

_____ 2. Smallest of major salivary glands

_____ 3. Tooth specialized for grinding

_____ 4. Chamber between tongue and palate

_____ 5. Projections on tongue surface

_____ 6. Cone-shaped projection of soft palate

_____ 7. Secretes the digestive enzymes in saliva

_____ 8. Attaches tooth to jaw

_____ 9. Chisel-shaped tooth

_____ 10. Roof of oral cavity

_____ 11. Space between the teeth, cheeks, and lips

_____ 12. Anchors tongue to floor of mouth

_____ 13. Lymphatic tissue in posterior wall of pharynx near auditory tubes

_____ 14. Portion of tooth projecting beyond gum

_____ 15. Splits starch into disaccharides

PART B

Complete the following:

1. The part of the pharynx superior to the soft palate is called the _____ .

2. The middle part of the pharynx is called the _____ .

3. The inferior portion of the pharynx is called the _____ .

4. The auditory tube opens through the wall of the _____ .

5. During swallowing, circular muscles called _____ pull the walls of the pharynx inward.

6. The esophagus passes through the mediastinum posterior to the _____ as it descends into the thorax.

7. _____ is the main secretion of the esophagus.

8. The esophagus penetrates the diaphragm through an opening called the _____ .

9. Summarize the functions of the esophagus. _____
_____
_____

PART C

Complete the following:

1. Name the four regions of the stomach. _____
_____

2. Name the valve that prevents regurgitation of food from the small intestine back into the stomach. _____
_____

3. Name three types of secretory cells found in gastric glands. _____
_____

4. Name the gastric cells that secrete digestive enzymes. _____

5. Name the gastric cells that secrete hydrochloric acid. _____

6. Name the most important digestive enzyme in gastric juice. _____

7. Name a substance needed for efficient absorption of vitamin B_{12}. _____

8. Name a hormone secreted by the stomach that stimulates gastric glands to secrete. _____

9. Name the semifluid paste of food particles and gastric juice. _____

10. Summarize the functions of the stomach. _____
_____
_____
_____

PART D

Match the terms in column A with the descriptions in column B. Place the letter of your choice in the space provided.

Column A		Column B
a. Amylase	_____ 1.	Activates protein-digesting enzyme trypsin
b. Bile salts	_____ 2.	Causes emulsification of fats
c. Cholecystokinin	_____ 3.	Carries on phagocytosis in liver
d. Enterokinase	_____ 4.	Iron-storing substance in liver
e. Ferritin		
f. Kupffer cells	_____ 5.	Carbohydrate-digesting enzyme
g. Lipase	_____ 6.	Fat-digesting enzyme
h. Nuclease	_____ 7.	Protein-digesting enzyme
i. Secretin	_____ 8.	Stimulates gallbladder to release bile
j. Somatostatin	_____ 9.	Stimulates pancreas to secrete fluids high in bicarbonate ions
k. Trypsin		
	_____ 10.	Nucleic acid-digesting enzyme
	_____ 11.	Inhibits secretion of acid by parietal cells

414

Complete the following:

1. Name the three portions of the small intestine. _____
   _____

2. Describe the function of the mesentery. _____
   _____
   _____

3. Name the lymphatic capillary found in an intestinal villus. _____

4. Name the tubular glands found between the bases of the intestinal villi. _____

5. Name five digestive enzymes secreted by the small intestinal mucosa. _____
   _____

6. Name the valve that controls movement of material between the small and large intestines. _____
   _____

7. Name the small projection that contains lymphatic tissue attached to the cecum. _____

8. Summarize the functions of the small intestine. _____
   _____
   _____
   _____

9. Summarize the functions of the large intestine. _____
   _____
   _____
   _____

CAT DISSECTION: DIGESTIVE SYSTEM

SAFETY

- Wear disposable gloves when working on the cat dissection.
- Dispose of tissue remnants and gloves as instructed.
- Wash the dissecting tray and instruments as instructed.
- Wash your laboratory table.
- Wash your hands before leaving the laboratory.

In this laboratory exercise, you will dissect the major digestive organs of an embalmed cat. As you observe these organs, compare them with those of the human by observing the parts of the human torso; however, keep in mind that the cat is a carnivore (flesh-eating mammal) and that the organs of its digestive system are adapted to capturing, holding, eating, and digesting the bodies of other animals. In contrast, humans are omnivores (eat both plant and animal substances). Because humans are adapted to eating and digesting a greater variety of foods, comparisons of the cat and the human digestive organs may not be precise.

PURPOSE OF THE EXERCISE

To examine the major digestive organs of the cat and to compare these organs with those of the human.

LEARNING OBJECTIVES

After completing this exercise, you should be able to

1. Locate and identify the major digestive organs of the embalmed cat.
2. Identify the corresponding organs in the human torso.
3. Compare the digestive system of the cat with that of the human.

PROCEDURE

1. Place the embalmed cat in the dissecting tray with its ventral surface down.
2. Locate the major salivary glands on one side of the head. To do this, follow these steps:
   a. Clear away any remaining fascia and other connective tissue from the region below the ear and near the joint of the mandible.
   b. Identify the *parotid gland,* a relatively large mass of glandular tissue just below the ear. Note the parotid duct (Stensen's duct) that passes over the surface of the masseter muscle and opens into the mouth.
   c. Look for the *submandibular gland* just below the parotid gland, near the angle of the jaw.
   d. Locate the *sublingual gland* that is adjacent and medial to the submandibular gland (fig. 50.1).
3. Open the oral cavity. To do this, follow these steps:
   a. Use scissors to cut through the soft tissues at the angle of the mouth.
   b. When you reach the bone of the jaw, use a bone cutter to cut through the bone, thus freeing the mandible.
   c. Open the mouth wide, and locate the following features:

   cheek

   lip

   vestibule

   palate

       hard palate

       soft palate

   palatine tonsils (small, rounded masses of glandular tissue in the lateral wall of the soft palate)

   tongue

       frenulum

       papillae (examine with a hand lens)

Figure 50.1 Major salivary glands of the cat.

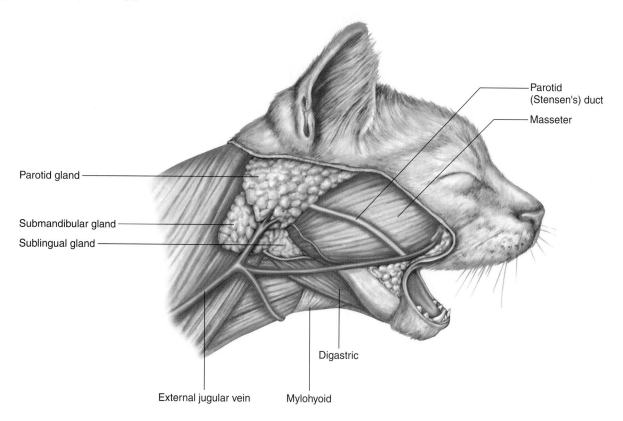

Parotid (Stensen's) duct

Masseter

Parotid gland

Submandibular gland

Sublingual gland

Digastric

External jugular vein

Mylohyoid

4. Examine the teeth of the maxilla. How many incisors, canines (cuspids), bicuspids, and molars are present?

_____

_____

5. Complete Part A of Laboratory Report 50.
6. Examine organs in the abdominal cavity with the cat positioned with its ventral surface up. Review the normal position and function of the greater omentum.
7. Examine the *liver,* which is located just beneath the diaphragm and is attached to the central portion of the diaphragm by the *falciform ligament.* Also, locate the *spleen,* which is posterior to the liver on the left side (fig. 50.2). The liver has five lobes— a right medial, right lateral (which is subdivided into two parts by a deep cleft), left medial, left lateral, and caudate lobe. The caudate lobe is the smallest lobe and is located in the median line, where it projects into the curvature of the stomach. The caudate lobe is covered by a sheet of mesentery, the *lesser omentum,* that connects the liver to the stomach (figs. 50.3 and 50.4). Find the greenish *gallbladder* on the inferior surface of the liver on the right side. Also note the *cystic duct,* by which the gallbladder is attached to the *common bile duct,* and the *hepatic duct,* which originates in the liver and attaches to the cystic duct. Trace the common bile duct to its connection with the duodenum.

8. Locate the *stomach* in the upper left side of the abdominal cavity. At its anterior end, note the union with the *esophagus,* which passes through the diaphragm. Identify the *cardiac, fundic, body,* and *pyloric regions* of the stomach. Use scissors to make an incision along the convex border of the stomach from the cardiac region to the pylorus. Note that the lining of the stomach has numerous folds (*rugae*). Examine the *pyloric sphincter,* which creates a constriction between the stomach and small intestine.

9. Locate the *pancreas.* It appears as a grayish, two-lobed elongated mass of glandular tissue. One lobe lies dorsal to the stomach and extends across to the duodenum. The other lobe is enclosed by the *mesentery,* which supports the duodenum (figs. 50.3 and 50.4).

Figure 50.2 Ventral view of abdominal organs of the cat.

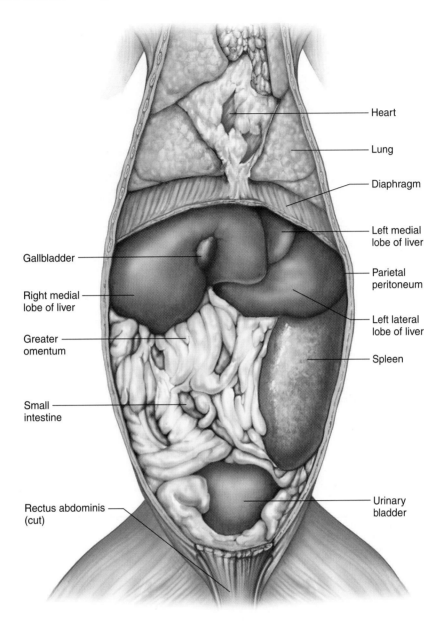

Heart

Lung

Diaphragm

Left medial lobe of liver

Parietal peritoneum

Left lateral lobe of liver

Spleen

Urinary bladder

Gallbladder

Right medial lobe of liver

Greater omentum

Small intestine

Rectus abdominis (cut)

10. Trace the *small intestine,* beginning at the pyloric sphincter. The first portion, the *duodenum,* travels posteriorly for several centimeters. Then it loops back around a lobe of the pancreas. The proximal half of the remaining portion of the small intestine is the *jejunum,* and the distal half is the *ileum.* Open the duodenum and note the velvety appearance of the villi. Note how the mesentery supports the small intestine from the dorsal body wall. The small intestine terminates on the right side, where it joins the large intestine.

11. Locate the *large intestine,* and identify the *cecum, ascending colon, transverse colon,* and *descending colon.* Also locate the *rectum,* which extends through the pelvic cavity to the *anus.* Make an incision at the junction between the ileum and cecum, and look for the *ileocecal sphincter.*

12. Complete Part B of the laboratory report.

Figure 50.3 Ventral view of the cat's abdominal organs with intestines reflected to the right side.

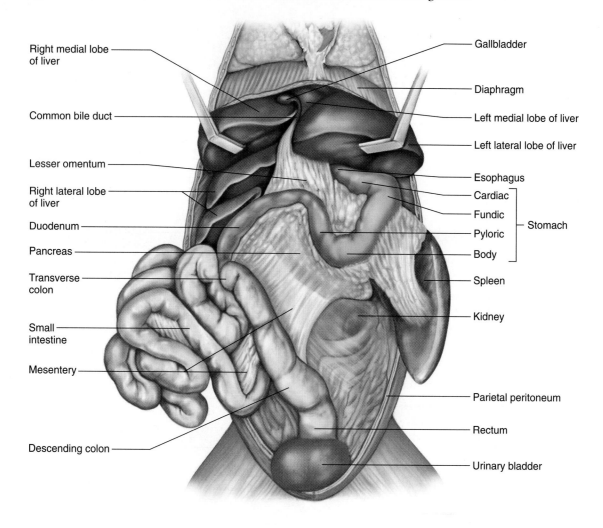

Right medial lobe of liver

Common bile duct

Lesser omentum

Right lateral lobe of liver

Duodenum

Pancreas

Transverse colon

Small intestine

Mesentery

Descending colon

Gallbladder

Diaphragm

Left medial lobe of liver

Left lateral lobe of liver

Esophagus

Cardiac
Fundic
Pyloric
Body

Stomach

Spleen

Kidney

Parietal peritoneum

Rectum

Urinary bladder

Figure 50.4 Abdominal organs of the cat with the greater omentum removed. The left liver lobes have been reflected to the right side.

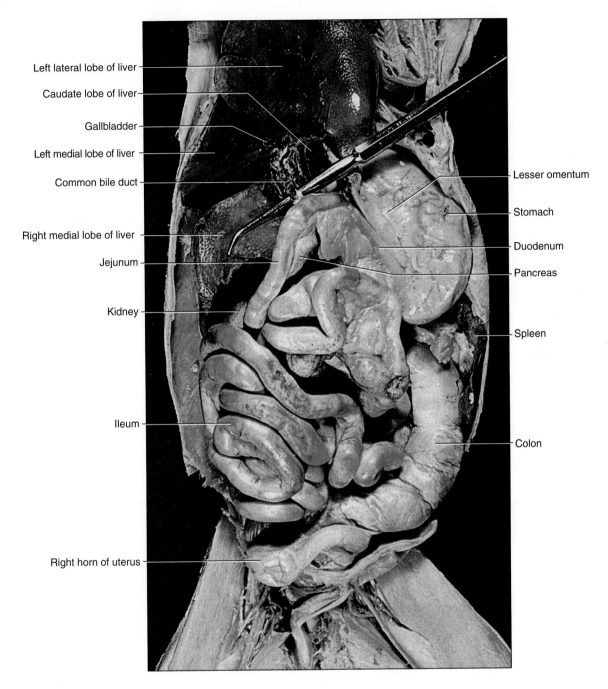

Left lateral lobe of liver

Caudate lobe of liver

Gallbladder

Left medial lobe of liver

Common bile duct

Right medial lobe of liver

Jejunum

Kidney

Ileum

Right horn of uterus

Lesser omentum

Stomach

Duodenum

Pancreas

Spleen

Colon

Name _____

Date _____

Section _____

CAT DISSECTION: DIGESTIVE SYSTEM

PART A

Complete the following:

1. Compare the locations of the major salivary glands of the human with those of the cat. _____
   _____
   _____
   _____
   _____

2. Compare the types and numbers of teeth present in the cat's maxilla with those of the human. _____
   _____
   _____
   _____

3. In what ways do the cat's teeth seem to be adapted to a special diet? _____
   _____
   _____
   _____
   _____

4. What part of the human soft palate is lacking in the cat? _____

5. What do you think is the function of the transverse ridges (rugae) in the hard palate of the cat? _____
   _____
   _____
   _____

6. How do the papillae on the surface of the human tongue compare with those of the cat? _____
   _____
   _____
   _____

PART B

Complete the following:

1. Describe how the peritoneum and mesenteries are associated with the organs in the abdominal cavity. _____

_____

_____

_____

_____

_____

_____

2. Describe the inner lining of the stomach. _____

_____

_____

_____

3. Compare the structure of the human liver with that of the cat. _____

_____

_____

_____

4. Compare the structure and location of the human pancreas with those of the cat. _____

_____

_____

_____

5. What feature of the human cecum is lacking in the cat? _____

_____

ACTION OF A DIGESTIVE ENZYME

SAFETY

- Use only a mechanical pipetting device (never your mouth). Use pipets with rubber bulbs or dropping pipets.
- Wear safety glasses when working with acids and when heating test tubes.
- Use test-tube clamps when handling hot test tubes.
- If an open flame is used for heating the test solutions, keep clothes and hair away from the flame.

The digestive enzyme in salivary secretions is called *amylase.* This enzyme splits starch molecules into sugar (disaccharide) molecules, which is the first step in the digestion of complex carbohydrates.

As in the case of other enzymes, amylase is a protein whose activity is affected by exposure to certain environmental factors, including excessive heat, radiation, electricity, and certain chemicals.

PURPOSE OF THE EXERCISE

To investigate the action of amylase and the effect of heat on its enzymatic activity.

LEARNING OBJECTIVES

After completing this exercise, you should be able to

1. Describe the action of amylase.
2. Test a solution for the presence of starch or the presence of sugar.
3. Test the effects of varying temperatures on the activity of amylase.

PROCEDURE A—AMYLASE ACTIVITY

1. Mark three clean test tubes as *tubes 1, 2, 3,* and prepare the tubes as follows:

    Tube 1: **Add 6 mL of amylase solution.**

    Tube 2: **Add 6 mL of starch solution.**

    Tube 3: **Add 5 mL of starch solution and 1 mL of amylase solution.**

2. Shake the tubes well to mix the contents, and place them in a warm water bath, 37°C (98.6°F), for 10 minutes.
3. At the end of the 10 minutes, test the contents of each tube for the presence of starch. To do this, follow these steps:
    a. Place 1 mL of the solution to be tested in a depression of a porcelain test plate.
    b. Next add one drop of iodine-potassium-iodide solution, and note the color of the mixture. If the solution becomes blue-black, starch is present.
    c. Record the results in Part A of Laboratory Report 51.
4. Test the contents of each tube for the presence of sugar (disaccharides in this instance). To do this, follow these steps:
    a. Place 1 mL of the solution to be tested in a clean test tube.
    b. Add 1 mL of Benedict's solution.
    c. Place the test tube with a test-tube clamp in a beaker of boiling water for 2 minutes.

d. Note the color of the liquid. If the solution becomes green, yellow, orange, or red, sugar is present. Blue indicates a negative test, whereas green indicates a positive test with the least amount of sugar, and red indicates the greatest amount of sugar present.

e. Record the results in Part A of the laboratory report.

5. Complete Part A of the laboratory report.

PROCEDURE B—EFFECT OF HEAT

1. Mark three clean test tubes as *tubes 4, 5,* and *6.*

2. Add 1 mL of amylase solution to each of the tubes, and expose each solution to a different test temperature for 3 minutes as follows:

Tube 4: **Place in beaker of ice water (about 0°C [32°F]).**

Tube 5: **Place in warm water bath (about 37°C [98.6°F]).**

Tube 6: **Place in beaker of boiling water (about 100°C [212°F]). Use a test-tube clamp.**

3. Add 5 mL of starch solution to each tube, shake to mix the contents, and return the tubes to their respective test temperatures for 10 minutes. It is important that the 5 mL of starch solution added to tube 4 be at ice-water temperature before it is added to the 1 mL of amylase solution.

4. At the end of the 10 minutes, test the contents of each tube for the presence of starch and the presence of sugar by following the directions in Procedure A.

5. Complete Part B of the laboratory report.

OPTIONAL ACTIVITY

Devise an experiment to test the effect of some other environmental factor on amylase activity. For example, you might test the effect of a strong acid by adding a few drops of concentrated hydrochloric acid to a mixture of starch and amylase solutions. Be sure to include a control in your experimental plan. That is, include a tube containing everything except the factor you are testing. Then you will have something with which to compare your results. *Carry out your experiment only if it has been approved by the laboratory instructor.*

ACTION OF A DIGESTIVE ENZYME

PART A—AMYLASE ACTIVITY

1. Test results:

Tube	Starch	Sugar
1 Amylase solution		
2 Starch solution		
3 Starch-amylase solution		

2. Complete the following:

   a. Explain the reason for including tube 1 in this experiment. _____

   _____

   _____

   _____

   b. What is the importance of tube 2? _____

   _____

   _____

   _____

   c. What do you conclude from the results of this experiment?_____

   _____

   _____

   _____

PART B—EFFECT OF HEAT

1. Test results:

Tube	Starch	Sugar
4 0°C (32°F)		
5 37°C (98.6°F)		
6 100°C (212°F)		

2. Complete the following:

   a. What do you conclude from the results of this experiment?_____

   _____

   _____

   _____

   _____

   _____

   b. If digestion failed to occur in one of the tubes in this experiment, how can you tell if the amylase was destroyed by the factor being tested or if the amylase activity was simply inhibited by the test treatment?

   _____

   _____

   _____

   _____

   _____

Critical Thinking Application

What test result would occur if the amylase you used contained sugar? _____
Would your results be valid? Explain your answer.

_____

_____

_____

_____

ORGANS OF THE RESPIRATORY SYSTEM

MATERIALS NEEDED

Textbook
Human skull (sagittal section)
Human torso
Larynx model
Thoracic organs model
Compound light microscope
Prepared microscope slides of the following:
    Trachea (cross section)
    Lung, human (normal)

For Demonstrations:

Animal lung with trachea (fresh or preserved)
Prepared microscope slides of the following:
    Lung tissue (smoker)
    Lung tissue (emphysema)

SAFETY

- Wear disposable gloves when working on the fresh or preserved animal lung demonstration.
- Wash your hands before leaving the laboratory.

The organs of the respiratory system include the nose, nasal cavity, sinuses, pharynx, larynx, trachea, bronchial tree, and lungs. They mainly function to process incoming air and to transport it to and from the atmosphere outside the body and the air sacs of the lungs.

In the air sacs, gas exchanges take place between the air and the blood of nearby capillaries. The blood, in turn, transports gases to and from the air sacs and the body cells. This entire process of transporting and exchanging gases between the atmosphere and the body cells is called *respiration.*

PURPOSE OF THE EXERCISE

To review the structure and function of the respiratory organs and to examine the tissues of some of these organs microscopically.

LEARNING OBJECTIVES

After completing this exercise, you should be able to

1. Locate the major organs of the respiratory system.
2. Describe the functions of these organs.
3. Recognize tissue sections of the trachea and lung.
4. Identify the major features of these tissue sections.

PROCEDURE A—RESPIRATORY ORGANS

1. Review the section entitled "Organs of the Respiratory System" in chapter 19 of the textbook.
2. As a review activity, label figures 52.1, 52.2, 52.3, and 52.4.
3. Examine the sagittal section of the human skull, and locate the following features:

nose

    nostrils (external nares)

nasal cavity

    nasal septum

    nasal conchae

    meatuses

sinuses

    maxillary sinus

    frontal sinus

    ethmoidal sinus

    sphenoidal sinus

Figure 52.1 Label the major features of the respiratory system.

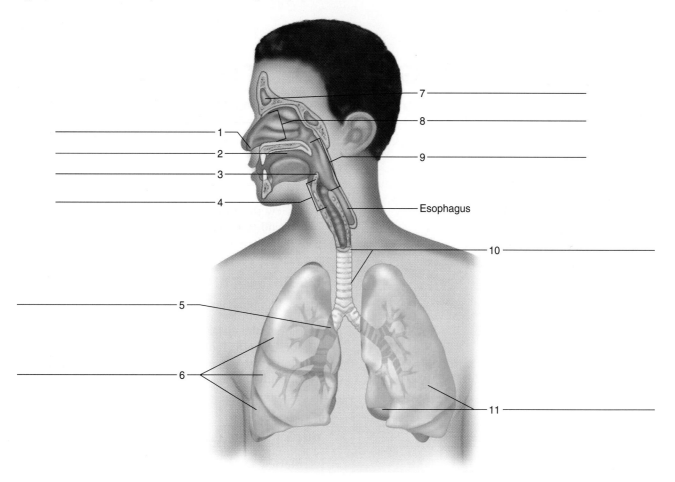

4. Observe the larynx model, the thoracic organs model, and the human torso. Locate the features listed in step 3. Also locate the following:

pharynx

nasopharynx

oropharynx

laryngopharynx

larynx (palpate your own larynx)

vocal cords

false vocal cords

true vocal cords

thyroid cartilage ("Adam's apple")

cricoid cartilage

epiglottis

epiglottic cartilage

arytenoid cartilages

corniculate cartilages

cuneiform cartilages

glottis

trachea (palpate your own trachea)

bronchi

primary bronchi

secondary bronchi

lung

hilum

lobes

superior lobe

middle lobe (right only)

inferior lobe

lobules

Figure 52.2 Label the features of this sagittal section of the upper respiratory tract.

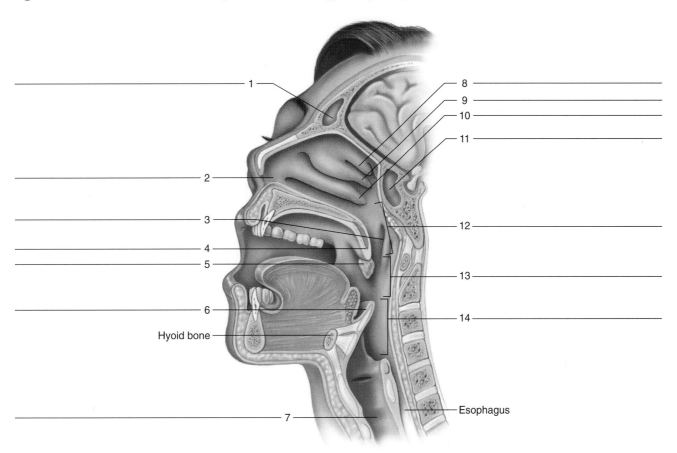

1 _____

2 _____

3 _____

4 _____

5 _____

6 _____

Hyoid bone

7 _____

8 _____

9 _____

10 _____

11 _____

12 _____

13 _____

14 _____

Esophagus

visceral pleura

parietal pleura

pleural cavity

5. Complete Part A of Laboratory Report 52.

DEMONSTRATION

Observe the animal lung and the attached trachea. Identify the larynx, major laryngeal cartilages, trachea, and the incomplete cartilaginous rings of the trachea. Open the larynx and locate the vocal folds. Examine the visceral pleura on the surface of a lung, and squeeze a portion of a lung between your fingers. How do you describe the texture of the lung?

_____

_____

_____

PROCEDURE B—RESPIRATORY TISSUES

1. Obtain a prepared microscope slide of a trachea, and use low-power magnification to examine it. Notice the inner lining of ciliated pseudostratified columnar epithelium and the deep layer of hyaline cartilage, which represents a portion of an incomplete (C-shaped) tracheal ring (fig. 52.5).

2. Use high-power magnification to observe the cilia on the free surface of the epithelial lining. Locate the wineglass-shaped goblet cells, which secrete a protective mucus, in the epithelium.

3. Prepare a labeled sketch of a representative portion of the tracheal wall in Part B of the laboratory report.

4. Obtain a prepared microscope slide of human lung tissue. Examine it, using low-power magnification, and note the many open spaces of the air sacs (alveoli). Look for a bronchiole—a tube with a relatively thick wall and a wavy inner lining. Locate the smooth muscle tissue in the wall of this tube (fig. 52.6). You also may see a section of cartilage as part of the bronchiole wall.

Figure 52.3 Label the major features of the larynx: (*a*) anterior view; (*b*) posterior view.

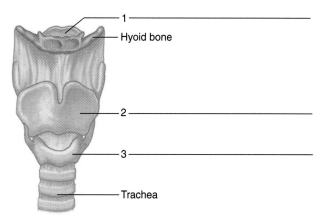

1 _____
Hyoid bone

2 _____

3 _____

Trachea

(a)

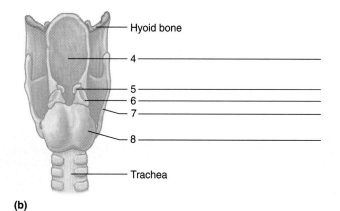

Hyoid bone

4 _____

5 _____
6 _____
7 _____

8 _____

Trachea

(b)

5. Use high-power magnification to examine the alveoli (fig. 52.7). Note that their walls are composed of simple squamous epithelium. You also may see sections of blood vessels filled with blood cells.
6. Prepare a labeled sketch of a representative portion of the lung in Part B of the laboratory report.
7. Complete Part C of the laboratory report.

DEMONSTRATION

Examine the prepared microscope slides of the lung tissue of a smoker and a person with emphysema, using low-power magnification. How does the smoker's lung tissue compare with that of the normal lung tissue that you examined previously?

_____
_____
_____

How does the emphysema patient's lung tissue compare with the normal lung tissue?

_____
_____
_____
_____

Web Quest

Review the structure and function of the respiratory system at
www.mhhe.com/shier10
Click on Student Edition. Click on Martin Lab Manual, Web Quest.

Figure 52.4 Label the features of the superior aspect of the larynx with the glottis closed.

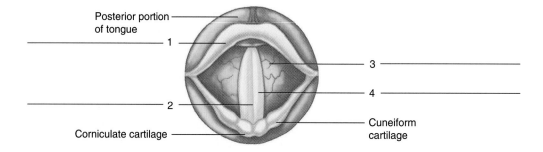

Posterior portion of tongue

1 _____

_____ 2

Corniculate cartilage

3 _____

4 _____

Cuneiform cartilage

Figure 52.5 Micrograph of a section of the tracheal wall (63×).

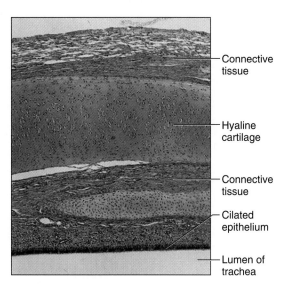

— Connective tissue

— Hyaline cartilage

— Connective tissue

— Cilated epithelium

— Lumen of trachea

Figure 52.6 Micrograph of human lung tissue (35×).

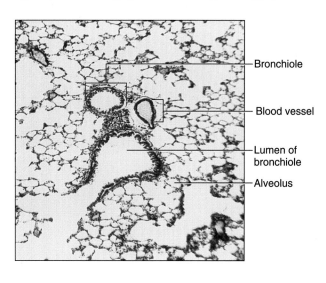

— Bronchiole

— Blood vessel

— Lumen of bronchiole

— Alveolus

Figure 52.7 Micrograph of human lung tissue (250×).

Capillary —

Simple squamous epithelial cells —

Alveolus —

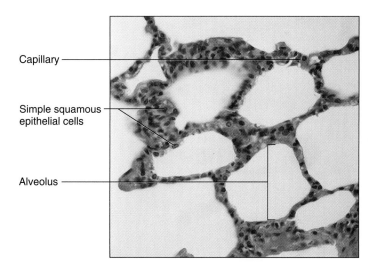

ORGANS OF THE RESPIRATORY SYSTEM

PART A

Match the terms in column A with the descriptions in column B. Place the letter of your choice in the space provided.

Column A	Column B
a. Alveolus	_____ 1. Potential space between visceral and parietal pleurae
b. Cricoid cartilage	
c. Epiglottis	_____ 2. Most inferior portion of larynx
d. Glottis	_____ 3. Serves as resonant chamber and reduces weight of skull
e. Lung	_____ 4. Microscopic air sac for gas exchange
f. Nasal concha	
g. Pharynx	_____ 5. Consists of large lobes
h. Pleural cavity	_____ 6. Opening between vocal cords
i. Sinus (paranasal sinus)	_____ 7. Fold of mucous membrane containing elastic fibers responsible for sounds
j. Vocal cord (true)	
	_____ 8. Increases surface area of nasal mucous membrane
	_____ 9. Passageway for air and food
	_____ 10. Partially covers opening of larynx during swallowing

PART B

1. Prepare a labeled sketch of a portion of the tracheal wall.

2. Prepare a labeled sketch of a portion of lung tissue.

PART C

Complete the following:

1. What is the function of the mucus secreted by the goblet cells? _____

_____

_____

2. Describe the function of the cilia in the respiratory tubes. _____

_____

_____

3. How is breathing affected if the smooth muscle of the bronchial tree relaxes? _____

_____

_____

Critical Thinking Application

Why are the alveolar walls so thin?

_____

_____

_____

_____

CAT DISSECTION: RESPIRATORY SYSTEM

MATERIALS NEEDED

Embalmed cat
Dissecting tray
Dissecting instruments
Disposable gloves
Human torso

SAFETY

- Wear disposable gloves when working on the cat dissection.
- Dispose of tissue remnants and gloves as instructed.
- Wash the dissecting tray and instruments as instructed.
- Wash your laboratory table.
- Wash your hands before leaving the laboratory.

In this laboratory exercise, you will dissect the major respiratory organs of an embalmed cat. As you observe these structures in the cat, compare them with those of the human torso.

PURPOSE OF THE EXERCISE

To examine the major respiratory organs of the cat and to compare these organs with those of the human.

LEARNING OBJECTIVES

After completing this exercise, you should be able to

1. Locate and identify the major respiratory organs of the embalmed cat.
2. Identify the corresponding organs in the human torso.
3. Compare the respiratory system of the human with that of the cat.

PROCEDURE

1. Place the embalmed cat in a dissecting tray with its ventral surface up.
2. Examine the *nostrils* (*external nares*) and *nasal septum.*
3. Open the mouth wide to expose the *soft palate.* Cut through the tissues of the soft palate, and observe the *nasopharynx* above it. Locate the small openings of the *auditory tubes* in the lateral walls of the nasopharynx. Insert a probe into an opening of an auditory tube.
4. Pull the tongue forward, and locate the *epiglottis* at its base. Also identify the *glottis,* which is the opening into the larynx. (In the human, the term *glottis* refers to the opening between the vocal folds within the larynx.) Locate the *esophagus,* which is dorsal to the larynx.
5. Open the thoracic cavity, and expose its contents.
6. Dissect the *trachea* in the neck, and expose the *larynx* at the anterior end near the base of the tongue. Note the *tracheal rings,* and locate the lobes of the *thyroid gland* on each side of the trachea, just posterior to the larynx. Also note the *thymus gland,* a rather diffuse mass of glandular tissue extending along the ventral surface of the trachea into the thorax (see fig. 47.3 and figs. 53.1 and 53.2).
7. Examine the larynx, and identify the *thyroid cartilage* and *cricoid cartilage* in its wall. Make a longitudinal incision through the ventral wall of the larynx, and locate the *vocal cords* inside.
8. Trace the trachea posteriorly to where it divides into *primary bronchi,* which pass into the lungs.
9. Examine the *lungs,* each of which is subdivided into three lobes—an *anterior,* a *middle,* and a *posterior lobe.* Notice the thin membrane, the *visceral pleura,* on the surface of each lung. Also notice the *parietal pleura,* which forms the inner lining of the thoracic wall, and locate the spaces of the *pleural cavities.*
10. Make an incision through a lobe of a lung, and examine its interior. Note the branches of the smaller air passages.
11. Examine the *diaphragm,* and note that its central portion is tendinous.
12. Locate the *phrenic nerve.* This nerve appears as a white thread passing along the side of the heart to the diaphragm.
13. Examine the organs located in the *mediastinum.* Note that the mediastinum separates the right and left lungs and pleural cavities within the thorax.
14. Complete Laboratory Report 53.

Figure 53.1 Ventral view of the cat's thoracic cavity.

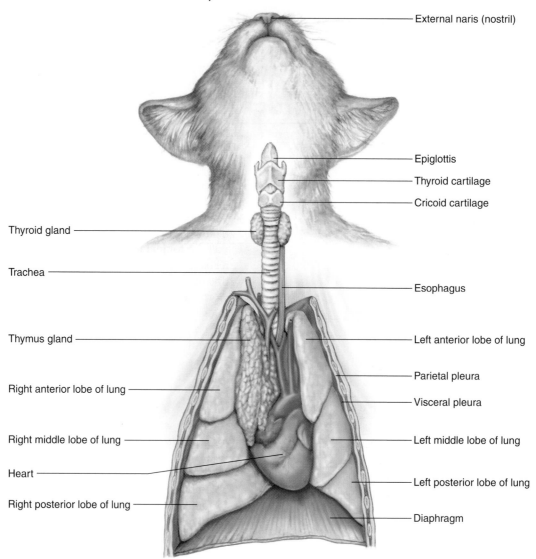

External naris (nostril)

Epiglottis

Thyroid cartilage

Cricoid cartilage

Thyroid gland

Trachea

Esophagus

Thymus gland

Left anterior lobe of lung

Parietal pleura

Right anterior lobe of lung

Visceral pleura

Right middle lobe of lung

Left middle lobe of lung

Heart

Left posterior lobe of lung

Right posterior lobe of lung

Diaphragm

Figure 53.2 Features of the cat's neck and oral cavity.

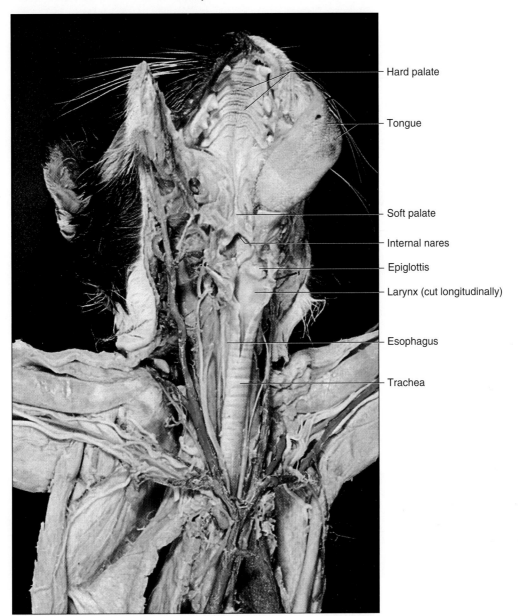

Hard palate

Tongue

Soft palate

Internal nares

Epiglottis

Larynx (cut longitudinally)

Esophagus

Trachea

Laboratory Report **53**

Name _____

Date _____

Section _____

Cat Dissection: Respiratory System

Complete the following:

1. What is the purpose of the auditory tubes opening into the nasopharynx? _____
   _____
   _____
   _____
   _____

2. Distinguish between the glottis and the epiglottis. _____
   _____
   _____
   _____
   _____

3. Are the tracheal rings of the cat complete or incomplete? _____ How does this feature
   compare with that of the human? _____
   _____
   _____

4. How does the structure of the primary bronchi compare with that of the trachea? _____
   _____
   _____

5. Compare the number of lobes in the human lungs with the number of lobes in the cat._____
   _____
   _____
   _____
   _____

6. Describe the attachments of the diaphragm. _____
   _____
   _____
   _____
   _____

7. What major structures are located within the mediastinum of the cat? _____
   _____ How does this compare to the human mediastinum?_____
   _____

BREATHING AND RESPIRATORY VOLUMES AND CAPACITIES

SAFETY

- Clean the spirometer with cotton moistened with 70% alcohol before each use.
- Place a new disposable mouthpiece on the stem of the spirometer before each use.
- Dispose of the cotton and mouthpieces according to your laboratory instructor's directions.

Breathing involves the movement of air from outside the body through the bronchial tree and into the alveoli and the reversal of this air movement. These movements are caused by changes in the size of the thoracic cavity that result from skeletal muscle contractions and from the elastic recoil of stretched tissues.

The volumes of air that move in and out of the lungs during various phases of breathing are called *respiratory air volumes* and *capacities.* These volumes can be measured by using an instrument called a *spirometer.* However, the values obtained vary with a person's age, sex, height, and weight. Various respiratory capacities can be calculated by combining two or more of the respiratory volumes.

PURPOSE OF THE EXERCISE

To review the mechanisms of breathing and to measure or calculate certain respiratory air volumes and respiratory capacities.

LEARNING OBJECTIVES

After completing this exercise, you should be able to

1. Describe the mechanisms responsible for inspiration and expiration.
2. Define the respiratory air volumes and respiratory capacities.
3. Measure or calculate the respiratory air volumes and capacities.

PROCEDURE A—BREATHING MECHANISMS

1. Review the sections entitled "Inspiration" and "Expiration" in chapter 19 of the textbook.
2. Complete Part A of Laboratory Report 54.

DEMONSTRATION

Observe the mechanical lung function model. Note that it consists of a heavy plastic bell jar with a rubber sheeting tied over its wide open end. Its narrow upper opening is plugged with a rubber stopper through which a glass Y tube is passed. Small rubber balloons are fastened to the arms of the Y (fig. 54.1). What happens to the balloons when the rubber sheeting is pulled downward?

_____

_____

What happens when the sheeting is pushed upward?

_____

_____

_____

How do you explain these changes?

_____

_____

What part of the respiratory system is represented by the rubber sheeting? _____
the bell jar?_____
the Y tube? _____
the balloons? _____

WARNING

If the subject begins to feel dizzy or lightheaded while performing Procedure B, stop the exercise and breathe normally.

Figure 54.1 A lung function model.

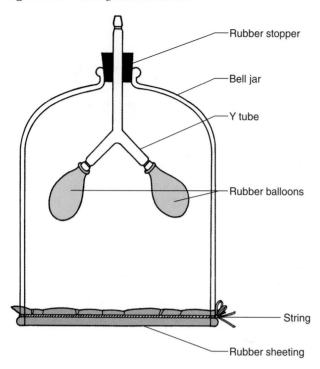

- Rubber stopper
- Bell jar
- Y tube
- Rubber balloons
- String
- Rubber sheeting

PROCEDURE B—RESPIRATORY AIR VOLUMES AND CAPACITIES

1. Review the section entitled "Respiratory Volumes and Capacities" in chapter 19 of the textbook.
2. Complete Part B of the laboratory report.
3. Obtain a handheld spirometer. Note that the needle can be set at zero by rotating the adjustable dial. Before using the instrument, clean it with cotton moistened with 70% alcohol and place a new disposable mouthpiece over its stem. The instrument should be held with the dial upward, and air should be blown into the disposable mouthpiece. Movement of the needle indicates the air volume that leaves the lungs (fig. 54.2).
4. *Tidal volume* is the volume of air that enters (or leaves) the lungs during a *respiratory cycle* (one inspiration plus the following expiration). *Resting tidal volume* is the volume of air that enters (or leaves) the lungs during normal, quiet breathing (fig. 54.3). To measure this volume, follow these steps:
   a. Sit quietly for a few moments.
   b. Position the spirometer dial so that the needle points to zero.
   c. Place the mouthpiece between your lips and exhale three ordinary expirations into it after inhaling through the nose each time. *Do not force air out of your lungs; exhale normally.*

Figure 54.2 A handheld spirometer can be used to measure respiratory air volumes.

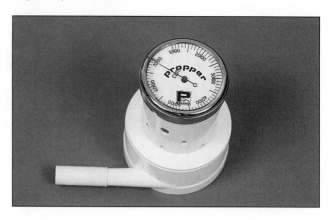

   d. Divide the total value indicated by the needle by 3, and record this amount as your resting tidal volume on the table in Part C of the laboratory report.
5. *Expiratory reserve volume* is the volume of air in addition to the tidal volume that leaves the lungs during forced expiration. To measure this volume, follow these steps:
   a. Breathe normally for a few moments. Set the needle to zero.
   b. At the end of an ordinary expiration, place the mouthpiece between your lips and exhale all of the air you can force from your lungs through the spirometer.
   c. Record the results as your expiratory reserve volume in Part C.
6. *Vital capacity* is the maximum volume of air that can be exhaled after taking the deepest breath possible. To measure this volume, follow these steps:
   a. Breathe normally for a few moments. Set the needle at zero.
   b. Breathe in and out deeply a couple of times, then take the deepest breath possible.
   c. Place the mouthpiece between your lips and exhale all the air out of your lungs, slowly and forcefully.
   d. Record the value as your vital capacity in Part C. Compare your result with that expected for a person of your sex, age, and height listed in tables 54.1 and 54.2. Use the meterstick to determine your height in centimeters if necessary or multiply your height in inches times 2.54 to calculate your height in centimeters. Considerable individual variations from the expected will be noted due to parameters other than sex, age, and height, which could include physical shape, health, medications, and others.

Table 54.1 Predicted Vital Capacities (in Milliliters) for Females

Age	Height in Centimeters																								
	146	148	150	152	154	156	158	160	162	164	166	168	170	172	174	176	178	180	182	184	186	188	190	192	194
16	2950	2990	3030	3070	3110	3150	3190	3230	3270	3310	3350	3390	3430	3470	3510	3550	3590	3630	3670	3715	3755	3800	3840	3880	3920
17	2935	2975	3015	3055	3095	3135	3175	3215	3255	3295	3335	3375	3415	3455	3495	3535	3575	3615	3655	3695	3740	3780	3820	3860	3900
18	2920	2960	3000	3040	3080	3120	3160	3200	3240	3280	3320	3360	3400	3440	3480	3520	3560	3600	3640	3680	3720	3760	3800	3840	3880
20	2890	2930	2970	3010	3050	3090	3130	3170	3210	3250	3290	3330	3370	3410	3450	3490	3525	3565	3605	3645	3695	3720	3760	3800	3840
22	2860	2900	2940	2980	3020	3060	3095	3135	3175	3215	3255	3290	3330	3370	3410	3450	3490	3530	3570	3610	3650	3685	3725	3765	3800
24	2830	2870	2910	2950	2985	3025	3065	3100	3140	3180	3220	3260	3300	3335	3375	3415	3455	3490	3530	3570	3610	3650	3685	3725	3765
26	2800	2840	2880	2920	2960	3000	3035	3070	3110	3150	3190	3230	3265	3300	3340	3380	3420	3455	3495	3530	3570	3610	3650	3685	3725
28	2775	2810	2850	2890	2930	2965	3000	3040	3070	3115	3155	3190	3230	3270	3305	3345	3380	3420	3460	3495	3535	3570	3610	3650	3685
30	2745	2780	2820	2860	2895	2935	2970	3010	3045	3085	3120	3160	3195	3235	3270	3310	3345	3385	3420	3460	3495	3535	3570	3610	3645
32	2715	2750	2790	2825	2865	2900	2940	2975	3015	3050	3090	3125	3160	3200	3235	3275	3310	3350	3385	3425	3460	3495	3535	3570	3610
34	2685	2725	2760	2795	2835	2870	2910	2945	2980	3020	3055	3090	3130	3165	3200	3240	3275	3310	3350	3385	3425	3460	3495	3535	3570
36	2655	2695	2730	2765	2805	2840	2875	2910	2950	2985	3020	3060	3095	3130	3165	3205	3240	3275	3310	3350	3385	3420	3460	3495	3530
38	2630	2665	2700	2735	2770	2810	2845	2880	2915	2950	2990	3025	3060	3095	3130	3170	3205	3240	3275	3310	3350	3385	3420	3455	3490
40	2600	2635	2670	2705	2740	2775	2810	2850	2885	2920	2955	2990	3025	3060	3095	3135	3170	3205	3240	3275	3310	3345	3380	3420	3455
42	2570	2605	2640	2675	2710	2745	2780	2815	2850	2885	2920	2955	2990	3025	3060	3100	3135	3170	3205	3240	3275	3310	3345	3380	3415
44	2540	2575	2610	2645	2680	2715	2750	2785	2820	2855	2890	2925	2960	2995	3030	3060	3095	3130	3165	3200	3235	3270	3305	3340	3375
46	2510	2545	2580	2615	2650	2685	2715	2750	2785	2820	2855	2890	2925	2960	2995	3030	3060	3095	3130	3165	3200	3235	3270	3305	3340
48	2480	2515	2550	2585	2620	2650	2685	2715	2750	2785	2820	2855	2890	2925	2960	2995	3030	3060	3095	3130	3160	3195	3230	3265	3300
50	2455	2485	2520	2555	2590	2625	2655	2690	2720	2755	2785	2820	2855	2890	2925	2955	2990	3025	3060	3090	3125	3155	3190	3225	3260
52	2425	2455	2490	2525	2555	2590	2625	2655	2690	2720	2755	2790	2820	2855	2890	2925	2955	2990	3020	3055	3090	3125	3155	3190	3220
54	2395	2425	2460	2495	2530	2560	2590	2625	2655	2690	2720	2755	2790	2820	2855	2885	2920	2950	2985	3020	3050	3085	3115	3150	3180
56	2365	2400	2430	2460	2495	2525	2560	2590	2625	2655	2690	2720	2755	2790	2820	2855	2885	2920	2950	2980	3015	3045	3080	3110	3145
58	2335	2370	2400	2430	2460	2495	2525	2560	2590	2625	2655	2690	2720	2750	2785	2815	2850	2880	2920	2945	2975	3010	3040	3075	3105
60	2305	2340	2370	2400	2430	2460	2495	2525	2560	2590	2625	2655	2685	2720	2750	2780	2810	2845	2875	2915	2940	2970	3000	3035	3065
62	2280	2310	2340	2370	2405	2435	2465	2495	2525	2560	2590	2620	2655	2685	2715	2745	2775	2810	2840	2870	2900	2935	2965	2995	3025
64	2250	2280	2310	2340	2370	2400	2430	2465	2495	2525	2555	2585	2620	2650	2680	2710	2740	2770	2805	2835	2865	2895	2920	2955	2990
66	2220	2250	2280	2310	2340	2370	2400	2430	2460	2490	2525	2555	2585	2615	2645	2675	2705	2735	2765	2800	2825	2860	2890	2920	2950
68	2190	2220	2250	2280	2310	2340	2370	2400	2430	2460	2490	2520	2550	2580	2610	2640	2670	2700	2730	2760	2795	2820	2850	2880	2910
70	2160	2190	2220	2250	2280	2310	2340	2370	2400	2425	2455	2485	2515	2545	2575	2605	2635	2665	2695	2725	2755	2780	2810	2840	2870
72	2130	2160	2190	2220	2250	2280	2310	2335	2365	2395	2425	2455	2480	2510	2540	2570	2600	2630	2660	2685	2715	2745	2775	2805	2830
74	2100	2130	2160	2190	2220	2245	2275	2305	2335	2360	2390	2420	2450	2475	2505	2535	2565	2590	2620	2650	2680	2710	2740	2765	2795

From E. DeF. Baldwin and E. W. Richards, Jr., "Pulmonary Insufficiency I, Physiologic Classification, Clinical Methods of Analysis, Standard Values in Normal Subjects" in Medicine 27:243. © William & Wilkins. Used by permission.

Table 54.2 Predicted Vital Capacities (in Milliliters) for Males

Age	Height in Centimeters																								
	146	148	150	152	154	156	158	160	162	164	166	168	170	172	174	176	178	180	182	184	186	188	190	192	194
16	3765	3820	3870	3920	3975	4025	4075	4130	4180	4230	4285	4335	4385	4440	4490	4540	4590	4645	4695	4745	4800	4850	4900	4955	5005
18	3740	3790	3840	3890	3940	3995	4045	4095	4145	4200	4250	4300	4350	4405	4455	4505	4555	4610	4660	4710	4760	4815	4865	4915	4965
20	3710	3760	3810	3860	3910	3960	4015	4065	4115	4165	4215	4265	4320	4370	4420	4470	4520	4570	4625	4675	4725	4775	4825	4875	4930
22	3680	3730	3780	3830	3880	3930	3980	4030	4080	4135	4185	4235	4285	4335	4385	4435	4485	4535	4585	4635	4685	4735	4790	4840	4890
24	3635	3685	3735	3785	3835	3885	3935	3985	4035	4085	4135	4185	4235	4285	4330	4380	4430	4480	4530	4580	4630	4680	4730	4780	4830
26	3605	3655	3705	3755	3805	3855	3905	3955	4000	4050	4100	4150	4200	4250	4300	4350	4395	4445	4495	4545	4595	4645	4695	4740	4790
28	3575	3625	3675	3725	3775	3820	3870	3920	3970	4020	4070	4115	4165	4215	4265	4310	4360	4410	4460	4510	4555	4605	4655	4705	4755
30	3550	3595	3645	3695	3740	3790	3840	3890	3935	3985	4035	4080	4130	4180	4230	4275	4325	4375	4425	4470	4520	4570	4615	4665	4715
32	3520	3565	3615	3665	3710	3760	3810	3855	3905	3950	4000	4050	4095	4145	4195	4240	4290	4340	4385	4435	4485	4530	4580	4625	4675
34	3475	3525	3570	3620	3665	3715	3760	3810	3855	3905	3950	4000	4045	4095	4140	4190	4225	4285	4330	4380	4425	4475	4520	4570	4615
36	3445	3495	3540	3585	3635	3680	3730	3775	3825	3870	3920	3965	4010	4060	4105	4155	4200	4250	4295	4340	4390	4435	4485	4530	4580
38	3415	3465	3510	3555	3605	3650	3695	3745	3790	3840	3885	3930	3980	4025	4070	4120	4165	4210	4260	4305	4350	4400	4445	4495	4540
40	3385	3435	3480	3525	3575	3620	3665	3710	3760	3805	3850	3900	3945	3990	4035	4085	4130	4175	4220	4270	4315	4360	4410	4455	4500
42	3360	3405	3450	3495	3540	3590	3635	3680	3725	3770	3820	3865	3910	3955	4000	4050	4095	4140	4185	4230	4280	4325	4370	4415	4460
44	3315	3360	3405	3450	3495	3540	3585	3630	3675	3725	3770	3815	3860	3905	3950	3995	4040	4085	4130	4175	4220	4270	4315	4360	4405
46	3285	3330	3375	3420	3465	3510	3555	3600	3645	3690	3735	3780	3825	3870	3915	3960	4005	4050	4095	4140	4185	4230	4275	4320	4365
48	3255	3300	3345	3390	3435	3480	3525	3570	3615	3655	3700	3745	3790	3835	3880	3925	3970	4015	4060	4105	4150	4190	4235	4280	4325
50	3210	3255	3300	3345	3390	3430	3475	3520	3565	3610	3650	3695	3740	3785	3830	3870	3915	3960	4005	4050	4090	4135	4180	4225	4270
52	3185	3225	3270	3315	3355	3400	3445	3490	3530	3575	3620	3660	3705	3750	3795	3835	3880	3925	3970	4010	4055	4100	4140	4185	4230
54	3155	3195	3240	3285	3325	3370	3415	3455	3500	3540	3585	3630	3670	3715	3760	3800	3845	3890	3930	3975	4020	4060	4105	4145	4190
56	3125	3165	3210	3255	3295	3340	3380	3425	3465	3510	3550	3595	3640	3680	3725	3765	3810	3850	3895	3940	3980	4025	4065	4110	4150
58	3080	3125	3165	3210	3250	3290	3335	3375	3420	3460	3500	3545	3585	3630	3670	3715	3755	3800	3840	3880	3925	3965	4010	4050	4095
60	3050	3095	3135	3175	3220	3260	3300	3345	3385	3430	3470	3500	3555	3595	3635	3680	3720	3760	3805	3845	3885	3930	3970	4015	4055
62	3020	3060	3110	3150	3190	3230	3270	3310	3350	3390	3440	3480	3520	3560	3600	3640	3680	3730	3770	3810	3850	3890	3930	3970	4020
64	2990	3030	3080	3120	3160	3200	3240	3280	3320	3360	3400	3440	3490	3530	3570	3610	3650	3690	3730	3770	3810	3850	3900	3940	3980
66	2950	2990	3030	3070	3110	3150	3190	3230	3270	3310	3350	3390	3430	3470	3510	3550	3600	3640	3680	3720	3760	3800	3840	3880	3920
68	2920	2960	3000	3040	3080	3120	3160	3200	3240	3280	3320	3360	3400	3440	3480	3520	3560	3600	3640	3680	3720	3760	3800	3840	3880
70	2890	2930	2970	3010	3050	3090	3130	3170	3210	3250	3290	3330	3370	3410	3450	3480	3520	3560	3600	3640	3680	3720	3760	3800	3840
72	2860	2900	2940	2980	3020	3060	3100	3140	3180	3210	3250	3290	3330	3370	3410	3450	3490	3530	3570	3610	3650	3680	3720	3760	3800
74	2820	2860	2900	2930	2970	3010	3050	3090	3130	3170	3200	3240	3280	3320	3360	3400	3440	3470	3510	3550	3590	3630	3670	3710	3740

From E. DeF. Baldwin and E. W. Richards, Jr., "Pulmonary Insufficiency 1, Physiologic Classification, Clinical Methods of Analysis, Standard Values in Normal Subjects" in Medicine 27:243, © by William & Wilkins. Used by permission.

Critical Thinking Application

It can be noted from the data in tables 54.1 and 54.2 that vital capacities gradually decrease with age. Propose an explanation for this normal correlation.

_____

_____

_____

_____

7. *Inspiratory reserve volume* (IRV) is the volume of air in addition to the tidal volume that enters the lungs during forced inspiration. Calculate your inspiratory reserve volume by subtracting your tidal volume (TV) and your expiratory reserve volume (ERV) from your vital capacity (VC):

$$IRV = VC - (TV + ERV)$$

8. *Inspiratory capacity* (IC) is the maximum volume of air a person can inhale following exhalation of the tidal volume. Calculate your inspiratory capacity by adding your tidal volume (TV) and your inspiratory reserve volume (IRV):

$$IC = TV + IRV$$

9. *Functional residual capacity* (FRC) is the volume of air that remains in the lungs following exhalation of the tidal volume. Calculate your functional residual capacity (FRC) by adding your expiratory reserve volume (ERV) and your residual volume (RV), which you can assume is 1,200 mL:

$$FRC = ERV + 1,200$$

10. Complete Part C of the laboratory report.

OPTIONAL ACTIVITY

Determine your *minute respiratory volume.* To do this, follow these steps:

1. Sit quietly for a while, and then to establish your breathing rate, count the number of times you breathe in 1 minute. This might be inaccurate because conscious awareness of breathing rate can alter the results. You might ask a laboratory partner to record your breathing rate at some time when you are not expecting it to be recorded.

2. Calculate your minute respiratory volume by multiplying your breathing rate by your tidal volume:

$$\underline{\hspace{2cm}} \times \underline{\hspace{2cm}} = \underline{\hspace{2cm}}$$
(breathing rate)   (tidal volume)   (minute respiratory volume)

3. This value indicates the total volume of air that moves into your respiratory passages during each minute of ordinary breathing.

Figure 54.3   Graphic representation of respiratory volumes and capacities.

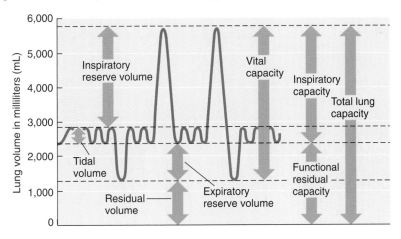

BREATHING AND RESPIRATORY VOLUMES AND CAPACITIES

PART A

Complete the following statements:

1. Breathing also can be called _____ .
2. The weight of air causes a force called _____ pressure.
3. The weight of air at sea level is sufficient to support a column of mercury within a tube _____ mm high.
4. If the pressure inside the lungs decreases, outside air is forced into the airways by _____ .
5. Nerve impulses are carried to the diaphragm by the _____ nerves.
6. When the diaphragm contracts, the size of the thoracic cavity _____ .
7. The ribs are raised by contraction of the _____ muscles, which increases the size of the thoracic cavity.
8. Only a thin film of lubricating serous fluid separates the parietal pleura from the _____ of a lung.
9. A mixture of lipoproteins, called _____ , acts to reduce the tendency of alveoli to collapse.
10. The force responsible for normal expiration comes from _____ of tissues and from surface tension.
11. The pressure between the pleural membranes is usually _____ than the atmospheric pressure.
12. Muscles that help to force out more than the normal volume of air by pulling the ribs downward and inward include the _____ .
13. The diaphragm can be forced to move higher than normal by contraction of the _____ muscles.

PART B

Match the air volumes in column A with the definitions in column B. Place the letter of your choice in the space provided.

Column A	Column B
a. Expiratory reserve volume	_____ 1. Volume in addition to tidal volume that leaves the lungs during forced expiration
b. Functional residual capacity	
c. Inspiratory capacity	_____ 2. Vital capacity plus residual volume
d. Inspiratory reserve volume	_____ 3. Volume that remains in lungs after the most forceful expiration
e. Residual volume	
f. Tidal volume	_____ 4. Volume that enters or leaves lungs during a respiratory cycle
g. Total lung capacity	_____ 5. Volume in addition to tidal volume that enters lungs during forced inspiration
h. Vital capacity	
	_____ 6. Maximum volume a person can exhale after taking the deepest possible breath
	_____ 7. Maximum volume a person can inhale following exhalation of the tidal volume
	_____ 8. Volume of air remaining in the lungs following exhalation of the tidal volume

449

PART C

1. Test results for respiratory air volumes and capacities:

Respiratory Volume or Capacity	Expected Value* (approximate)	Test Results	Percent of Expected Value (test result/expected value × 100)
Tidal volume (resting) (TV)	500 mL		
Expiratory reserve volume (ERV)	1,100 mL		
Vital capacity (VC)	4,600 mL (or enter yours from the table)		
Inspiratory reserve volume (IRV)	3,000 mL		
Inspiratory capacity (IC)	3,500 mL		
Functional residual capacity (FRC)	2,300 mL		

*The values listed are most characteristic for a tall, young adult.

2. Complete the following:

   a. How do your test results compare with the expected values? _____

   _____

   _____

   _____

   _____

   b. How does your vital capacity compare with the average value for a person of your sex, age, and height?

   _____

   _____

   c. What measurement in addition to vital capacity is needed before you can calculate your total lung capacity?

   _____

3. If your experimental results are considerably different than the predicted vital capacities, propose reasons for the differences. As you write this paragraph, consider factors such as smoking, exercise, respiratory disorders, and medications. (Your instructor might have you make some class correlations from class data.)

   _____

   _____

   _____

   _____

   _____

   _____

   _____

   _____

CONTROL OF BREATHING

Normal breathing is controlled from regions of the brainstem called the *respiratory center.* This center initiates and regulates nerve impulses that travel to various breathing muscles, causing rhythmic breathing movements. The respiratory center includes the *medullary rhythmicity area* and the *pneumotaxic area.*

Various factors can influence the respiratory center and thus affect the rate and depth of breathing. These factors include stretch of the lung tissues, emotional state, and the presence in the blood of certain chemicals, such as carbon dioxide, hydrogen ions, and oxygen. For example, the breathing rate increases as the blood concentration of carbon dioxide or hydrogen ions increases or as the concentration of oxygen decreases.

PURPOSE OF THE EXERCISE

To review the mechanisms that control breathing and to investigate some of the factors that affect the rate and depth of breathing.

LEARNING OBJECTIVES

After completing this exercise, you should be able to

1. Locate the breathing centers in the brainstem.
2. Describe the mechanisms that control normal breathing.
3. List several factors that influence the breathing center.
4. Test the effect of various factors on the rate and depth of breathing.

PROCEDURE A—CONTROL OF BREATHING

1. Review the section entitled "Control of Breathing" in chapter 19 of the textbook.
2. Complete Part A of Laboratory Report 55.

DEMONSTRATION

When a solution of calcium hydroxide is exposed to carbon dioxide, a chemical reaction occurs and a white precipitate of calcium carbonate is formed as indicated by the following reaction:

$$Ca(OH)_2 + CO_2 \rightarrow CaCO_2 + H_2O$$

Thus, a clear water solution of calcium hydroxide (limewater) can be used to detect the presence of carbon dioxide because the solution becomes cloudy if this gas is bubbled through it.

The laboratory instructor will demonstrate this test for carbon dioxide by drawing some air through limewater in an apparatus such as that shown in figure 55.1. Then the instructor will blow an equal volume of expired air through a similar apparatus. (*Note:* A new sterile mouthpiece should be used each time the apparatus is demonstrated.) Watch for the appearance of a precipitate that causes the limewater to become cloudy. Was there any carbon dioxide in the atmospheric air drawn through the limewater?

_____

If so, how did the amount of carbon dioxide in the atmospheric air compare with the amount in the expired air?

_____
_____
_____

PROCEDURE B—FACTORS AFFECTING BREATHING

Perform each of the following tests, using your laboratory partner as a test subject.

1. *Normal breathing.* To determine the subject's normal breathing rate and depth, follow these steps:
   a. Have the subject sit quietly for a few minutes.
   b. After the rest period, ask the subject to count backwards mentally, beginning with five hundred.

Figure 55.1 Apparatus used to demonstrate the presence of carbon dioxide in air: (*a*) atmospheric air is drawn through limewater; (*b*) expired air is blown through limewater.

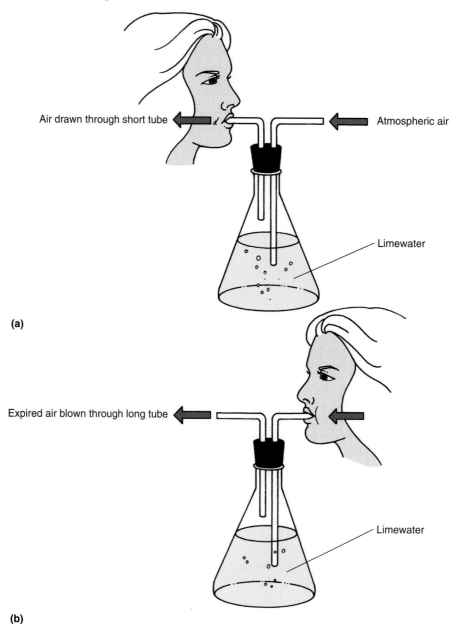

Air drawn through short tube ← | → Atmospheric air

Limewater

(a)

Expired air blown through long tube ← | ←

Limewater

(b)

c. While the subject is distracted by counting, watch the subject's chest movements, and count the breaths taken in a minute. Use this value as the normal breathing rate (breaths per minute).

d. Note the relative depth of the breathing movements.

e. Record your observations in the table in Part B of the laboratory report.

2. *Effect of hyperventilation.* To test the effect of hyperventilation on breathing, follow these steps:

a. Seat the subject and *guard to prevent the possibility of the subject falling over.*

b. Have the subject breathe rapidly and deeply for a maximum of 1 minute. *If the subject begins to feel dizzy, the hyperventilation should be halted immediately to prevent the subject from fainting from complications of alkalosis. The increased blood pH causes vasoconstriction of cerebral arterioles, which decreases circulation and oxygen to the brain.*

c. After the period of hyperventilation, determine the subject's breathing rate, and judge the breathing depth as before.

d. Record the results in Part B of the laboratory report.

3. *Effect of rebreathing air.* To test the effect of rebreathing air on breathing, follow these steps:
   a. Have the subject sit quietly (approximately 5 minutes) until the breathing rate returns to normal.
   b. Have the subject breathe deeply into a small paper bag that is held tightly over the nose and mouth. *If the subject begins to feel light-headed or like fainting, the rebreathing air should be halted immediately to prevent further acidosis and fainting.*
   c. After 2 minutes of rebreathing air, determine the subject's breathing rate, and judge the depth of breathing.
   d. Record the results in Part B of the laboratory report.
4. *Effect of breath holding.* To test the effect of breath holding on breathing, follow these steps:
   a. Have the subject sit quietly (approximately 5 minutes) until the breathing rate returns to normal.
   b. Have the subject hold his or her breath as long as possible. *If the subject begins to feel light-headed or like fainting, breath holding should be halted immediately to prevent further acidosis and fainting.*
   c. As the subject begins to breathe again, determine the rate of breathing, and judge the depth of breathing.
   d. Record the results in Part B of the laboratory report.
5. *Effect of exercise.* To test the effect of exercise on breathing, follow these steps:
   a. Have the subject sit quietly (approximately 5 minutes) until breathing rate returns to normal.
   b. Have the subject exercise by rapidly running in place for 3–5 minutes. *This exercise should be avoided by anyone with health risks.*
   c. After the exercise, determine the breathing rate, and judge the depth of breathing.
   d. Record the results in Part B of the laboratory report.
6. Complete Part B of the laboratory report.

OPTIONAL ACTIVITY

A *pneumograph* is a device that can be used together with some type of recording apparatus to record breathing movements. The laboratory instructor will demonstrate the use of this equipment to record various movements, such as those that accompany coughing, laughing, yawning, and speaking.

Devise an experiment to test the effect of some factor, such as hyperventilation, rebreathing air, or exercise, on the length of time a person can hold the breath. *After the laboratory instructor has approved your plan,* carry out the experiment, using the pneumograph and recording equipment. What conclusion can you draw from the results of your experiment?

_____

_____

_____

_____

Laboratory Report **55**

Name _____

Date _____

Section _____

CONTROL OF BREATHING

PART A

Complete the following statements:

1. The respiratory center is widely scattered throughout the _____ and _____ of the brainstem.

2. The two major components of the respiratory center are the _____ and _____ .

3. The _____ group establishes the basic rhythm of breathing.

4. The _____ group functions during forceful breathing.

5. The _____ regulates the duration of inspiratory bursts and controls the breathing rate.

6. Chemosensitive areas of the respiratory center are located in the ventral portion of the _____ .

7. These chemosensitive areas are stimulated by changes in the blood concentrations of _____ and _____ .

8. As the blood concentration of carbon dioxide increases, the breathing rate _____ .

9. _____ combines with water to form carbonic acid.

10. When carbonic acid dissociates, _____ and hydrogen ions are released.

11. As a result of increased breathing, the blood concentration of carbon dioxide is _____ .

12. The chemoreceptors that sense low blood oxygen concentration are located in the _____ and _____ .

13. As the blood concentration of oxygen decreases, the breathing rate _____ .

14. The inflation reflex involves _____ receptors located in the lungs.

15. The _____ nerve carries sensory impulses from the lungs to the respiratory center.

16. As a result of hyperventilation, breath-holding time is _____ .

PART B

1. Record the results of your breathing tests in the table.

Factor Tested	Breathing Rate (breaths/min)	Breathing Depth (+, ++, +++)
Normal		
Hyperventilation		
Rebreathing air		
Breath holding		
Exercise		

2. Briefly explain the reason for the changes in breathing that occurred in each of the following cases:

  a. Hyperventilation

_____

_____

_____

_____

  b. Rebreathing air

_____

_____

_____

_____

  c. Breath holding

_____

_____

_____

_____

  d. Exercise

_____

_____

_____

_____

3. Complete the following:

   a. Why is it important to distract a person when you are determining the normal rate of breathing?

   _____

   _____

   b. How can the depth of breathing be measured accurately?

   _____

   _____

   _____

Critical Thinking Application

Why is it dangerous for a swimmer to hyperventilate in order to hold the breath for a longer period of time?

_____

_____

_____

_____

STRUCTURE OF THE KIDNEY

MATERIALS NEEDED

Textbook
Human torso
Kidney model
Preserved pig (or sheep) kidney
Dissecting tray
Dissecting instruments
Long knife
Compound light microscope
Prepared microscope slide of a kidney section

SAFETY

- Wear disposable gloves when working on the kidney dissection.
- Dispose of the kidney and gloves as instructed.
- Wash the dissecting tray and instruments as instructed.
- Wash your laboratory table.
- Wash your hands before leaving the laboratory.

The two kidneys are the primary organs of the urinary system. They are located in the upper quadrants of the abdominal cavity, against the posterior wall and behind the parietal peritoneum. They perform a variety of complex activities that lead to the production of urine.

The other organs of the urinary system include the ureters, which transport urine away from the kidneys; the urinary bladder, which stores urine; and the urethra, which conveys urine to the outside of the body.

PURPOSE OF THE EXERCISE

To review the structure of the kidney, to dissect a kidney, and to observe the major structures of a nephron microscopically.

LEARNING OBJECTIVES

After completing this exercise, you should be able to

1. Describe the location of the kidneys.
2. Locate and identify the major structures of a kidney.
3. Identify the major structures of a nephron.
4. Trace the path of filtrate through a renal tubule.
5. Trace the path of blood through the renal blood vessels.

PROCEDURE A—KIDNEY STRUCTURE

1. Review the section entitled "Kidney Structure" in chapter 20 of the textbook.
2. As a review activity, label figures 56.1 and 56.2.
3. Complete Part A of Laboratory Report 56.
4. Observe the human torso and the kidney model. Locate the following:

   kidneys

   ureters

   urinary bladder

   urethra

   renal sinus

   renal pelvis

      major calyces

      minor calyces

   renal papillae

   renal medulla

      renal pyramids

   renal cortex

      renal columns (extensions of renal cortex
       between renal pyramids)

   nephrons

5. To observe the structure of a kidney, follow these steps:
   a. Obtain a pig or sheep kidney and rinse it with water to remove as much of the preserving fluid as possible.
   b. Carefully remove any adipose tissue from the surface of the specimen.
   c. Locate the following features:

   renal capsule

   hilum

   renal artery

   renal vein

   ureter

Figure 56.1 Label the major structures of the urinary system.

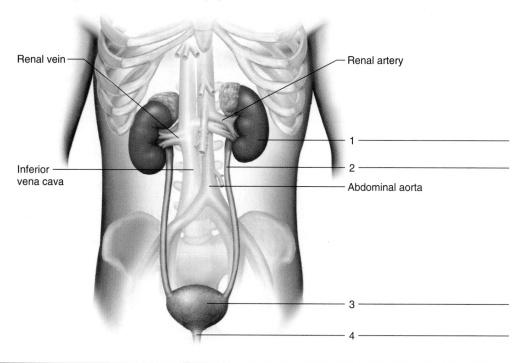

Renal vein

Renal artery

1

Inferior
vena cava

2

Abdominal aorta

3

4

Figure 56.2 Label the major structures in the longitudinal section of a kidney.

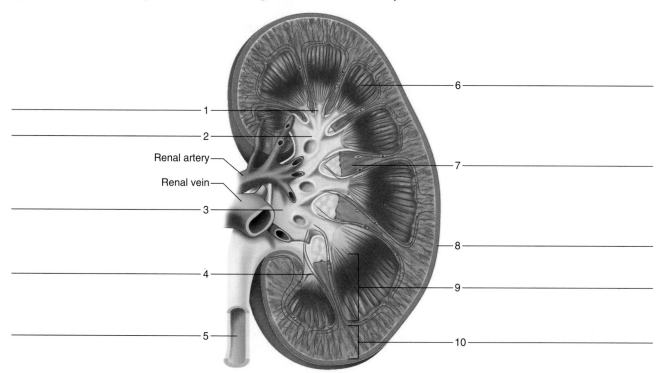

1

2

Renal artery

Renal vein

3

4

5

6

7

8

9

10

d. Use a long knife to cut the kidney in half
   longitudinally along the coronal plane,
   beginning on the convex border.

e. Rinse the interior of the kidney with water, and
   using figure 56.3 as a reference, locate the
   following:

renal pelvis

   major calyces

   minor calyces

renal cortex

   renal columns (extensions of renal cortex
   between renal pyramids)

renal medulla

   renal pyramids

Figure 56.3 Longitudinal section of a pig kidney that has a triple injection of latex (*red* in the renal artery, *blue* in the renal vein and *yellow* in the ureter and renal pelvis).

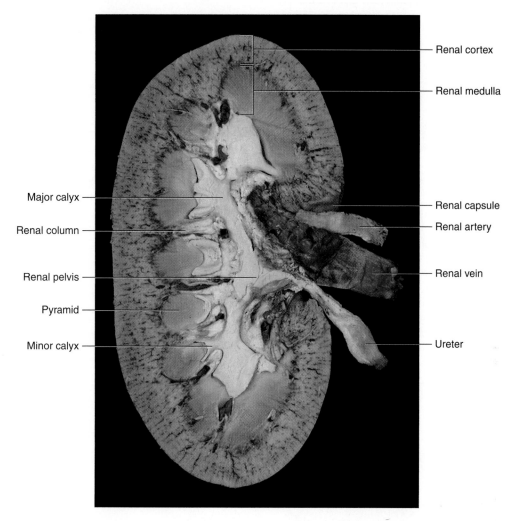

Renal cortex

Renal medulla

Major calyx

Renal column

Renal pelvis

Pyramid

Minor calyx

Renal capsule

Renal artery

Renal vein

Ureter

PROCEDURE B—THE RENAL BLOOD VESSELS AND NEPHRONS

1. Review the sections entitled "Renal Blood Vessels" and "Nephrons" in chapter 20 of the textbook.
2. As a review activity, label figure 56.4.
3. Complete Part B of the laboratory report.
4. Obtain a microscope slide of a kidney section and examine it, using low-power magnification. Locate the *renal capsule,* the *renal cortex* (which appears somewhat granular and may be more darkly stained than the other renal tissues,) and the *renal medulla* (fig. 56.5).
5. Examine the renal cortex, using high-power magnification. Locate a *renal corpuscle.* These structures appear as isolated circular areas. Identify the *glomerulus,* which is the capillary cluster inside the corpuscle, and the *glomerular capsule,* which appears as a clear area surrounding the glomerulus. Also note the numerous sections of renal tubules that occupy the spaces between renal corpuscles (fig. 56.5a).

6. Prepare a labeled sketch of a representative section of renal cortex in Part C of the laboratory report.
7. Examine the renal medulla, using high-power magnification. Identify longitudinal and cross sections of various collecting ducts. Note that these ducts are lined with simple epithelial cells, which vary in shape from squamous to cuboidal (fig. 56.5b).
8. Prepare a labeled sketch of a representative section of renal medulla in Part D of the laboratory report.

Web Quest

Review the structure and function of the urinary system at
www.mhhe.com/shier10

Click on Student Edition. Click on Martin Lab Manual, Web Quest.

Figure 56.4 Label the major structures of the nephron and the blood vessels associated with it, using the terms provided.

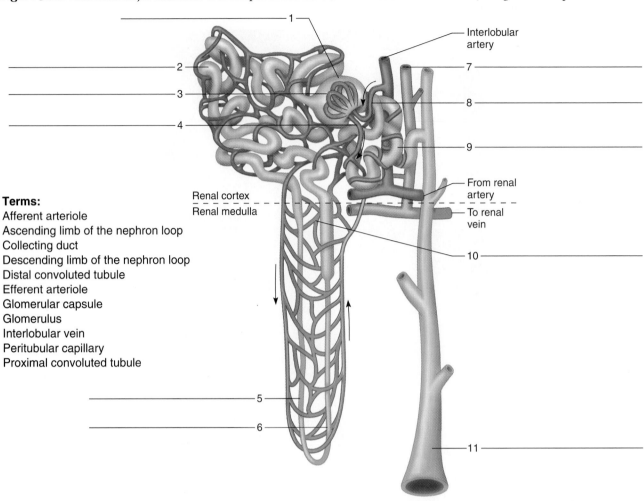

Interlobular artery

Renal cortex
Renal medulla

From renal artery

To renal vein

Terms:

Afferent arteriole
Ascending limb of the nephron loop
Collecting duct
Descending limb of the nephron loop
Distal convoluted tubule
Efferent arteriole
Glomerular capsule
Glomerulus
Interlobular vein
Peritubular capillary
Proximal convoluted tubule

Figure 56.5 (*a*) Micrograph of a section of the renal cortex (220×). (*b*) Micrograph of a section of the renal medulla (80× micrograph enlarged to 200×).

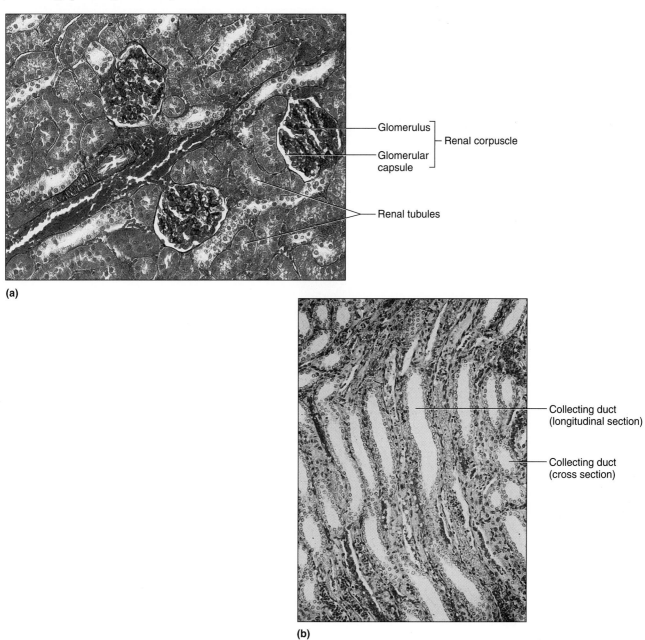

(a)

Glomerulus — Renal corpuscle
Glomerular capsule

Renal tubules

Collecting duct (longitudinal section)

Collecting duct (cross section)

(b)

Laboratory Report 56

Name _____

Date _____

Section _____

STRUCTURE OF THE KIDNEY

PART A

Match the terms in column A with the descriptions in column B. Place the letter of your choice in the space provided.

Column A	Column B
a. Calyces	_____ 1. Shell around the renal medulla
b. Hilum	_____ 2. Branches of renal pelvis to renal papillae
c. Nephron	_____ 3. Conical mass of tissue within renal medulla
d. Renal column	_____ 4. Projection with tiny openings into a minor calyx
e. Renal cortex	_____ 5. Hollow chamber within kidney
f. Renal papilla	_____ 6. Microscopic functional unit of kidney
g. Renal pelvis	_____ 7. Cortical tissue between renal pyramids
h. Renal pyramid	_____ 8. Superior funnel-shaped end of ureter inside the renal sinus
i. Renal sinus	_____ 9. Medial depression for blood vessels and ureter to enter kidney chamber

PART B

Complete the following:

1. Distinguish between a renal corpuscle and a renal tubule. _____

_____

_____

_____

_____

2. Number the following structures to indicate their respective positions in relation to the nephron. Assign the number 1 to the structure attached to the glomerular capsule.

_____ Ascending limb of nephron loop

_____ Collecting duct

_____ Descending limb of nephron loop

_____ Distal convoluted tubule

_____ Proximal convoluted tubule

_____ Renal papilla

465

3. Number the following structures to indicate their respective positions in the blood pathway within the kidney. Assign the number 1 to the vessel nearest the renal artery.

_____ Afferent arteriole

_____ Efferent arteriole

_____ Glomerulus

_____ Peritubular capillary

_____ Renal vein

4. Explain how the blood vessels associated with the renal corpuscle help to maintain relatively high blood pressure within the glomerulus. _____

_____

_____

_____

_____

_____

_____

5. Describe the *juxtaglomerular apparatus.* _____

_____

_____

_____

_____

_____

PART C

Prepare a sketch of a representative section of renal cortex. Label the glomerulus, glomerular capsule, and sections of renal tubules.

PART D

Prepare a sketch of a representative section of renal medulla. Label a longitudinal section and a cross section of a collecting duct.

URINALYSIS

MATERIALS NEEDED

Disposable urine-collecting container
Paper towel
Urinometer cylinder
Urinometer hydrometer
Laboratory thermometer
pH test paper
Reagent strips (individual or combination strips such as
    Chemstrip or Multistix) to test for the presence of
    the following:
    glucose
    protein
    ketones
    bilirubin
    hemoglobin/occult blood
Compound light microscope
Microscope slide
Coverslip
Centrifuge
Centrifuge tube
Graduated cylinder, 10 mL
Medicine dropper
Sedi-stain
Normal and abnormal simulated urine specimens
    (optional)

SAFETY

- Wear disposable gloves when working with body fluids.
- Work only with your own urine sample.
- Use an appropriate disinfectant to wash the laboratory table before and after the procedures.
- Place glassware in a disinfectant when finished.
- Dispose of contaminated items as instructed.
- Wash your hands before leaving the laboratory.

Urine is the product of kidney functions, which include the removal of various waste substances from the blood and the maintenance of body fluid and electrolyte balances. Consequently, the composition of urine varies from time to time because of differences in dietary intake and physical activity. Also, the volume of urine produced by the kidneys varies with such factors as fluid intake, environmental temperature, relative humidity, respiratory rate, and body temperature.

An analysis of urine composition and volume often is used to evaluate the functions of the kidneys and other organs. This procedure, called *urinalysis,* involves observing the physical characteristics of a urine sample, testing for the presence of certain organic and inorganic substances, and examining the microscopic solids present in the sample.

PURPOSE OF THE EXERCISE

To perform the observations and tests commonly used to analyze the characteristics and composition of urine.

LEARNING OBJECTIVES

After completing this exercise, you should be able to

1. Evaluate the color, transparency, and specific gravity of a urine sample.
2. Determine the pH of a urine sample.
3. Test a urine sample for the presence of glucose, protein, ketones, bilirubin, and hemoglobin.
4. Perform a microscopic study of urine sediment.
5. Evaluate the results of these observations and tests.

WARNING

While performing the following tests, you should wear disposable latex gloves so that skin contact with urine is avoided. Observe all safety procedures listed for this lab. (Normal and abnormal simulated urine specimens could be used instead of real urine for this lab.)

PROCEDURE

1. Proceed to the restroom with a clean disposable container. The first small volume of urine should not be collected because it contains abnormally high levels of microorganisms from the urethra. Collect a midstream sample of about 50 mL of urine. The best collections are the first specimen in the morning or one taken 3 hours after a meal. Refrigerate samples if they are not used immediately.
2. Place a sample of urine in a clean, transparent container. Describe the *color* of the urine. Normal urine varies from light yellow to amber, depending on the presence of urochromes, which are end-product pigments produced during the

decomposition of hemoglobin. Dark urine indicates a high concentration of pigments.

Abnormal urine colors include yellow-brown or green, due to elevated concentrations of bile pigments, and red to dark brown, due to the presence of blood. Certain foods, such as beets or carrots, and various drug substances also may cause color changes in urine, but in such cases the colors have no clinical significance. Enter the results of this and the following tests in Part A of Laboratory Report 57.

3. Evaluate the *transparency* of the urine sample (that is, judge whether the urine is clear, slightly cloudy, or very cloudy). Normal urine is clear enough to see through. You can read newsprint through slightly cloudy urine; you can no longer read newsprint through cloudy urine. Cloudy urine indicates the presence of various substances that may include mucus, bacteria, epithelial cells, fat droplets, or inorganic salts.

4. Determine the *specific gravity* of the urine sample. Specific gravity is the ratio of the weight of something to the weight of an equal volume of pure water. For example, mercury (at 15°C) weighs 13.6 times as much as an equal volume of water; thus, it has a specific gravity of 13.6. Although urine is mostly water, it has substances dissolved in it and is slightly heavier than an equal volume of water. Thus, urine has a specific gravity of more than 1.000. Actually, the specific gravity of normal urine varies from 1.003 to 1.035.

To determine the specific gravity of a urine sample, follow these steps:

a. Pour enough urine into a clean urinometer cylinder to fill it about three-fourths full. Any foam that appears should be removed with a paper towel.

b. Use a laboratory thermometer to measure the temperature of the urine.

c. Gently place the urinometer hydrometer into the urine, and *make sure that the float is not touching the sides or the bottom of the cylinder* (fig. 57.1).

d. Position your eye at the level of the urine surface. Determine which line on the stem of the hydrometer intersects the lowest level of the concave surface (meniscus) of the urine.

e. Because liquids tend to contract and become denser as they are cooled, or to expand and become less dense as they are heated, it may be necessary to make a temperature correction to obtain an accurate specific gravity measurement. To do this, add 0.001 to the hydrometer reading for each 3 degrees of urine temperature above 25°C or subtract 0.001 for each 3 degrees below 25°C. Enter this calculated value in the table of the laboratory report as the test result.

468

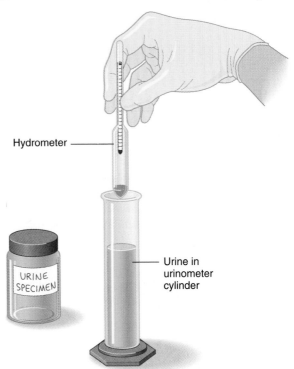

Figure 57.1 Float the hydrometer in the urine, making sure that it does not touch the sides or the bottom of the cylinder.

Hydrometer

Urine in urinometer cylinder

URINE SPECIMEN

5. Reagent strips can be used to perform a variety of urine tests. In each case, directions for using the strips are found on the strip container. *Be sure to read them.*

To perform each test, follow these steps:

a. Obtain a urine sample and the proper reagent strip.

b. Read the directions on the strip container.

c. Dip the strip in the urine sample.

d. Remove the strip at an angle and let it touch the inside rim of the urine container to remove any excess liquid.

e. Wait for the length of time indicated by the directions on the container before you compare the color of the test strip with the standard color scale on the side of the container. The value or amount represented by the matching color should be used as the test result and recorded in Part A of the laboratory report.

ALTERNATIVE PROCEDURE

If combination reagent test strips (Multistix or Chem-strip) are being used, locate the appropriate color chart for each test being evaluated. Wait the designated time for each color reaction before the comparison is made to the standard color scale.

6. Perform the *pH test.* The pH of normal urine varies from 4.6 to 8.0, but most commonly, it is near 6.0 (slightly acidic). The pH of urine may decrease as a result of a diet high in protein, or it may increase

with a vegetarian diet. Significant daily variations within the broad normal range are results of concentrations of excesses from variable diets.

7. Perform the *glucose test.* Normally, there is no glucose in urine. However, glucose may appear in the urine temporarily following a meal high in carbohydrates. Glucose also may appear in the urine as a result of uncontrolled diabetes mellitus.

8. Perform the *protein test.* Normally, proteins of large molecular size are not present in urine. However, those of small molecular sizes, such as albumins, may appear in trace amounts, particularly following strenuous exercise. Increased amounts of proteins also may appear as a result of kidney diseases in which the glomeruli are damaged or as a result of high blood pressure.

9. Perform the *ketone test.* Ketones are products of fat metabolism. Usually they are not present in urine. However, they may appear in the urine if the diet fails to provide adequate carbohydrate, as in the case of prolonged fasting or starvation, or as a result of insulin deficiency (diabetes mellitus).

10. Perform the *bilirubin test.* Bilirubin, which results from hemoglobin decomposition in the liver, normally is absent in urine. It may appear, however, as a result of liver disorders that cause obstructions of the biliary tract. Urochrome, a normal yellow component of urine, is a result of additional breakdown of bilirubin.

11. Perform the *hemoglobin/occult blood test.* Hemoglobin occurs in the red blood cells, and because such cells normally do not pass into the renal tubules, hemoglobin is not found in normal urine. Its presence in urine usually indicates a disease process, a transfusion reaction, or an injury to the urinary organs.

12. Complete Part A of the laboratory report.

13. A urinalysis usually includes a study of urine sediment—the microscopic solids present in a urine sample. This sediment normally includes mucus, certain crystals, and a variety of cells, such as the epithelial cells that line the urinary tubes and an occasional white blood cell. Other types of solids, such as casts or red blood cells, may indicate a disease or injury if they are present in excess. (Casts are cylindrical masses of cells or other substances that form in the renal tubules and are flushed out by the flow of urine.)

To observe urine sediment, follow these steps:

a. Thoroughly stir or shake a urine sample to suspend the sediment, which tends to settle to the bottom of the container.

b. Pour 10 mL of urine into a clean centrifuge tube and centrifuge it for 5 minutes at slow speed (1,500 rpm). Be sure to balance the centrifuge with an even number of tubes filled to the same levels.

c. Carefully decant 9 mL (leave 1 mL) of the liquid from the sediment in the bottom of the centrifuge tube, as directed by your laboratory instructor. Resuspend the 1 mL of sediment.

d. Use a medicine dropper to remove some of the sediment and place it on a clean microscope slide.

e. Add a drop of Sedi-stain to the sample, and add a coverslip.

f. Examine the sediment with low-power (reduce the light when using low power) and high-power magnifications.

g. Identify the kinds of solids present with the aid of figure 57.2.

h. In Part B of the laboratory report, make a sketch of each type of sediment that you observed.

14. Complete Part B of the laboratory report.

Figure 57.2 Types of urine sediment. Healthy individuals lack many of these sediments and possess only occasional to trace amounts of others.

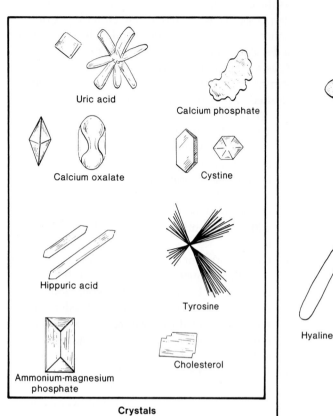

Crystals

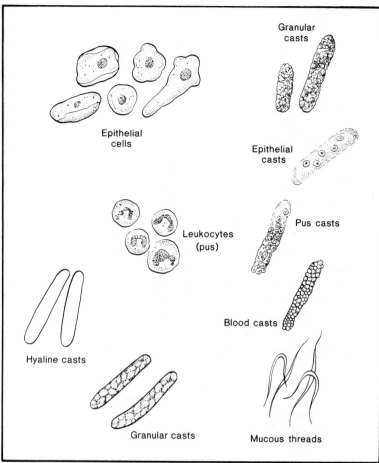

Cells and casts

Laboratory Report 57

Name _____

Date _____

Section _____

URINALYSIS

PART A

1. Enter your observations, test results, and evaluations in the following table:

Urine Characteristics	Observations and Test Results	Normal Values	Evaluations
Color		Light yellow to amber	
Transparency		Clear	
Specific gravity (corrected for temperature)		1.003–1.035	
pH		4.6–8.0	
Glucose		0 (negative)	
Protein		0 to trace	
Ketones		0	
Bilirubin		0	
Hemoglobin/occult blood		0	
(Other)			
(Other)			

2. Summarize the results of the urinalysis. _____

_____

_____

_____

_____

_____

_____

Critical Thinking Application

Why do you think it is important to refrigerate a urine sample if an analysis cannot be performed immediately after collecting it?

_____

_____

_____

_____

PART B

1. Make a sketch for each type of sediment you observed. Label any from those shown in figure 57.2.

2. Summarize the results of the urine sediment study. _____

_____

_____

_____

_____

_____

_____

CAT DISSECTION: URINARY SYSTEM

In this laboratory exercise, you will dissect the urinary organs of the embalmed cat. As you observe these structures, compare them with the corresponding human organs by observing the parts of the human torso.

PURPOSE OF THE EXERCISE

To examine the urinary organs of the cat and to compare them with those of the human.

LEARNING OBJECTIVES

After completing this exercise, you should be able to

1. Locate and identify the urinary organs of the cat.
2. Identify the corresponding organs in the human torso.
3. Compare the urinary organs of the cat with those of the human.

PROCEDURE

1. Place the embalmed cat in a dissecting tray with its ventral surface up.
2. Open the abdominal cavity, and remove the liver, stomach, and spleen.
3. Push the intestines to one side, and locate the *kidneys* in the dorsal abdominal wall on either side of the vertebral column. Note that the kidneys are located dorsal to the *parietal peritoneum* (retroperitoneal).

4. Carefully remove the parietal peritoneum and the adipose tissue surrounding the kidneys. Locate the following, using figures 47.8 and 58.1 as guides:

   ureters

   renal arteries

   renal veins

5. Locate the *adrenal glands,* which lie medially and above the kidneys and are surrounded by connective tissues. Note that the adrenal glands are attached to blood vessels and are separated by the abdominal aorta and inferior vena cava.
6. Expose the ureters by cleaning away the connective tissues along their lengths. Note that they enter the *urinary bladder* on the dorsal surface.
7. Examine the urinary bladder, and note how it is attached to the abdominal wall by folds of peritoneum that form ligaments. Locate the *medial ligament* on the ventral surface of the bladder that connects it with the linea alba and the *two lateral ligaments* on the dorsal surface.
8. Remove the connective tissue from around the urinary bladder. Use a sharp scalpel to open the bladder, and examine its interior. Locate the openings of the ureters and the urethra on the inside.
9. Expose the *urethra* at the posterior of the urinary bladder.
10. Remove one kidney, and section it longitudinally along the coronal plane (figs. 58.2 and 58.3). Identify the following features:

    renal capsule

    renal cortex

    renal medulla

    renal pyramid

    renal papilla

    renal sinus

    renal pelvis

11. Discard the organs and tissues that were removed from the cat, as directed by the laboratory instructor.
12. Complete Laboratory Report 58.

Figure 58.1 The urinary organs of a female cat.

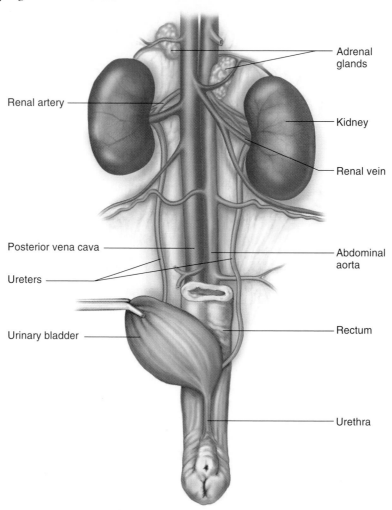

Adrenal
glands

Renal artery

Kidney

Renal vein

Posterior vena cava

Abdominal
aorta

Ureters

Urinary bladder

Rectum

Urethra

Figure 58.2 Longitudinal section of a cat's kidney.

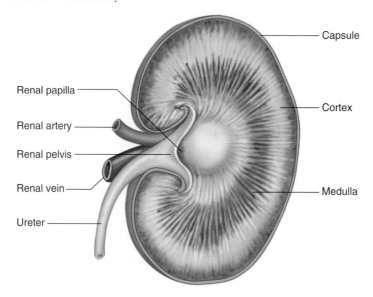

Capsule

Renal papilla

Renal artery

Cortex

Renal pelvis

Renal vein

Ureter

Medulla

Figure 58.3 Female urinary and reproductive systems of a pregnant cat.

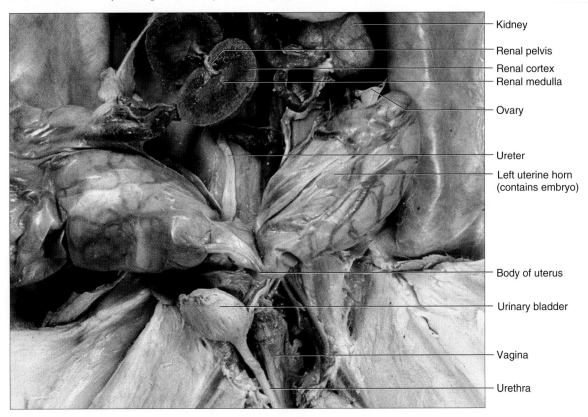

- Kidney
- Renal pelvis
- Renal cortex
- Renal medulla
- Ovary
- Ureter
- Left uterine horn (contains embryo)
- Body of uterus
- Urinary bladder
- Vagina
- Urethra

Name _____

Date _____

Section _____

CAT DISSECTION: URINARY SYSTEM

Complete the following:

1. Compare the positions of the kidneys in the cat with those in the human. _____
   _____
   _____
   _____

2. Compare the locations of the adrenal glands in the cat with those in the human. _____
   _____
   _____
   _____
   _____
   _____

3. What structures of the cat urinary system are retroperitoneal? _____
   _____
   _____
   _____

4. Describe the wall of the urinary bladder of the cat. _____
   _____
   _____
   _____

5. Compare the renal pyramids and renal papillae of the human kidney with those of the cat. _____
   _____
   _____
   _____
   _____

MALE REPRODUCTIVE SYSTEM

MATERIALS NEEDED

Textbook
Human torso
Model of the male reproductive system
Anatomical chart of the male reproductive system
Compound light microscope
Prepared microscope slides of the following:
  Testis section
  Epididymis, cross section
  Penis, cross section

The organs of the male reproductive system are specialized to produce and maintain the male sex cells, to transport these cells together with supporting fluids to the female reproductive tract, and to produce and secrete male sex hormones.

These organs include the testes, in which sperm cells and male sex hormones are produced, and sets of internal and external accessory organs. The internal organs include various tubes and glands, whereas the external structures are the scrotum and the penis.

PURPOSE OF THE EXERCISE

To review the structure and functions of the male reproductive organs and to examine some of these organs microscopically.

LEARNING OBJECTIVES

After completing this exercise, you should be able to

1. Locate and identify the organs of the male reproductive system.
2. Describe the functions of these organs.
3. Recognize sections of the testis, epididymis, and penis microscopically.
4. Identify the major features of these microscopic sections.

PROCEDURE A—MALE REPRODUCTIVE ORGANS

1. Review the sections entitled "Testes," "Male Internal Accessory Organs," and "Male External Reproductive Organs" in chapter 22 of the textbook.
2. As a review activity, label figures 59.1 and 59.2.

3. Observe the human torso, the model, and anatomical chart of the male reproductive system. Locate the following features:

   testes

      seminiferous tubules

      interstitial cells

   inguinal canal

   spermatic cord

   epididymis

   vas deferens

   ejaculatory duct

   seminal vesicles

   prostate gland

   bulbourethral glands

   scrotum

   penis

      corpora cavernosa

      corpus spongiosum

      tunica albuginea

      glans penis

      external urethral orifice

      prepuce

      crura

      bulb

4. Complete Part A of Laboratory Report 59.

PROCEDURE B—MICROSCOPIC ANATOMY

1. Obtain a microscope slide of human testis section and examine it, using low-power magnification (fig. 59.3). Locate the thick *fibrous capsule* (tunica albuginea) on the surface and the numerous sections of *seminiferous tubules* inside.
2. Focus on some of the seminiferous tubules, using high-power magnification (fig. 59.4). Locate the

479

Figure 59.1 Label the major structures of the male reproductive system in this sagittal view.

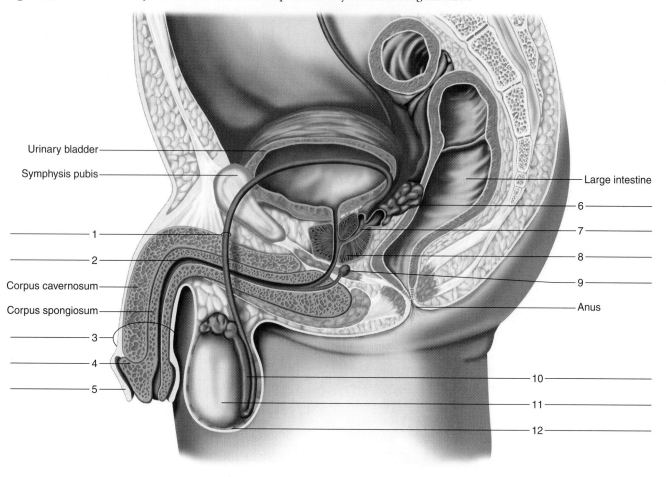

Urinary bladder

Symphysis pubis

Corpus cavernosum

Corpus spongiosum

Large intestine

Anus

1
2
3
4
5
6
7
8
9
10
11
12

Figure 59.2 Label the diagram of (*a*) the sagittal section of a testis and (*b*) a cross section of a seminiferous tubule by placing the correct numbers in the spaces provided.

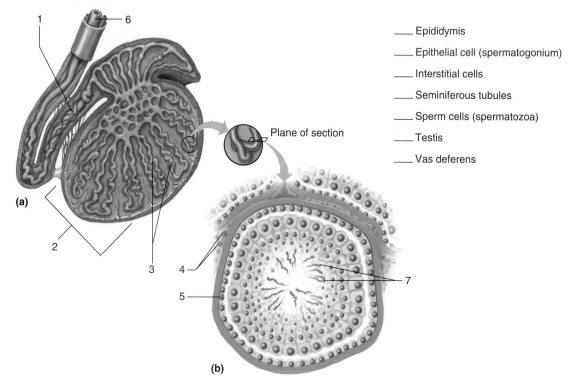

_____ Epididymis

_____ Epithelial cell (spermatogonium)

_____ Interstitial cells

_____ Seminiferous tubules

_____ Sperm cells (spermatozoa)

_____ Testis

_____ Vas deferens

Plane of section

(a)

(b)

Figure 59.3 Micrograph of a human testis (1.7×).

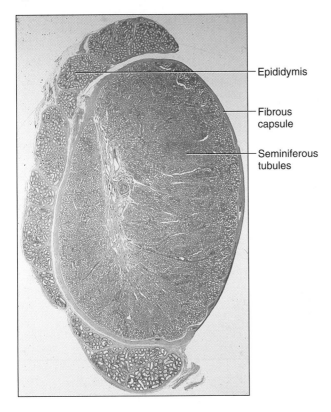

- Epididymis
- Fibrous capsule
- Seminiferous tubules

Figure 59.4 Micrograph of seminiferous tubules (50× micrograph enlarged to 135×).

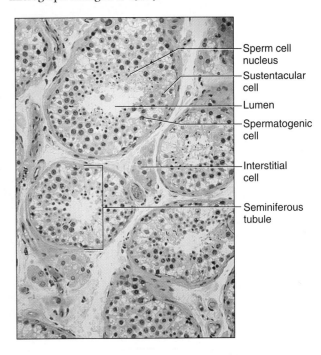

- Sperm cell nucleus
- Sustentacular cell
- Lumen
- Spermatogenic cell
- Interstitial cell
- Seminiferous tubule

epithelium, which forms the inner lining of each tube. Within this epithelium, identify some *sustentacular cells* or supporting cells (Sertoli's cells), which have pale, oval-shaped nuclei, and some *spermatogenic cells,* which have smaller, round nuclei. Near the lumen of the tube, find some darkly stained, elongated heads of developing sperm cells. In the spaces between adjacent seminiferous tubules, locate some isolated *interstitial cells* (cells of Leydig).

3. Prepare a labeled sketch of a representative section of the testis in Part B of the laboratory report.
4. Obtain a microscope slide of a cross section of *epididymis* (fig. 59.5). Examine its wall, using high-power magnification. Note the elongated, *pseudostratified columnar epithelial cells* that comprise most of the inner lining. These cells have nonmotile stereocilia (microvilli) on their free surfaces. Also note the thin layer of smooth muscle and connective tissue surrounding the tube.
5. Prepare a labeled sketch of the epididymis wall in Part B of the laboratory report.

6. Obtain a microscope slide of a *penis* cross section, and examine it with low-power magnification (fig. 59.6). Identify the following features:

corpora cavernosa

corpus spongiosum

tunica albuginea

urethra

skin

7. Prepare a labeled sketch of a penis cross section in Part B of the laboratory report.
8. Complete Part B of the laboratory report.

Web Quest

Review the structures and functions of the male reproductive system at www.mhhe.com/shier10
Click on Student Edition. Click on Martin Lab Manual, Web Quest.

Figure 59.5   Micrograph of a cross section of a human epididymis (50× micrograph enlarged to 145×).

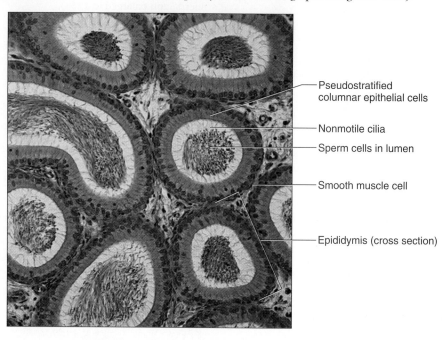

Pseudostratified columnar epithelial cells

Nonmotile cilia

Sperm cells in lumen

Smooth muscle cell

Epididymis (cross section)

Figure 59.6   Micrograph of a cross section of the body of the penis (5×).

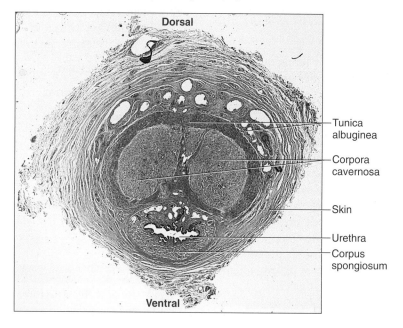

Dorsal

Tunica albuginea

Corpora cavernosa

Skin

Urethra

Corpus spongiosum

Ventral

Laboratory Report **59**

Name _____

Date _____

Section _____

MALE REPRODUCTIVE SYSTEM

PART A

Complete the following statements:

1. Each testis is suspended by a (an) _____ within the scrotum.

2. The descent of the testes from near the developing kidneys is stimulated by the male sex hormone _____ .

3. The descent of the testes is aided by a fibromuscular cord called the _____ .

4. As the testes descend, they pass through the _____ of the abdominal wall into the scrotum.

5. Connective tissue subdivides a testis into many _____ , which contain seminiferous tubules.

6. The _____ is a highly coiled tube on the surface of the testis.

7. The _____ cells of the epithelium that line the seminiferous tubule give rise to sperm cells.

8. _____ is the process by which sperm cells are formed.

9. Immature sperm cells formed from secondary spermatocytes during meiosis are called _____ .

10. The number of chromosomes normally present in a human sperm cell is _____ .

11. The anterior end of a sperm head, called the _____, contains enzymes that aid the penetration of an egg cell at the time of fertilization.

12. Sperm cells undergo maturation while they are stored in the _____ .

13. The secretion of the seminal vesicles is rich in the monosaccharide called _____ .

14. Prostate gland secretion helps neutralize seminal fluid because the prostate secretion is _____ .

15. The secretion of the _____ glands lubricates the end of the penis in preparation for sexual intercourse.

16. The pH of seminal fluid is slightly _____ .

17. The dartos muscle is in the subcutaneous tissue in the wall of the _____ .

18. The sensitive, cone-shaped end of the penis is called the _____ .

19. _____ is the movement of sperm cells and various glandular secretions into the urethra.

20. _____ is the process by which semen is forced out through the urethra.

PART B

1. Prepare a labeled sketch of a representative section of the testis.

2. Prepare a labeled sketch of a region of the epididymis.

3. Prepare a labeled sketch of a penis cross section.

4. Briefly describe the function of each of the following:
    a. sustentacular cell (supporting cell)

    _____
    _____
    _____

    b. spermatogenic cell

    _____
    _____
    _____

    c. interstitial cell

    _____
    _____
    _____

    d. epididymis

    _____
    _____
    _____

    e. corpora cavernosa and corpus spongiosum

    _____
    _____
    _____

484

FEMALE REPRODUCTIVE SYSTEM

MATERIALS NEEDED

Textbook
Human torso
Model of the female reproductive system
Anatomical chart of the female reproductive system
Compound light microscope
Prepared microscope slides of the following:
    Ovary section with maturing follicles
    Uterine tube, cross section
    Uterine wall section

For Demonstration:

Prepared microscope slides of the following:
    Uterine wall, early proliferative phase
    Uterine wall, secretory phase
    Uterine wall, early menstrual phase

The organs of the female reproductive system are specialized to produce and maintain the female sex cells, to transport these cells to the site of fertilization, to provide a favorable environment for a developing offspring, to move the offspring to the outside, and to produce female sex hormones.

These organs include the ovaries, which produce the egg cells and female sex hormones, and sets of internal and external accessory organs. The internal accessory organs include the uterine tubes, uterus, and vagina. The external organs are the labia majora, labia minora, clitoris, and vestibular glands.

PURPOSE OF THE EXERCISE

To review the structure and functions of the female reproductive organs and to examine some of their features microscopically.

LEARNING OBJECTIVES

After completing this exercise, you should be able to

1. Locate and identify the organs of the female reproductive system.
2. Describe the functions of these organs.
3. Recognize sections of the ovary, uterine tube, and uterine wall microscopically.
4. Identify the major features of these microscopic sections.

PROCEDURE A—FEMALE REPRODUCTIVE ORGANS

1. Review the sections entitled, "Ovaries," "Female Internal Accessory Organs," "Female External Reproductive Organs," and "Mammary Glands" in chapter 22 of the textbook.
2. As a review activity, label figures 60.1, 60.2, 60.3, 60.4, and 60.5
3. Observe the human torso, the model, and the anatomical chart of the female reproductive system. Locate the following features:

ovaries
    medulla
    cortex
ligaments
    broad ligament
    suspensory ligament
    ovarian ligament
    round ligament
uterine tubes (oviducts; fallopian tubes)
    infundibulum
    fimbriae
uterus
    fundus
    body
    cervix
    cervical orifice
    endometrium
    myometrium
    perimetrium (serous membrane)
rectouterine pouch
vagina
    fornices
    vaginal orifice
    hymen
    mucosal layer
    muscular layer
    fibrous layer

Figure 60.1 Label the ligaments and other structures associated with the female internal reproductive organs.

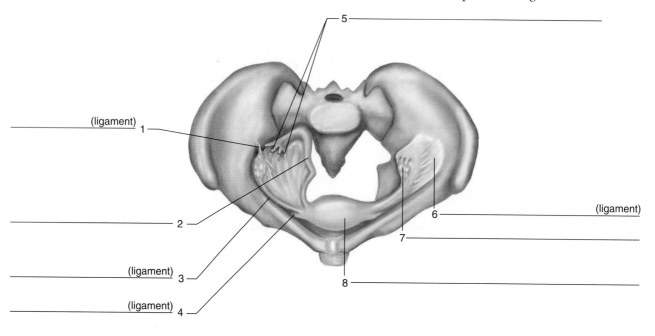

(ligament) 1 _____

5 _____

(ligament) _____

6 _____

7 _____

2 _____

(ligament) 3 _____

8 _____

(ligament) 4 _____

Figure 60.2 Label the structures of the female reproductive system in this sagittal view.

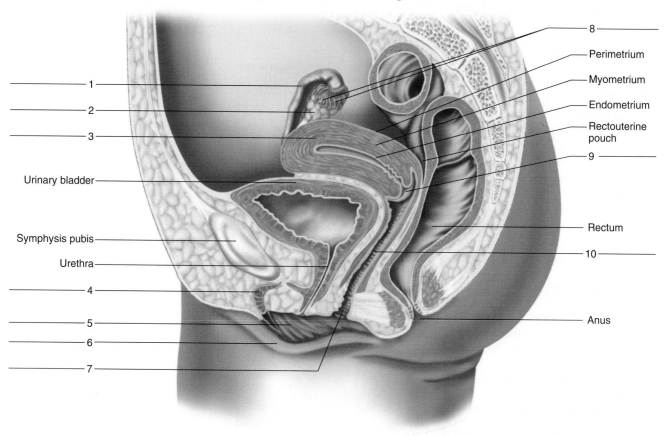

1 _____

2 _____

3 _____

Urinary bladder _____

Symphysis pubis _____

Urethra _____

4 _____

5 _____

6 _____

7 _____

8 _____

Perimetrium

Myometrium

Endometrium

Rectouterine pouch

9 _____

Rectum

10 _____

Anus

Figure 60.3 Label the female external reproductive organs and associated structures.

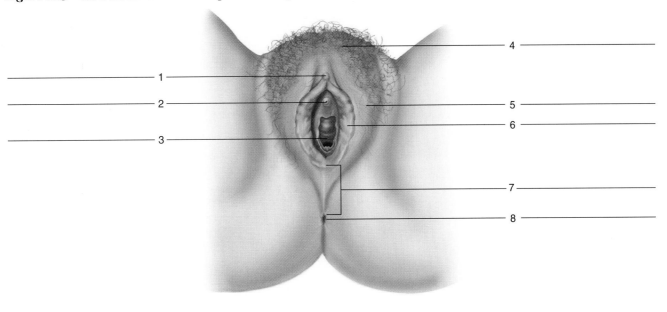

Figure 60.4 Label this ovary by placing the correct numbers in the spaces provided.

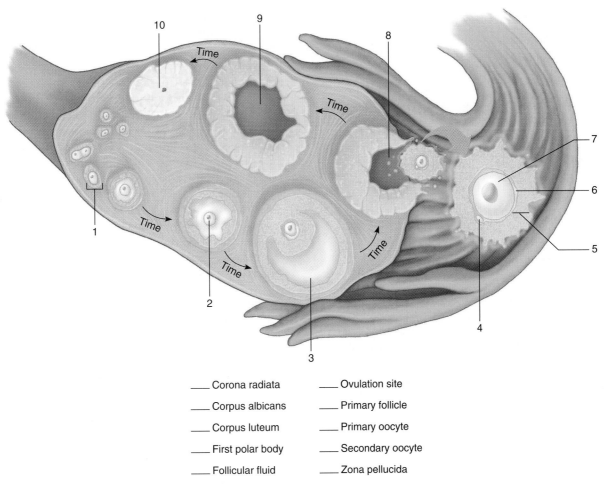

___ Corona radiata	___ Ovulation site
___ Corpus albicans	___ Primary follicle
___ Corpus luteum	___ Primary oocyte
___ First polar body	___ Secondary oocyte
___ Follicular fluid	___ Zona pellucida

Figure 60.5 Label the structures of the breast (anterior view), using the terms provided.

Terms:
Adipose tissue
Alveolar glands
Areola
Lactiferous duct
Nipple

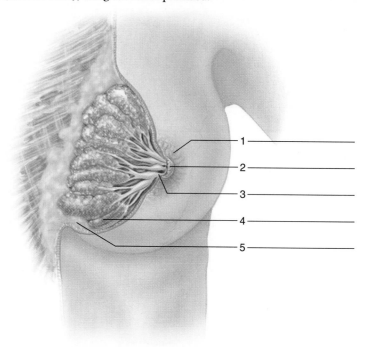

1 —————
2 —————
3 —————
4 —————
5 —————

mons pubis

vulva (external accessory organs)

   labia majora

   labia minora

   vestibular glands

   clitoris

      corpora cavernosa

      glans

vestibule

vestibular bulbs

breasts

   nipple

   areola

   alveolar glands (compose mammary glands)

   lactiferous ducts

   adipose tissue

4. Complete Part A of Laboratory Report 60.

PROCEDURE B—MICROSCOPIC ANATOMY

1. Obtain a microscope slide of an ovary section with maturing follicles, and examine it with low-power magnification (fig. 60.6). Locate the outer layer, or *cortex,* which is composed of densely packed cells, and the inner layer, or *medulla,* which largely consists of loose connective tissue.

2. Focus on the cortex of the ovary, using high-power magnification (fig. 60.7). Note the thin layer of small cuboidal cells on the free surface. These cells comprise the *germinal epithelium.* Also locate some *primordial follicles* just beneath the germinal epithelium. Note that each follicle consists of a single, relatively large *primary oocyte* with a prominent nucleus and a covering of *follicular cells.*

3. Prepare a labeled sketch of the ovarian cortex in Part B of the laboratory report.

4. Use low-power magnification to search the ovarian cortex for maturing follicles in various stages of development. Prepare three labeled sketches in Part B of the laboratory report to illustrate the changes that occur in a follicle as it matures.

5. Obtain a microscope slide of a cross section of a uterine tube. Examine it, using low-power magnification (fig. 60.8). Note that the shape of the lumen is very irregular.

6. Focus on the inner lining of the uterine tube, using high-power magnification. Note that the lining is composed of *simple columnar epithelium* and that some of the epithelial cells are ciliated on their free surfaces.

7. Prepare a labeled sketch of a representative region of the wall of the uterine tube in Part B of the laboratory report.

8. Obtain a microscope slide of the uterine wall section. Examine it, using low-power magnification (fig. 60.9), and locate the following:

endometrium (inner mucosal layer)

myometrium (middle, thick muscular layer)

perimetrium (outer serosal layer)

Figure 60.6 Micrograph of the ovary (30× micrograph enlarged to 80×).

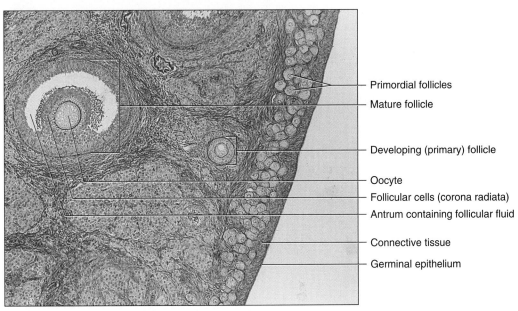

- Primordial follicles
- Mature follicle
- Developing (primary) follicle
- Oocyte
- Follicular cells (corona radiata)
- Antrum containing follicular fluid
- Connective tissue
- Germinal epithelium

Figure 60.7 Micrograph of the ovarian cortex (100× micrograph enlarged to 200×).

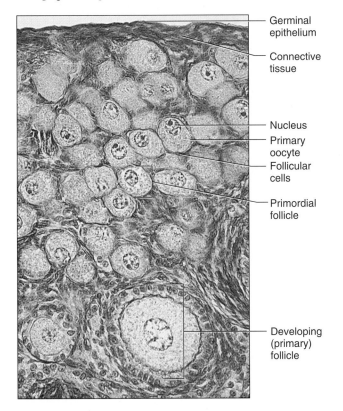

- Germinal epithelium
- Connective tissue
- Nucleus
- Primary oocyte
- Follicular cells
- Primordial follicle
- Developing (primary) follicle

9. Prepare a labeled sketch of a representative section of the uterine wall in Part B of the laboratory report.
10. Complete Part B of the laboratory report.

DEMONSTRATION

Observe the slides in the demonstration microscopes. Each slide contains a section of uterine mucosa taken during a different phase in the reproductive cycle. In the *early proliferative phase*, note the simple columnar epithelium on the free surface of the mucosa and the many sections of tubular uterine glands in the tissues beneath the epithelium. In the *secretory phase*, note that the endometrium is thicker and that the uterine glands appear more extensive and that they are coiled. In the *early menstrual phase*, note that the endometrium is thinner because its surface layer has been lost. Also note that the uterine glands are less apparent and that the spaces between the glands contain many leukocytes. What is the significance of these changes?

_____

_____

_____

Web Quest

Review the structures and functions of the female reproductive system at

www.mhhe.com/shier10

Click on Student Edition. Click on Martin Lab Manual, Web Quest.

Figure 60.8   Micrograph of a cross section of the uterine tube (8×).

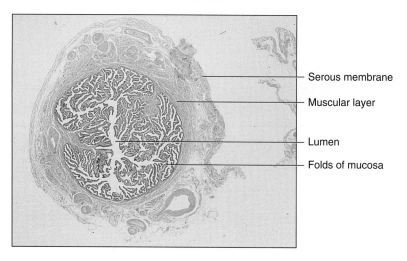

Serous membrane

Muscular layer

Lumen

Folds of mucosa

Figure 60.9   Micrograph of a section of the uterine wall (10× micrograph enlarged to 35×).

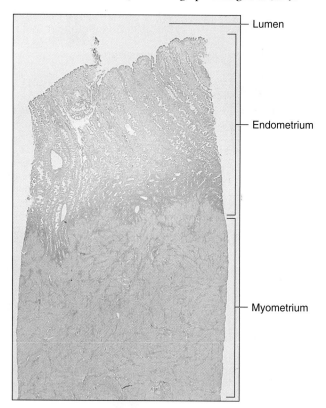

Lumen

Endometrium

Myometrium

FEMALE REPRODUCTIVE SYSTEM

PART A

Complete the following statements:

1. The ovaries are located in the lateral wall of the _____ cavity.
2. The largest of the ovarian attachments is called the _____ ligament.
3. The ovarian cortex appears granular because of the presence of _____ .
4. The meiosis of egg formation is called _____ .
5. A primary oocyte is closely surrounded by flattened epithelial cells called _____ cells.
6. When a primary oocyte divides, a secondary oocyte and a(an) _____ are produced.
7. Primordial follicles are stimulated to develop into primary follicles by the hormone called _____ .
8. _____ is the process by which a secondary oocyte is released from the ovary.
9. Uterine tubes also are called _____ .
10. The _____ is the funnel-shaped expansion at the end of a uterine tube.
11. The _____ ligament is a band of tissue within the broad ligament that helps to hold the uterus in proper position.
12. A portion of the uterus called the _____ extends downward into the upper portion of the vagina.
13. The inner mucosal lining of the uterus is called the _____ .
14. The myometrium is largely composed of _____ tissue.
15. The vaginal orifice is partially closed by a thin membrane called the _____ .
16. The group of external accessory organs that surround the openings of the urethra and vagina comprise the _____ .
17. The rounded mass of fatty tissue overlying the symphysis pubis of the female is called the _____ .
18. The female organ that corresponds to the male penis is the _____ .
19. The _____ of the female corresponds to the bubourethral glands of the male.
20. Orgasm is accompanied by a series of reflexes involving the lumbar and _____ regions of the spinal cord.

PART B

1. Prepare a labeled sketch of a representative region of the ovarian cortex.

491

2. Prepare a series of three labeled sketches to illustrate follicular maturation.

3. Prepare a labeled sketch of a representative section of the wall of a uterine tube.

4. Prepare a labeled sketch of a representative section of uterine wall.

5. Complete the following:
    a. Describe the fate of a mature follicle. _____
    _____
    _____

    b. Describe the function of the cilia in the lining of the uterine tube. _____
    _____
    _____

    c. Briefly describe the changes that occur in the uterine lining during a reproductive cycle._____
    _____
    _____

LABORATORY EXERCISE 61

CAT DISSECTION: REPRODUCTIVE SYSTEMS

SAFETY

- Wear disposable gloves when working on the cat dissection.
- Dispose of tissue remnants and gloves as instructed.
- Wash the dissecting tray and instruments as instructed.
- Wash your laboratory table.
- Wash your hands before leaving the laboratory.

In this laboratory exercise, you will dissect the reproductive system of the embalmed cat. If you have a female cat, begin with Procedure A. If you have a male cat, begin with Procedure B. After completing the dissection, exchange cats with someone who has dissected one of the opposite sex, and examine its reproductive organs.

As you observe the cat reproductive organs, compare them with the corresponding human organs by examining the models of the human reproductive systems.

PURPOSE OF THE EXERCISE

To examine the reproductive organs of the embalmed cat and to compare them with the corresponding organs of the human.

LEARNING OBJECTIVES

After completing this exercise, you should be able to

1. Locate and identify the reproductive organs of a cat.
2. Identify the corresponding organs in models of the human reproductive systems.
3. Compare the reproductive organs of the cat with those of the human.

PROCEDURE A—FEMALE REPRODUCTIVE SYSTEM

1. Place the embalmed female cat in a dissecting tray with its ventral surface up, and open its abdominal cavity.
2. Locate the small oval ovaries just posterior to the kidneys (figs. 58.3 and 61.1).
3. Examine an ovary, and note that it is suspended from the dorsal body wall by a fold of peritoneum called the *mesovarium.* Another attachment, the *ovarian ligament,* connects the ovary to the tubular *uterine horn* (figs. 58.3 and 61.1).
4. Near the anterior end of the ovary, locate the funnel-shaped *infundibulum,* which is at the end of the *uterine tube,* or *oviduct.* Note the tiny projections, or *fimbriae,* that create a fringe around the edge of the infundibulum. Trace the uterine tube around the ovary to its connection with the uterine horn near the attachment of the ovarian ligament.
5. Examine the uterine horn. Note that it is suspended from the body wall by a fold of peritoneum, the *broad ligament,* that is continuous with the mesovarium of the ovary. Also note the fibrous *round ligament,* which extends from the uterine horn laterally and posteriorly to the body wall (figs. 58.3 and 61.1).
6. To observe the remaining organs of the reproductive system more easily, remove the left hindlimb by severing it near the hip with bone cutters. Then use the bone cutters to remove the left anterior portion of the pelvic girdle. Also remove the necessary muscles and connective tissue to expose the structures shown in figure 61.2.
7. Trace the uterine horns posteriorly, and note that they unite to form the *uterine body,* which is located between the urethra and rectum. The uterine body is continuous with the *vagina,* which leads to the outside.
8. Trace the *urethra* from the urinary bladder posteriorly, and note that it and the vagina open into a common chamber, called the *urogenital sinus.* The opening of this chamber, which is ventral to the anus, is called the *urogenital aperture.* Locate the *labia majora,* which are folds of skin on either side of the urogenital aperture.

493

Figure 61.1   Ventral view of the female cat's reproductive system.

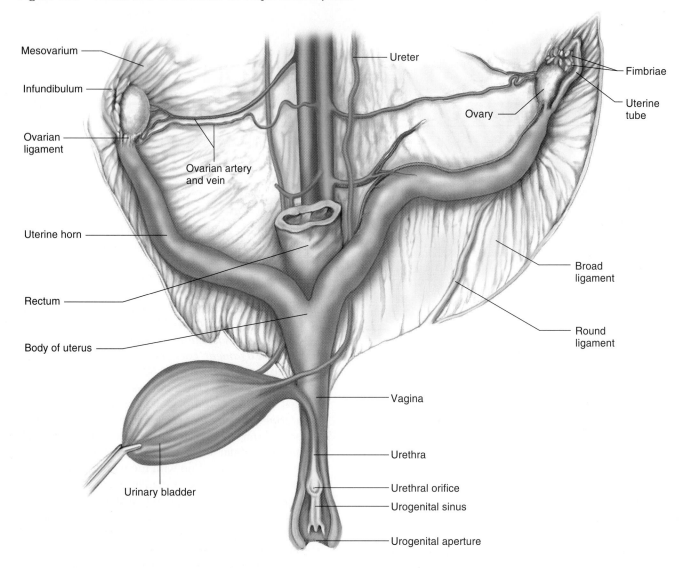

9. Use scissors to open the vagina along its lateral wall, beginning at the urogenital aperture and continuing to the body of the uterus. Note the *urethral orifice* in the ventral wall of the urogenital sinus, and locate the small, rounded *cervix* of the uterus, which projects into the vagina at its deep end. The *clitoris* is located in the ventral wall of the urogenital sinus near its opening to the outside.

10. Complete Part A of Laboratory Report 61.

PROCEDURE B—MALE REPRODUCTIVE SYSTEM

1. Place the embalmed male cat in a dissecting tray with its ventral surface up.

2. Locate the *scrotum* between the hindlimbs. Make an incision along the midline of the scrotum, and expose the *testes* inside (fig. 61.3). Note that the testes are separated by a septum and that each testis is enclosed in a sheath of connective tissue.

3. Remove the sheath surrounding the testes, and locate the convoluted *epididymis* on the dorsal surface of each testis (fig. 61.4).

4. Locate the *spermatic cord* on the right side, leading away from the testis. This cord contains the *vas deferens,* which is continuous with the epididymis as well as with the nerves and blood vessels that supply the testis on that side. Trace the spermatic cord to the body wall, where its contents pass through the *inguinal canal* and enter the pelvic cavity (fig. 61.4).

5. Locate the *penis,* and identify the *prepuce,* which forms a sheath around the penis. Make an incision through the skin of the prepuce, and expose the *glans* of the penis. Note that the glans has minute spines on its surface.

6. To observe the remaining structures of the reproductive system more easily, remove the left hindlimb by severing it near the hip with bone cutters. Then use the bone cutters to

Figure 61.2 Lateral view of the female cat's reproductive system.

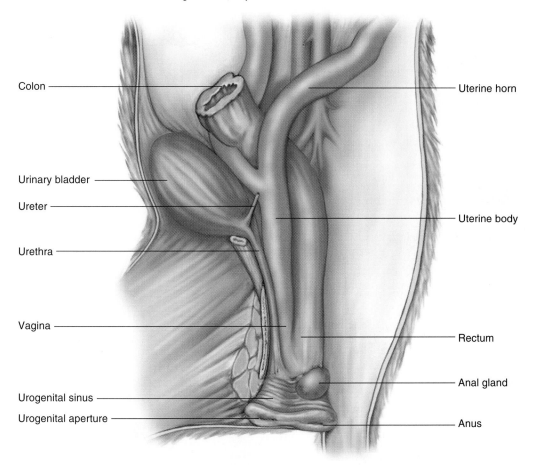

Colon

Urinary bladder

Ureter

Urethra

Vagina

Urogenital sinus

Urogenital aperture

Uterine horn

Uterine body

Rectum

Anal gland

Anus

remove the left anterior portion of the pelvic girdle. Also remove the necessary muscles and connective tissues to expose the organs shown in figure 61.5.

7. Trace the vas deferens from the inguinal canal to the penis. Note that the vas deferens loops over the ureter within the pelvic cavity and passes downward behind the urinary bladder to join the urethra. Locate the *prostate gland,* which appears as an enlargement at the junction of the vas deferens and urethra.

8. Trace the urethra to the penis. Locate the *bulbourethral glands,* which form small swellings on either side at the proximal end of the penis.

9. Use a sharp scalpel to cut a transverse section of the penis. Identify the *corpora cavernosa,* each of which contains many blood spaces surrounded by a sheath of connective tissue.

10. Complete Part B of the laboratory report.

Figure 61.3 Male reproductive system of the cat.

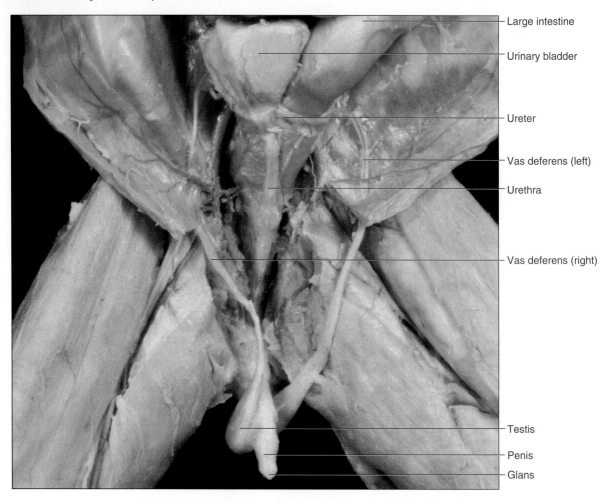

Large intestine

Urinary bladder

Ureter

Vas deferens (left)

Urethra

Vas deferens (right)

Testis

Penis

Glans

Figure 61.4 Ventral view of the male cat's reproductive system.

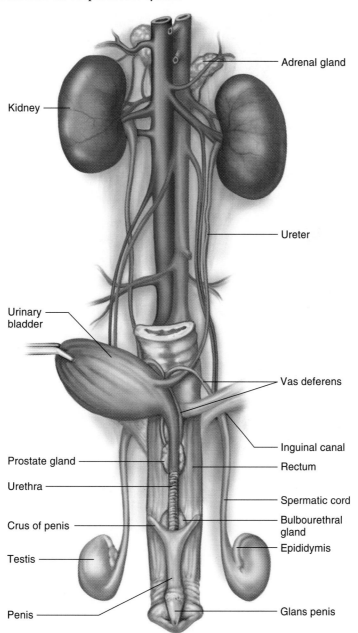

Kidney

Adrenal gland

Ureter

Urinary bladder

Vas deferens

Inguinal canal

Prostate gland

Rectum

Urethra

Spermatic cord

Crus of penis

Bulbourethral gland

Testis

Epididymis

Penis

Glans penis

Figure 61.5   Lateral view of the male cat's reproductive sytem.

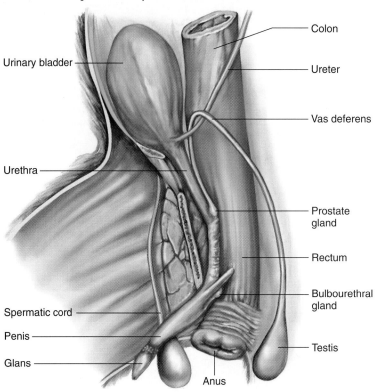

Urinary bladder

Urethra

Spermatic cord

Penis

Glans

Colon

Ureter

Vas deferens

Prostate
gland

Rectum

Bulbourethral
gland

Testis

Anus

CAT DISSECTION: REPRODUCTIVE SYSTEMS

PART A

Complete the following:

1. Compare the relative lengths and paths of the uterine tubes (oviducts) of the cat and the human. _____

_____

_____

_____

_____

_____

2. How do the shape and structure of the uterus of the cat compare with that of the human? _____

_____

_____

_____

_____

_____

3. Assess the function of the uterine horns of the cat. _____

_____

_____

_____

_____

_____

4. Compare the relationship of the urethra and the vagina in the cat and in the human. _____

_____

_____

_____

_____

_____

PART B

Complete the following:

1. Compare the glans of the penis in the cat and in the human. _____
   _____
   _____
   _____
   _____
   _____

2. How do the location and the relative size of the prostate gland of the cat compare with that of the human? _____
   _____
   _____
   _____
   _____
   _____

3. What glands associated with the human vas deferens are missing in the cat? _____
   _____
   _____

4. Compare the prepuce in the cat and in the human. _____
   _____
   _____
   _____

FERTILIZATION AND EARLY DEVELOPMENT

MATERIALS NEEDED

Textbook
Sea urchin egg suspension*
Sea urchin sperm suspension*
Compound light microscope
Depression microscope slide
Coverslip
Medicine droppers
Prepared microscope slide of the following:
    Sea urchin embryos (early and late cleavage)
Models of human embryos

For Optional Activity:

Vaseline
Toothpick

For Demonstration:

Preserved mammalian embryos

*See the Instructor's Manual for a source of materials.

Fertilization is the process by which the nuclei of an egg and a sperm cell come together and combine their chromosomes, forming a single cell called a zygote.

Ordinarily, before fertilization can occur in a human female, an egg cell must be released from the ovary and must be carried into a uterine tube. Also, semen containing sperm cells must be deposited in the vagina; some of these sperm cells must travel through the uterus and into the uterine tubes. Although many sperm cells may reach an egg cell, only one will participate in the fertilization of the egg.

Shortly after fertilization, the zygote undergoes division (mitosis) to form two cells. These two cells become four, they in turn divide into eight, and so forth. The resulting mass of cells continues to grow and undergoes developmental changes and growth that give rise to an offspring.

PURPOSE OF THE EXERCISE

To review the process of fertilization, to observe sea urchin eggs being fertilized, and to examine embryos in early stages of development.

LEARNING OBJECTIVES

After completing this exercise, you should be able to

1. Describe the process of fertilization.
2. Describe the early developmental stages of a sea urchin.
3. Describe the early developmental stages of a human.
4. Identify the major features of human embryo models.

PROCEDURE A—FERTILIZATION

1. Review the sections entitled "Fertilization" and "Period of Cleavage" in chapter 23 of the textbook.
2. As a review activity, label figure 62.1.
3. Complete Part A of Laboratory Report 62.
4. Although it is difficult to observe fertilization in animals since the process occurs internally, it is possible to view forms of external fertilization. For example, egg and sperm cells can be collected from sea urchins, and the process of fertilization can be observed microscopically. To make this observation, follow these steps:
   a. Place a drop of sea urchin egg-cell suspension in the chamber of a depression slide, and add a coverslip.
   b. Examine the egg cells, using low-power magnification.
   c. Focus on a single egg cell with high-power magnification, and sketch the cell in Part B of the laboratory report.
   d. Remove the coverslip and add a drop of sea urchin sperm-cell suspension to the depression slide. Replace the coverslip, and observe the sperm cells with high-power magnification as they cluster around the egg cells. This attraction is stimulated by gamete secretions.
   e. Observe the egg cells with low-power magnification once again, and watch for the appearance of *fertilization membranes.* Such a membrane forms as soon as an egg cell is penetrated by a sperm cell; it looks like a clear halo surrounding the egg cell.
   f. Focus on a single fertilized egg cell, and sketch it in Part B of the laboratory report.

Figure 62.1 Label the diagram of the stages of early human development.

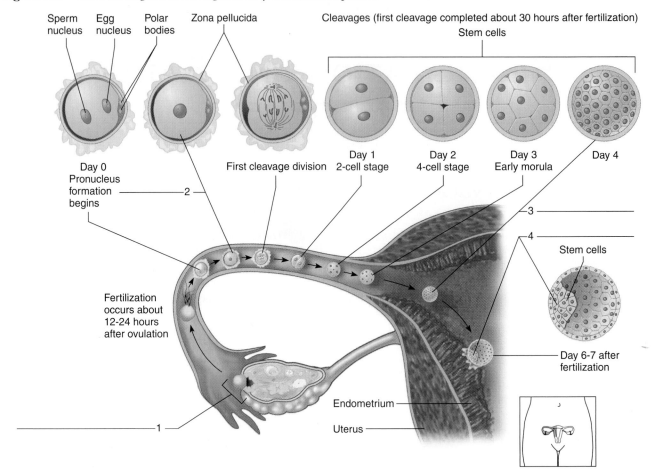

PROCEDURE B—SEA URCHIN EARLY DEVELOPMENT

1. Obtain a prepared microscope slide of developing sea urchin embryos. This slide contains embryos in various stages of cleavage. Search the slide, using low-power magnification, and locate embryos in two-, four-, eight-, and sixteen-cell stages. Observe that cleavage results in an increase of cell numbers; however, the cells get progressively smaller.

2. Prepare a sketch of each stage in Part C of the laboratory report.

PROCEDURE C—HUMAN EARLY DEVELOPMENT

1. Review the sections entitled "Period of Cleavage" and "Embryonic Stage" in chapter 23 of the textbook.
2. As a review activity, study and label figures 62.2 and 62.3.
3. Complete Parts D and E of the laboratory report.
4. Observe the models of human embryos, and identify the following features:

blastomeres

morula

blastocyst

Figure 62.2 Label the major features of the early embryo and the structures associated with it.

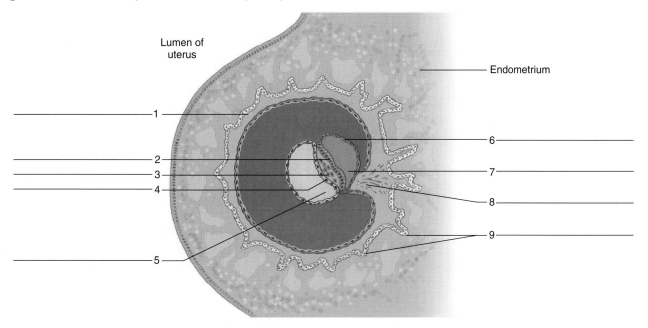

inner cell mass

trophoblast

embryonic disc

primary germ layers

    ectoderm

    endoderm

    mesoderm

connecting stalk

chorion

chorionic villi (from trophoblast)

lacunae

amnion

amniotic fluid

umbilical cord

    umbilical arteries

    umbilical vein

yolk sac

allantois

placenta

DEMONSTRATION

Observe the preserved mammalian embryos that are on display. In addition to observing the developing external body structures, identify such features as the chorion, chorionic villi, amnion, yolk sac, umbilical cord, and placenta. What special features provide clues to the type of mammal these embryos represent?

_____

_____

_____

Web Quest

Examine topics of infertility, fetal development, pregnancy, birth, and many more.
www.mhhe.com/shier10
Click on Student Edition. Click on Martin Lab Manual, Web Quest.

Figure 62.3 Label the structures associated with this seven-week-old embryo.

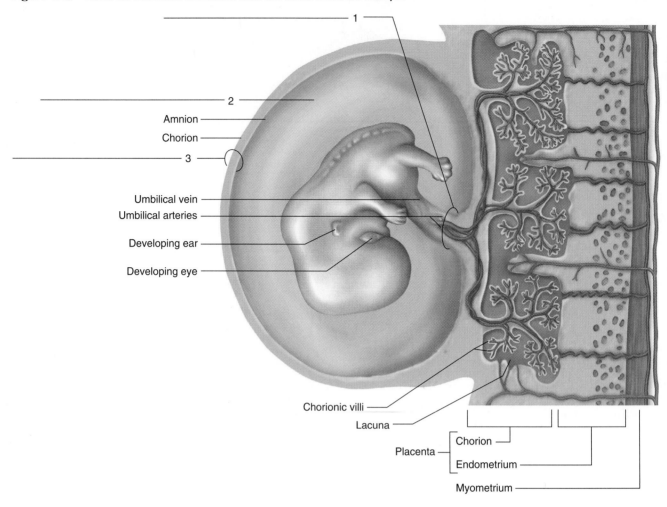

1

2

Amnion

Chorion

3

Umbilical vein

Umbilical arteries

Developing ear

Developing eye

Chorionic villi

Lacuna

Placenta

Chorion

Endometrium

Myometrium

FERTILIZATION AND EARLY DEVELOPMENT

PART A

Complete the following statements:

1. The zona pellucida surrounds the cell membrane of a(an) _____ .
2. Enzymes secreted by the _____ of a sperm cell help it to penetrate the zona pellucida.
3. The cell resulting from fertilization is called a(an) _____ .
4. The cell resulting from fertilization divides by the process of _____ .
5. _____ is the phase of development during which cellular divisions result in smaller and smaller cells.
6. _____ on the inner lining of the uterine tube aid in moving a developing embryo.
7. The trip from the site of fertilization to the uterine cavity takes about _____ days.
8. The hollow ball of cells formed early in development is called a(an) _____ .
9. A human offspring is called a(an) _____ until the end of the eighth week of development.
10. After the eighth week, a developing human is called a(an) _____ until the time of birth.

PART B

Prepare sketches of the following:

Single sea urchin egg (____×)

Fertilized sea urchin egg (____×)

PART C

Prepare sketches of the following:

Two-cell sea urchin embryo (___×)

Four-cell sea urchin embryo (___×)

Eight-cell sea urchin embryo (___×)

Sixteen-cell sea urchin embryo (___×)

PART D

Match the terms in column A with the descriptions in column B. Place the letter of your choice in the space provided.

Column A		Column B
a. Blastocyst	_____ 1.	Germ layer that gives rise to muscle and bone tissues
b. Chorionic villi	_____ 2.	Blastocyst cells that give rise to embryo proper
c. Connecting stalk	_____ 3.	Spaces around chorionic villi that become filled with maternal blood
d. Ectoderm		
e. Endoderm	_____ 4.	Hollow ball of cells
f. Inner cell mass	_____ 5.	Cells forming wall of blastocyst
g. Lacunae		
h. Mesoderm	_____ 6.	Solid ball of about sixteen cells
i. Morula	_____ 7.	Inner germ layer of embryonic disc
j. Trophoblast	_____ 8.	Slender extensions that grow out from the trophoblast
	_____ 9.	Tissue connecting embryo proper to developing placenta
	_____ 10.	Outer germ layer of embryonic disc

PART E

Complete the following statements:

1. The thin membrane that separates the embryonic blood from the maternal blood is called the _____ .

2. The embryonic membrane that is attached to the edge of the embryonic disc and surrounds the developing body is called the _____ .

3. The umbilical cord contains three blood vessels, two of which are _____ .

4. Eventually the amniotic cavity is surrounded by a double-layered membrane called the _____ .

5. The _____ and allantois function to form blood cells during the early stages of development.

6. The _____ gives rise to the cells that later become sex cells.

7. The _____ gives rise to the umbilical blood vessels.

8. The embryonic stage is completed by the end of the _____ week of development.

9. All essential external and internal body parts are formed during the _____ stage of development.

10. _____ fluid protects the embryo from jarred movements and provides a watery environment for development.

PREPARATION OF SOLUTIONS

Amylase solution, 0.5%
Place 0.5 g of bacterial amylase in a graduated cylinder or volumetric flask. Add distilled water to the 100-mL level. Stir until dissolved. See the Instructor's Manual for a supplier of amylase that is free of sugar. (Store amylase powder in a freezer until mixing this solution.)

Benedict's solution
Prepared solution is available from various suppliers.

Caffeine, 0.2%
Place 0.2 g of caffeine in a graduated cylinder or volumetric flask. Add distilled water to the 100-mL level. Stir until dissolved.

Calcium chloride, 2.0%
Place 2.0 g of calcium chloride in a graduated cylinder or volumetric flask. Add distilled water to the 100-mL level. Stir until dissolved.

Calcium hydroxide solution (limewater)
Add an excess of calcium hydroxide to 1 L of distilled water. Stopper the bottle and shake thoroughly. Allow the solution to stand for 24 hours. Pour the supernatant fluid through a filter. Store the clear filtrate in a stoppered container.

Epsom salt solution, 0.1%
Place 0.5 g of Epsom salt in a graduated cylinder or volumetric flask. Add distilled water to the 500-mL level. Stir until dissolved.

Glucose, 1%
Place 1 g of glucose in a graduated cylinder or volumetric flask. Add distilled water to the 100-mL level. Stir until dissolved.

Iodine-potassium-iodide (IKI solution)
Add 20 g of potassium iodide to 1 L of distilled water, and stir until dissolved. Then add 4.0 g of iodine, and stir again until dissolved. Solution should be stored in a dark stoppered bottle.

Methylene blue
Dissolve 0.3 g of methylene blue powder in 30 mL of 95% ethyl alcohol. In a separate container, dissolve 0.01 g of potassium hydroxide in 100 mL of distilled water. Mix the two solutions. (Prepared solution is available from various suppliers.)

Physiological saline solution
Place 0.9 g of sodium chloride in a graduated cylinder or volumetric flask. Add distilled water to the 100-mL level. Stir until dissolved.

Potassium chloride, 5%
Place 5.0 g of potassium chloride in a graduated cylinder or volumetric flask. Add distilled water to the 100-mL level. Stir until dissolved.

Quinine sulfate, 0.5%
Place 0.5 g of quinine sulfate in a graduated cylinder or volumetric flask. Add distilled water to the 100-mL level. Stir until dissolved.

Ringer's solution (frog)
Dissolve the following salts in 1 L of distilled water:

> 6.50 g sodium chloride
> 0.20 g sodium bicarbonate
> 0.14 g potassium chloride
> 0.12 g calcium chloride

Sodium chloride solutions
1. *0.9% solution.* Place 0.9 g of sodium chloride in a graduated cylinder or volumetric flask. Add distilled water to the 100-mL level. Stir until dissolved.
2. *1.0% solution.* Place 1.0 g of sodium chloride in a graduated cylinder or volumetric flask. Add distilled water to the 100-mL level. Stir until dissolved.
3. *3.0% solution.* Place 3.0 g of sodium chloride in a graduated cylinder or volumetric flask. Add distilled water to the 100-mL level. Stir until dissolved.
4. *5.0% solution.* Place 5.0 g of sodium chloride in a graduated cylinder or volumetric flask. Add distilled water to the 100-mL level. Stir until dissolved.

Starch solutions
1. *0.5% solution.* Add 5 g of cornstarch to 1 L of distilled water. Heat until the mixture boils. Cool the liquid, and pour it through a filter. Store the filtrate in a refrigerator.
2. *1.0% solution.* Add 10 g of cornstarch to 1 L of distilled water. Heat until the mixture boils. Cool the liquid, and pour it through a filter. Store the filtrate in a refrigerator.

Sucrose, 5% solution
Place 5.0 g of sucrose in a graduated cylinder or volumetric flask. Add distilled water to the 100-mL level. Stir until dissolved.

Wright's stain
Prepared solution is available from various suppliers.

ASSESSMENTS OF LABORATORY REPORTS

Many assessment models can be used for laboratory reports. A rubric, which can be used for performance assessments, contains a description of the elements (requirements or criteria) of success to various degrees. The term *rubric* originated from *rubrica terra,* which is Latin for the application of red earth to indicate anything of importance. A rubric used for assessment contains elements for judging student performance, with points awarded for varying degrees of success in meeting the objectives. The content and the quality level necessary to attain certain points are indicated in the rubric. It is effective if the assessment tool is shared with the students before the laboratory exercise is performed.

Following are two sample rubrics that could easily be modified to meet the needs of a specific course. Some of the elements for these sample rubrics may not be necessary for every laboratory exercise. The generalized rubric needs to contain the possible assessment points that correspond to objectives for a specific course. The point value for each element may vary. The specific rubric example contains performance levels for laboratory reports. The elements and the point values could easily be altered to meet the value placed on laboratory reports for a specific course.

ASSESSMENT: GENERALIZED LABORATORY REPORT RUBRIC

Element	Assessment Points Possible	Assessment Points Earned
1. Figures are completely and accurately labeled.	_____	_____
2. Sketches are accurate, contain proper labels, and are of sufficient detail.	_____	_____
3. Colored pencils were used extensively to differentiate structures on illustrations.	_____	_____
4. Matching and fill-in-the-blank answers are completed and accurate.	_____	_____
5. Short-answer/discussion questions contain complete, thorough, and accurate answers. Some elaboration is evident for some answers.	_____	_____
6. Data collected are complete, accurately displayed, and contain a valid explanation.	_____	_____
Total Points	_____	_____

ASSESSMENT: SPECIFIC LABORATORY REPORT RUBRIC

Element	Excellent Performance (4 points)	Proficient Performance (3 points)	Marginal Performance (2 points)	Novice Performance (1 point)	Points Earned
Figure labels	Labels completed with ≥ 90% accuracy.	Labels completed with 80%–89% accuracy.	Labels completed with 70%–79% accuracy.	Labels < 70% accurate.	
Sketches	Accurate use of scale, details illustrated, and all structures labeled accurately.	Minor errors in sketches. Missing or inaccurate labels on one or more structures.	Sketch is not realistic. Missing or inaccurate labels on two or more structures.	Several missing or inaccurate labels.	
Matching and fill-in-the-blanks	All completed and accurate.	One to two errors or omissions.	Three to four errors or omissions.	Five or more errors or omissions.	
Short-answer and discussion questions	Answers are complete, valid, and contain some elaboration. No misinterpretations are noted.	Answers are generally complete and valid. Only minor inaccuracies were noted. Minimal elaboration exists.	Marginal answers to the questions and contains inaccurate information.	Many answers are incorrect or fail to address the topic. There may be misinterpretations.	
Data collection and analysis	Data are complete and displayed with a valid interpretation.	Only minor data missing or a slight misinterpretation exists.	Some omissions. Not displayed or interpreted accurately.	Data are incomplete or show serious misinterpretations.	

Total Points Earned _____

CREDITS

INDEX